THE SQUARES

INSIDE TECHNOLOGY

Edited by Wiebe E. Bijker, Trevor J. Pinch, and Rebecca Slayton

A list of books in the series appears at the back of the book.

THE SQUARES

US Physical and Engineering Scientists in the Long 1970s

CYRUS C. M. MODY

The MIT Press
Cambridge, Massachusetts
London, England

The open access edition of this book was made possible by generous funding from Arcadia—a charitable fund of Lisbet Rausing and Peter Baldwin.

ARCADIA
A charitable fund of Lisbet Rausing and Peter Baldwin

The MIT Press would like to thank the anonymous peer reviewers who provided comments on drafts of this book. The generous work of academic experts is essential for establishing the authority and quality of our publications. We acknowledge with gratitude the contributions of these otherwise uncredited readers.

This book was set in Scala and ScalaSans by New Best-set Typesetters Ltd. Printed and bound in the United States of America.

Library of Congress Cataloging-in-Publication Data

Names: Mody, Cyrus C. M. (Cyrus Cawas Maneck), 1974– author.
Title: The squares : US physical and engineering scientists in the long 1970s / Cyrus C. M. Mody.
Description: Cambridge, Massachusetts : The MIT Press, [2022] | Series: Inside technology | Includes bibliographical references and index. | Identifiers: LCCN 2021035172 | ISBN 9780262543613 (paperback)
Subjects: LCSH: Science—United States. | Scientists—United States. | Engineers—United States. | Physical scientists—United States.
Classification: LCC Q149.U5 M63 2022 | DDC 502.373—dc23
LC record available at https://lccn.loc.gov/2021035172

10 9 8 7 6 5 4 3 2 1

For Daria Lara Mody: may she rise calmly

CONTENTS

PREFACE AND ACKNOWLEDGMENTS

I doubt that I am unusual in writing books partly for myself: for the pleasure of writing, of solving the puzzle of how to organize a chapter, of discovering what I actually know by putting my thoughts into words. But any book that is published is also written for others: family, friends, colleagues, and hopefully wider, unknown audiences. The latter, especially, deserve some thanks for taking a chance on a stranger's book. So if you are reading this and we don't know each other, well, consider yourself acknowledged! I hope to thank you personally someday. Similarly, if you're reading this and we're related, well, I can't thank you enough. Karen and Daria, in particular—this book is just one small souvenir from our adventures together. Whatever is hopeful and optimistic in what follows owes everything to you.

So much for family and strangers. That just leaves friends and colleagues. Any list would be incomplete and unsatisfactory, so I won't aim to be comprehensive. Instead, let me try to thank a few people and organizations for their specific contributions to each of the chapters. Unfortunately, word counts and memories are finite, so I will neglect some of you. If you are one of those not explicitly acknowledged here, please know that you are one of those for whom this book was written and without whom it could not have been written. Also, some of you are only thanked once, whereas you should know that I am grateful for too many occasions of your assistance and intervention to count.

The overview of the book's argument that appears in chapters 1 and 7 was rehearsed earlier in an article in *Centaurus* that benefited from advice from Matthias Heymann, Ida Stamhuis, Mike Egan, and anonymous referees; a white paper for the Washington Center for Equitable Growth commissioned by Jonathan Moreno and Ed Paisley and discussed at a workshop attended by Heidi Voskuhl, Chris Jones, and several of those thanked here; a presentation at the MIT-Wharton National Science Foundation (NSF) Conference for which I was suggested by Sonali Shah; and a plenary address to the Society for the Studies of New and Emerging Technologies (S.NET) courtesy of Harro van Lente and Tsjalling Swierstra. Audiences at Cornell University, The University of Texas at Dallas, University of Notre Dame, and Seoul National University also saw talks that summarized the book as a whole. In addition, my thinking on the long, lean 1970s benefited from conversations with colleagues at Rice University (especially Randal Hall and Carl Caldwell), at the University of Houston (especially Julie Cohn, Ioanna Semendeferi, Kairn Klieman, and Jimmy Schafer), and at Maastricht University (especially Dani Shanley, Darryl Cressman, Geert Somsen, Vincent Lagendijk, and Tom Turnbull).

A semipopular version of chapter 2 appeared in *Physics Today* at the invitation of Charles Day after he saw me on a panel at the 2013 American Physical Society (APS) March Meeting in Baltimore that Joe Martin organized. Versions with more footnotes have also appeared in two very different edited volumes: *Groovy Science*, edited by David Kaiser and Patrick McCray, and *Public Universities and Regional Growth*, edited by Martin Kenney and David Mowery. My thanks to all four of them, plus the other authors in those volumes. At Anna Guagnini's suggestion I also supplied a paper based on this chapter for a workshop at the Max Planck Institute for the History of Science organized by Alexander Schwerin, Jaromír Balcar, and Florian Schmaltz. In addition, I presented this material to a group that influenced this book in many ways, the Maastricht University Science, Technology, and Society Studies (MUSTS) research program. This chapter would not have been possible without the generous help of the University of California, Santa Barbara (UCSB) Physics Department, Philip Wyatt, Virgil Elings, Mike Buchin, and Linda and Dan Phillips.

A version of chapter 3 was originally slated for *Groovy Science* and therefore received feedback from participants in the workshop that led to that volume. Particular thanks go to Michael Gordin (in this instance as host, but for much else besides), Matt Wisnioski, and Laura Stark (for her insightful comments on a very "draft-y" paper). As it turned out, that paper was instead published in *Investigating Interdisciplinary Research*, edited by Scott Frickel, Barbara Prainsack, and Mathieu Albert. Some elements of this chapter were developed in collaboration with Mara Mills and in a separate collaboration with Andrew Nelson that resulted in an article in the *Osiris* volume edited by Julia Kursell, Alix Hui, and Myles Jackson (with thanks to the participants in that volume's "Catfish Workshop"). Other pieces of this chapter appear in the *Oxford Handbook of Sound Studies*, edited by Trevor Pinch and Karin Bijsterveld. Audiences at UCSB, Rice University, The University of Texas at Austin, the Society for the History of Technology, and the Society for Social Studies of Science gave useful feedback. So did the members of the Descartes Center at Utrecht University, where I was invited to speak by Bert Theunissen at the suggestion of Frank Huisman.

That talk then led to my presenting a version of chapter 4 for the Science, Technology, and Policy Studies (STePS) Program at University of Twente at the invitation of Lissa Roberts. I also spoke about NASA and socially relevant research at Carnegie Mellon University thanks to David Hounshell, Erica Fuchs, and Hassan Khan. Members of the MUSTS group read a version for our Summer Harvest organized by Alexandra Supper, as did participants in the NASA and the 'Long' Civil Rights Movement Conference organized by Brian Odom. Matt Hersch, Asif Siddiqi, and Peter Westwick offered helpful advice and encouragement as I made my first foray into space history.

Chapter 5 was inspired by an archival visit that wouldn't have happened but for Jessica Martucci's request that I moderate a panel at the Southern Association for Women Historians meeting. Later, Dominic Boyer and Cymene Howe asked me to talk about Jack Kilby's solar energy work at the annual symposium of Rice's Center for Energy and Environmental Research in the Human Sciences (CENHS). I also talked about Kilby at the 2015 APS March meeting (at Bruce Hunt's invitation) and at a workshop organized by Christian Forstner and Falk Müller. Finally, with Jean Kumagai and Rachel Courtland's skillful editing I wrote an article about Kilby for *IEEE Spectrum*. That magazine has been a wonderful supporter of the small community of nano- and semiconductor historians, many of whom also helped with chapters 5 and 6—especially David Brock, Amy Slaton, Hyungsub Choi, Ben Gross, Matt Eisler, Jorijn van Duijn, Christophe Lécuyer, and the much-missed Ann Johnson.

Finally, a word of thanks to our sponsors, especially the American Council of Learned Societies and the NSF—not only for the monetary assistance but also for the connections they facilitated. I particularly thank Mike Gorman for his time at NSF, and a variety of NSF-funded organizations that supported the research in this book: the Center for Biological and Environmental Nanotechnology at Rice University, the National Nanotechnology Infrastructure Network (particularly Don Tennant and other members of the Cornell NanoScale Science and Technology Facility and the Stanford Nanofabrication Facility), and the Center for Nanotechnology in Society at UCSB. This book is in part the finale to a trilogy on the history of nanotechnology made possible by nearly continuous NSF assistance from 1997 to 2016 and by collaborations facilitated by that assistance, particularly with Patrick, Hyungsub, Ann, and Dave. This book *also* marks my transition from nanotechnology to energy history: chapter 5, in particular, incorporates material from the Managing Scarcity and Sustainability project VI.C.191.067 supported by the Dutch Research Council (NWO).

LIST OF FIGURES

1

SOCIETY IN DISARRAY? PROFESSIONAL SOCIETIES AND PHYSICS IN THE 1970S

Most of the abstracts for the January 1970 meeting of the American Physical Society (APS) were conventional, if esoteric, summaries of findings about Fermi levels, peaks in spectroscopic curves, and the like. But one abstract contained a pictorial element that caught many readers' eyes: a peace symbol (figure 1.1). The paper to which the symbol belonged was by Charles Schwartz of Berkeley and was entitled "I Pledge That I Will Not Participate in War Research or Weapons Production; I Further Pledge to Counsel My Students and Urge My Colleagues to Do the Same."[1]

Crucially, the APS left a line from the abstract that Schwartz had submitted out of the published version: "Work supported by the US Air Force." The APS executive secretary, Bill Havens, had excised that acknowledgment after the Air Force informed him that the "support" in question consisted of travel expenses and an honorarium for Schwartz's invited participation in the Sixth Annual Symposium on Science, Philosophy, and Religion held at Kirtland Air Force Base in October 1969.[2] Schwartz didn't dispute that that was the form of support he was referring to. Havens, however, thought that Schwartz was playing a rhetorical trick, because most APS members would read "work supported by" as

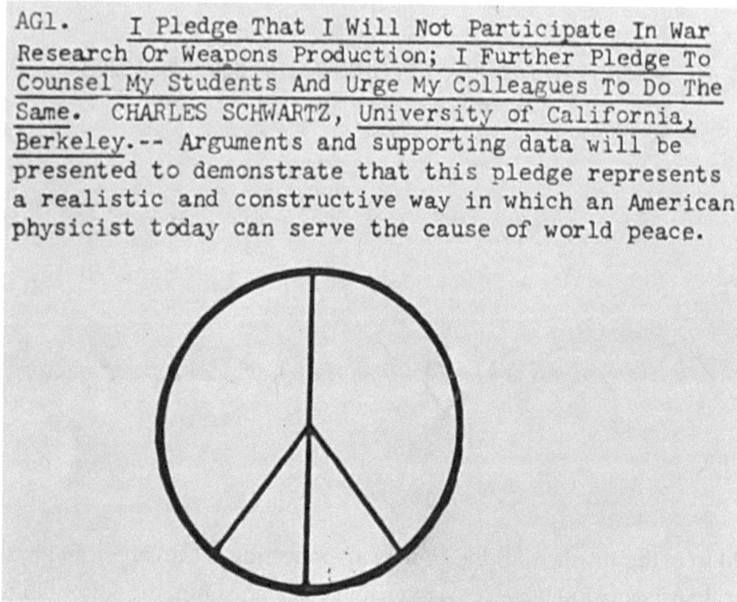

AG1. I Pledge That I Will Not Participate In War Research Or Weapons Production; I Further Pledge To Counsel My Students And Urge My Colleagues To Do The Same. CHARLES SCHWARTZ, University of California, Berkeley.-- Arguments and supporting data will be presented to demonstrate that this pledge represents a realistic and constructive way in which an American physicist today can serve the cause of world peace.

Figure 1.1

Charles Schwartz's abstract for the 1970 APS meeting. *Source*: *Bulletin of the American Physical Society* 15.1 (1970): 28. Permission granted by the American Physical Society. For the APS's own look back at this episode see "Consciousness Raising," *APS News* 9.1 (January 2000). https://www.aps.org/publications/apsnews/200001/knowledge.cfm.

meaning that the Air Force had given Schwartz a grant to perform the research presented in his paper. Schwartz, in turn, took the excision of the last line of his abstract as an opportunity to try to force the APS either to reprimand Havens for "censoring the abstract submitted by a member" or to denounce the Air Force for "unwarranted interference in the internal affairs of this society."[3]

This was neither the first nor the last time that Schwartz collided with APS officialdom. In 1968 he had proposed an amendment to the APS constitution that would have facilitated APS members in offering resolutions of political and societal concern for debate by their fellow members.[4] Later in 1970 he circulated a petition proposing to amend Article II of the APS constitution ("The object of the Society shall be the advancement and diffusion of the knowledge of physics") by adding "and concern for the application of this knowledge to the welfare of mankind."[5] In 1974 he again put forward changes to the APS constitution, this time to require candidates for APS office to disclose "relevant interests and affiliations which might reveal to the APS membership the views of their representatives on any given issue"—in particular, affiliations with the nuclear power industry that Schwartz believed might influence the Society's pronouncements on the safety of nuclear energy.[6] And so on; these examples are only a sampling of Schwartz's attempts to force the APS to debate the political controversies of the day.

None of these actions were taken alone. Some of Schwartz's allies, such as Martin Perl of Stanford, worked closely with him, for instance in forming Scientists and Engineers for Social and Political Action (better known as Science for the People).[7] Others, such as Jay Orear of Cornell, organized their own, parallel challenges to the political inertia of the APS leadership.[8] With Schwartz's moral support, for instance, Orear led a campaign to convince the APS to move the 1969 meeting away from Chicago in the wake of Mayor Richard Daley's violent crackdown on protestors at the 1968 Democratic Party convention.[9]

Schwartz also made many enemies. Although none of his proposals to amend the APS constitution succeeded, each of his attempts to put the Vietnam War and other injustices in front of the APS membership did prompt a stream of angry letters to the society's leadership. To give just a sampling of the vitriol: one Los Alamos researcher informed Havens that

> I note with pain, sorrow and great rage the appearance of abstract AG1 . . . for the 1970 Annual Meeting. . . . Many of us in the "silent majority" had hoped that the divisive and destructive influences appearing in our scientific society would die away with minimum untoward effects especially after the overwhelming vote on the Schwartz amendment. It seems that the biting of the flea cannot be simply ignored and eventually a slap must be administered.[10]

Others insinuated that the APS leadership might also be targeted for "slaps" because they had provided Schwartz with a platform. In response to the peace symbol abstract, for instance, a lawyer for the Atomic Energy Commission (AEC) requested that Luis Alvarez, then president of the APS, "please be good enough to let me know the name and office of the person who approved the inclusion in the program of item AG-1 by Charles Schwartz."[11] The same abstract occasioned four physicists from Baylor University to write Alvarez to

> register a complaint for Schwartz's abstract because we suspect that it was inserted for political reasons, either because the editor concurred with his views or because the editor was afraid to oppose Schwartz's views. If most of us quiet, unassuming "low-brow" physicists had submitted an abstract of the caliber of Schwartz's, we are sure the abstract would have been tossed in the waste can.[12]

Clearly, as this correspondence shows, Schwartz's actions provoked deeply felt reactions from colleagues who saw his behavior as demeaning and immoral.

Schwartz, of course, thought that it was colleagues such as these who were themselves defying morality by refusing to condemn the injustices of the Vietnam War. The years around 1970 were, therefore, an era in which affectively powerful expectations—moralized expectations—about how scientists should behave were contested and changing. Some, like Schwartz, believed that they had to do what they could to accelerate that change. Others believed just as strongly that they had a responsibility to hinder it.

Both camps employed extreme or disingenuous rhetoric in opposing the other and in pressuring the APS leadership to do their bidding. Schwartz's trick regarding Air Force "support" is one example. From the other side, letters such as one from a prewar émigré from Germany (then working at the University of Michigan) show how casually some of Schwartz's detractors poisoned the well with references to the twentieth century's bleak history: "the inclusion of the so-called peace symbol is in the worst taste. Is it a diagram? Or a mathematical symbol, [as] contained in the font of AIP [American Institute of Physics]? No, it is a political symbol, a shibboleth, of the same kind as the swastika of Hitler Germany."[13]

Given such reckless rhetoric, this was a flummoxing time for those like Havens and Alvarez whose loyalty to the institution meant that they had to navigate between polarized views. The APS leadership moved cautiously, conceding the partisans' points enough to show concern for their views while doing nothing that they thought would endanger the long-term viability of the organization and its profession. As Alvarez put it in his standard reply to the letters mentioned earlier,

> Charles Schwartz is a member in good standing of the Society, and although I disagree strongly with many of the things he believes in, I think it would hurt our Society much more to dismiss arbitrarily his abstracts than it would be to let him give them. He thinks his points are important and directly concerned with the future of physics. The object of the Society is the advancement and diffusion of knowledge of physics, and Charles Schwartz takes the point of view that if we continue to play footsie with the Pentagon, physics as we know it will disappear, and there will no longer be any knowledge of physics for the Society to diffuse. I do not believe it would be possible to write a restriction that would keep such an abstract from being published in our bulletin, without at the same time opening the doors to wide censorship of abstracts of a type we should protect, in my opinion.[14]

A spirited defense! Yet in some versions, Alvarez ended his letter with a wearier plaint: "Everyone has problems, and my problem this year has been Charles Schwartz. I will be most happy to turn the affairs of the Society over to" the next APS president.

This is not a book about Schwartz and his allies, nor is it a book about Schwartz's foes. This book is, in part, about those such as Alvarez and Havens who may have disagreed with Schwartz but who thought that shutting him out would burn down the house of science. It is, even more, a book about those rank-and-file applied physicists and engineering scientists who were ambivalent about Schwartz's views, and even more so about his tactics, but who were not immune to the argument that "if we continue to play footsie with the Pentagon, physics as we know it will disappear."

The views of this class were succinctly expressed by a Harvard physicist who wrote to the APS in reaction to the era's events:

> I would like to explain why I am voting to change the 1970 meeting site from Chicago, even though I voted against the Schwartz amendment. While I am against formal political positions on the part of the Society, I believe the membership still has the right to decide where to hold their conventions, for maximum convenience and comfort. I personally would be very uncomfortable attending a meeting in Chicago; I couldn't pass near a Chicago policeman without feeling the bitterness of 1968 recur.[15]

Yet the central characters of this book also resemble another letter writer, from the Illinois Institute of Technology, whose views were as ambivalent as, but *inverted*, his Harvard colleague's:

> If some of the APS members wish to use the allocation of Meeting sites as a club against political incompetency and police brutality, why was there not an outcry against holding the 1969 Annual Meeting in New York City[?] Are our memories so short that we can no longer conjure up the bloody incidents of last spring at Columbia University or examples of racial bigotry and violence condoned or handled ineffectively by Mayor Lindsay this past fall? . . . When it can be showed that APS members are likely to experience physical abuse or indignity at the hands of Mayor Daley or the Chicago police while attending to their professional interests at a Meeting, I shall be in the vanguard of those fleeing the city. Until then I trust the majority of APS members will continue to show good judgement in exercising their concern for social and political justice outside the framework of the APS.[16]

Between these two letters we can glimpse the ambivalences of the era. Both expressed "concern for social and political justice," in distinct contrast to Schwartz's antagonists. Yet, like most APS members, neither correspondent was willing to take his concern as far as Schwartz. Moreover, despite the similarity of their views on social and political justice, those views led them to diametrically opposite positions on the question of whether the APS should hold its meeting in Chicago.

Mostly, though, this is a book about those who constituted a majority so silent that they could move themselves neither to instigate for reform like Schwartz, nor to write rage-filled rejoinders like that of their self-proclaimed "low-brow" colleagues at Baylor or even to pen ambivalent missives like the ones I've just quoted. Specifically, it is about those who were not moved to protest but who were in a position to use their *technical* knowledge to do something about the injustices Schwartz and his allies decried. As we've seen, Schwartz called for his colleagues to become less dependent on the military; left somewhat tacit in those calls was the implication that American physicists would therefore have to shift toward civilian funders and toward topics more oriented to the concerns of civil society. Yet as a theorist noted mainly for calculating electron scattering coefficients, Schwartz was not particularly well placed to practice a civilian-oriented "science for the people" directed at, say, designing better public housing or mass transit.

In contrast, practitioners of applied physics, materials science, and a variety of engineering fields as well those fields' more academically oriented "engineering science" wings had the capacity to design such technologies without much retooling. In the years when Schwartz and Science for the People were most active, a significant number of their peers *did* move away from defense work and toward topics of immediate, applied relevance to civil society. Some of those who did so were sympathetic to Schwartz and were equally vocal in expressing their politics. There were also many who moved in that direction yet were unsympathetic or ambivalent toward Schwartz or did not draw attention to their politics; *those* are the people this book is about. Thus, this book asks: what led some politically ambivalent or reticent physical and engineering scientists (but not others) to practice socially relevant forms of research and development in the 1970s, despite their reluctance to engage in avowed activism of the kind practiced by Schwartz and Science for the People?

THE EXCLUDED MIDDLE

There are good reasons and bad ones for examining this reticent group. A poor motivation per se, but a necessary step toward better ones, is that these people have not been studied before, at least not as a group squeezed on all sides in the years during and just after large-scale American participation in the Vietnam War. The groups doing the squeezing, however, *have* received attention from historians, sociologists, and science and technology studies (STS) scholars. Kelly Moore, Sarah Bridger, Sigrid Schmalzer, and others have written about Schwartz, Science for the People, and other people and organizations that confronted American scientists with their complicity in war, inequality, and injustice; while Alice Bell and Simone Turchetti have examined the transnational networks to which US activist scientists were connected.[17] Their opponents have received less attention, but there is a growing literature on where some of the most prominent of them ended up: as denialists continuing (or so they believed) the fight against communism by undermining scientific evidence concerning the dangers of ozone layer depletion, anti-ballistic missile defense, nuclear winter, and climate change.[18]

There is also a voluminous literature on "groovy" scientists and engineers—those who embraced the counterculture but not necessarily New Left politics and who brought their interest in communes, parapsychology, psychotropic drugs,. and New Age spirituality into their scientific and technological activities.[19] A handful of the people who appear in this book were involved in such pursuits, but in a (literally) buttoned-down manner that was connected to yet distinct from the mushroom tea–powered Fundamental Fysiks of Berkeley and the hot tubs of Esalen. Most of the figures in this book, however, were uninterested in or put off by the youth counterculture, and many lumped the commune dwellers together with the New Left as exemplars of a general antiscientific "revolt against reason."[20]

One of the arguments of works on groovy science is that its proponents differed from the New Left in that they cultivated ties to the military–industrial complex, as evidenced by their interest in the Pentagon's Advanced Research Projects Agency Network (ARPANET, precursor of the internet) and their later admiration for nuclear power, space shuttles, and space "colonies." Indeed, one of the unremarked curiosities of some books on groovy science is that they end with an admission that one or more of the protagonists later embraced deeply conservative political movements.[21] Such transformations are perhaps not entirely surprising because these were fluid times, in which many people transited the political spectrum. An example related to chapter 3 is the Stanford Artificial Intelligence Laboratory, where the director, John McCarthy, started as a red-diaper baby and ended as a belligerent, racially inflaming and misogynistic conservative—whereas his deputy, Les Earnest, started out on a "bigoted path until after I voted for Richard Nixon over John F. Kennedy, then figured out that I was on the wrong track and eventually evolved into a radical socialist who aims to destroy Wall Street and similar piracy organizations around the world."[22]

McCarthy and Earnest are perhaps extreme examples who don't accurately represent the wider science and engineering community. The more usual direction of travel in this era, according to Sarah Bridger, was for younger scientists to move leftward and more established figures to the right.[23] Yet there were many—represented by the people featured in this book—who did *not* shift far politically, at least not visibly. Indeed, that lack of visibility is my point: we have a good understanding of the motivations and trajectories of scientists and engineers who publicly stated their views on the war, both pro and con, but we don't know much about those whose views were more ambivalent or less public.

As a shorthand, I will refer to this group as *square scientists*. As should become clear over the course of the book, I do not use the term pejoratively, but rather to underline the distinction between them and their more-studied "groovy" peers.[24] In the next section I will define the term more carefully. For now I simply want to show that this group is conspicuous in its absence from histories of 1970s American science and engineering. We do know a little about some individual squares, of course. For instance, Luis Alvarez, the APS president who defended but also tired of Charles Schwartz, is famous both for his Nobel Prize–winning work in high energy physics and for his later proposal (with his son Walter) that an asteroid impact killed the dinosaurs.[25] His predecessor as APS president, John Bardeen, won *two* Nobel Prizes yet was so square that his reticence and lack of charisma are the central enigmas of Lillian Hoddeson and Vicki Daitch's biography of him.[26]

As a class, though, the square scientists occupy an excluded middle that is hard to locate in the literature. Perhaps the closest they have come to the spotlight is Matthew Wisnioski's *Engineers for Change: Competing Visions of Technology in 1960s America*.[27] Yet Wisnioski's intellectual history focuses largely on the public statements of those who were willing to make public statements. This book will look instead at what less-vocal square scientists and engineers *did* in their laboratories, garages, and classrooms.

There have been a handful of works that have examined square scientists outside applied physics and engineering science, though without explicit reference to squares as a class. I would particularly highlight the small but simmering debate as to whether the dislocations of the 1970s had any effect on molecular biology and the emergence of the biotechnology industry. On one side, Eric Vettel argues that campus protests at Berkeley, Stanford, and the University of California, San Francisco stimulated reforms even among politically reticent and ambivalent scientists and thereby led to the founding of one of the first biotech firms, Cetus.[28] On the other side, Nicolas Rasmussen argues that the "spirit of '68" had little to do with biotech—the industry was simply a natural evolution of longstanding trends internal to the discipline.[29] My view is that both Vettel and Rasmussen somewhat overstate their cases but that their debate is an important one for understanding how politically ambivalent, nonactivist scientists positioned themselves with respect to the social movements of the era. This book attempts to extend that understanding to the physical and engineering sciences. That extension is necessary because from the 1940s onward the physical and engineering sciences became almost existentially dependent on the military—more so than the life sciences, which could more easily turn to civilian patrons such as the National Institutes of Health. Hence, the debate that became unavoidable in the late 1960s—namely, whether and how to become less dependent on the Pentagon—was more fraught for American physical and engineering scientists than for their colleagues in biomedical fields.

THE SQUARES' ROOTS

So square physical and engineering scientists constitute a proverbial gap in the literature. But such gaps don't require us to fill them. There is no sense in which the squares *deserve* to have their story told, no way in which their tribulations merit our attention nor their triumphs over adversity our praise. Their general reticence regarding the war and the social movements of the day was not exactly heroic. Their lives were, in the main, comfortable. In fact, many scientists and engineers entered those fields in the 1950s and early 1960s precisely because those fields were seen as a ticket to an affluent suburban lifestyle.[30] That strategy worked out nicely for some of the people in this book, though less so for others; but in general, this book is about people who were reasonably well off before 1970, and who—despite occasional "problems" posed by Schwartz and his foes and allies—prospered after 1970 as well.

We can, cautiously, make a few more generalizations about politically reticent and ambivalent scientists and engineers in this period. For one thing, the people for whom a science or engineering career seemed a path to middle-class comfort were overwhelmingly white men. The APS Committee on the Status of Women in Physics estimated in the early 1970s that only 3.7 percent of US physicists were women, and the APS Committee on Minorities in Physics estimated that less than 0.4 percent were from an Indigenous, Latinx, or African-American background (to use today's terminology).[31] Other physical and engineering sciences were even less representative of the population as a whole. There were, to be sure, politically ambivalent or reticent scientists and engineers who identified as women and/or as African-American, or who were subaltern or minoritized in other ways, but they were a small—if growing—minority in the organizations we will look at.

For the people featured in this book, there was also *some* correlation between political reticence or ambivalence and a "small-c" conservative presentation of self. Figure 2.1 shows the typical uniform of the physical scientist circa 1970, at least for square men: suit or sport coat, white shirt, tie (generally plain, but with exceptions). We'll see that for some squares, wearing this kind of uniform—perhaps along with a crew cut or other identifier—was meant to tacitly signal their conservative politics. Unsurprisingly, some of these squares complained occasionally about "the kids with the long hair" and other people who weren't square in their appearance, lifestyle, or politics.[32] For other squares, though, wearing a suit and tie was simply how you went along to get along; we can read their habitus as a signifier of their reticence but not of their ideological leanings. Where possible, I will try to distinguish between these modes of being square and between the more political and cultural dimensions of squareness.

Along both dimensions, squareness should be seen as a relational category. Many of the actors I will discuss were square relative to some of their peers and flamboyantly countercultural relative to other colleagues. Those are exactly the people whom historians of 1970s science have not examined yet, at least not in an explicit way. There are a number of studies of physical and engineering scientists who were groovier than almost everyone, or to the left or right of almost everyone—but we don't have studies that examine the complexities involved in being groovier than some and squarer than others, more hawkish than some and more dovish than others.

I also view squareness as context dependent. We'll see in chapter 6, for instance, that carousing, heavy drinking, and even theft were standard, almost mandatory, practice among square (white, male, middle-class, professional) engineers working in the microelectronics industry. In other contexts, such behavior would have been seen as rather more transgressive. Similarly, in chapter 3, we'll see that an interest in parapsychology could pass for square science in the California of the long 1970s, but less so in other times and places (and not uncontested even in that time and place). Crucially, the context dependence of squareness changed over time: a male scientist who wore a suit and tie every day in 1970 *didn't* stick out, whereas the same costume in 1980 was further from the median.

If square is a relational and context-dependent category it is also—mostly—an analyst's category. I'm doubtful many squares would've used that label themselves, though dichotomies like hip/square and hawk/dove were certainly part of the discourse of the times. Where possible, I will defer to their evaluations of themselves as "plain vanilla" or their labeling of nonsquares as, for example, having a "messiah complex." Because I define the squares by their public reticence, though, such characterizations were rare and generally private. In a few places I will show the gap between squares' public and private views. Neither their public nor private personae were more "real" than the other; rather, I am highlighting that some squares were willing to express strong views in private even if their public selves were harder to pin down.

But why characterize public political reticence and ambivalence as "squareness?" Squareness implies adherence to expected behavior. So why use public political reticence as the yardstick of orthodoxy? Here we need to acknowledge the importance of history, and of the particular developmental path of the physical and engineering sciences in the United States. Edwin Layton's 1971 classic *The Revolt of the Engineers: Social Responsibility and the American Engineering Profession* showed that from the beginning, engineers tended to come from the middling, square stratum of American society: middle-class, white, native-born men who grew up in medium-sized towns in the Midwest.[33] Those origins didn't necessarily equate to political quiescence, however; as Layton explained, in the early twentieth century the profession was roiled by debates over whether engineers were natural allies of labor or management. Allies of management won the argument, and engineers who wanted their peers to distance themselves from capital—or even to form unions of their own—were purged.[34] As Matthew Wisnioski has argued, Layton made this point to push engineers of the 1970s to rediscover their radical forebears. The "social responsibility" in Layton's subtitle gives away that he operated in the milieu examined in this book.

With plenty of exceptions, American applied physicists mostly came from similar places and a similar stratum of society as their engineering peers. For instance, one of the most frequently mentioned organizations on the resumés of American applied physicists was Bell Labs, which had a policy of preferentially hiring white, male Protestants from the Midwest.[35] Nor was Bell Labs atypical. Jews and other ethnosectarian minorities had somewhat more success in theoretical subfields, though that success led to further dangers. For, as David Kaiser has shown, in the McCarthy era witch hunters drew extensively on antisemitic tropes.[36] The cultural nonsquareness of, say, Robert Oppenheimer—his ethnicity, bodily "ethereality," and love affairs—made it easier for his enemies to cast doubts on his loyalty to the United States. More generally, as Jessica Wang argues, McCarthyism showed American scientists that if they stuck their necks out politically they were likely to be culled.[37] The result was that those who remained in the physical and engineering sciences tended to be quietists.

Note that I don't say that the squares were apolitical. The squares' choice not to broadcast their political views was a political choice. As Audra Wolfe has argued, cultivating a publicly apolitical science was a means to further the highly politicized ends of the national security state.[38] One could serve those politicized ends by, for instance, snitching on colleagues, as the chemist Kenneth Pitzer did to Oppenheimer. (Pitzer's ignominious deserts appear in chapter 3.) But many squares served those ends instead by keeping their heads down and conforming to what was expected of them. Until the late 1960s that mainly meant pursuing research questions set by their disciplinary peers or by the national security state, for which squares were amply rewarded with research grants, praise from colleagues, and a comfortable lifestyle.

In the late 1960s, though, expectations for American scientists and engineers shifted rapidly. Working for the recognition of one's disciplinary peers and the national security state suddenly drew widespread criticism while no longer guaranteeing job security. The era's social movements and economic conditions muddied square scientists' and engineers' expectations for themselves and others. This upending of convention opened the door (a little) for minoritized scientists and engineers to advance: if no one knew quite what was expected of scientists and engineers, then the expectation that they be white men became less stringent. As the literature on groovy science has shown, the abeyance of expectations also fostered greater diversity in the *content* of science: with the military no longer calling the shots, physicists could think deeply about quantum paradoxes and even parapsychology.[39]

Yet the upending of postwar expectations also represented a potential threat to those who had benefited most from those expectations. We will see that some squares—as in the letters quoted earlier—fiercely resisted any change in their status. #NotAllSquares, of course: many who were discomfited by events around 1970 responded by trying to make their profession more—in the buzzwords of the day—"societally relevant" and "socially responsible." We needn't celebrate such moves as magnanimous noblesse oblige, though, because they were in many cases self-interested responses to an environment in flux. The two APS committees mentioned earlier, for instance, would not have been set up without pressure applied by the civil rights movement and second-wave feminism. Nor should we be rosy about the results of such reforms. Attempts to make science and engineering more inclusive have continued to the present but have made very slow progress.[40] Other forays into societal relevance were, as we will repeatedly see, abandoned after just a few years. Often, I will argue, socially responsible projects were put aside precisely because the squares didn't see the era's social movements as possible partners—and, to some extent, vice versa. The middle has been excluded in today's historiography in large part because it was excluded at the time.

In writing about that excluded middle, I am moving somewhat counter to the direction that historians have been moving since the 1970s. Because of the very social changes that form a background for this study, historians since then have increasingly emphasized the agency of those whom previous generations of historians regarded as lacking agency: women, people of color, proletarians, subalterns in colonial regimes, and increasingly nonhuman organisms and machines.[41] There are signs that this tide is ebbing in some subfields, but it is still an urgent project far from completion in the history of science and technology.[42]

This book works the other side of the fence, though with a common purpose. Square scientists and engineers are the type of people whose agency is often taken for granted. That is, in this period the American physical and engineering sciences were overwhelmingly composed of square white men, the group that faced the fewest legal and cultural constraints in American society. There were limits to this group's agency even before the late 1960s, of course; for example, as noted, limits imposed by the chilling effect of McCarthyism. Yet I will use the episodes that follow to argue that the contributions of this group became more oriented to aiding civil society in exactly the period when their agency was most challenged by civil society.

Given their overrepresentation in their profession, square scientists hold the keys to revealing some general features of how science and engineering fields deal with such pressure for reform. We of course need more histories of those who exerted that pressure. But if we want science and engineering to become more representative, inclusive, and socially responsible then we also need histories of those who were included and well represented to begin with, and especially histories of what those people did in eras when their privileges became less secure.

FROM SOCIAL RESPONSIBILITY TO RESPONSIBLE INNOVATION

The question of how to effect change in science and engineering and how to open up zones of cooperation among science, engineering, and their (potential) societal partners is still very much alive. Since the early 1990s there have been a series of official initiatives, especially in North America and Western Europe, aiming to facilitate scientists' and engineers' engagement with various "stakeholders." Those initiatives have usually been branded with variations on similar themes: Ethical, Legal, and Social Implications/Issues/Aspects (ELSI/ELSA); Societal and Ethical Issues/Implications (SEI); Responsible Development; Responsible Research and Innovation (RRI); and so on.[43] As Danielle Shanley has shown, several red threads connect ELSI/ELSA, SEI, and today's

RRI to the quandaries that square scientists and engineers faced in the 1970s.[44]

We therefore now have a very long history of both top-down and grassroots efforts to encourage scientists and engineers to practice more responsibly. Curiously, though, each framework for reform reinvents itself partly by forgetting what came before, so we know much less than we could about that history and about what kinds of encouragement succeed. This book attempts to draws out those lessons. I will try to avoid undue presentism—the 1970s were different from today in many ways—yet I will keep an eye on what the past tells us about how we got to today and what we should bear in mind going forward.

In particular, I hope that my case studies cast light on the most recent such framework, RRI, and its main tenets. Since about 2010, "responsible innovation" has achieved widespread currency as a "transparent, interactive process by which societal actors and innovators become mutually responsive to each other with a view to the (ethical) acceptability, sustainability, and societal desirability of the innovation process and its marketable products."[45] In Europe, it has been adopted at the highest reaches of science policy in the United Kingdom, the Netherlands, and other countries as well as by the European Union. In North America the state has not adopted RRI so forthrightly, but government agencies such as the National Science Foundation (NSF) have supported efforts to give RRI some institutional presence, such as the Society for the Studies of New and Emerging Technologies (S.NET) and the *Journal of Responsible Innovation* (JRI).

Note that the S.NET was originally the Society for the Studies of *Nanoscience* and Emerging Technologies, and that both it and the JRI are closely associated with the Center for Nanotechnology in Society at Arizona State University as well as with veterans of NanoNed, the Dutch nanotechnology initiative. As the editorial in the first issue of JRI put it,

> self-conscious socio-technical integration, with an aim toward better innovations for a better society would wait decades [after 1950] to emerge. It was not until the dawn of the new millennium that the terms "responsible innovation" (RI),

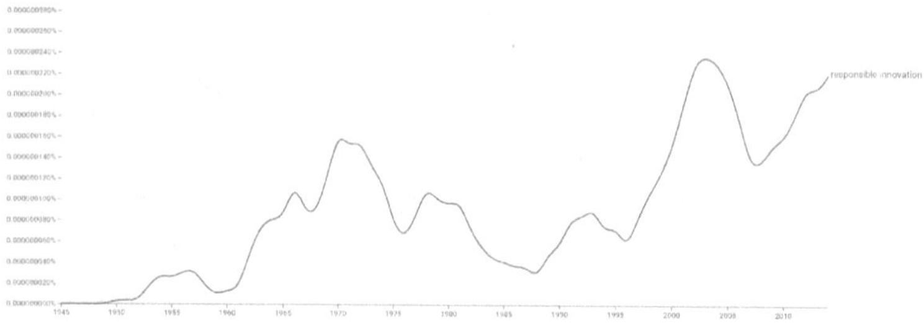

Figure 1.2

Responsible innovation's forgotten peak in 1970. The peak in 2002 probably relates to the National Nanotechnology Initiative. The curve rises again from 2008 onward thanks to adoption of responsible innovation in EU research funding and other arenas. I cut off the Ngram at 2014 because since then the term has become popular enough to dwarf earlier eras. My point here is that the 1970 peak was not definitively outpaced until 2010. *Source*: Google's Ngram service, American English, smoothing of 2.

> "responsible research and innovation" (RRI) and "responsible development" fully emerged. . . . In particular, the language of RI and its cognates began to take off along with large-scale national programs to conduct R&D on nanotechnology.[46]

In two previous books I have examined some of the historical roots of the natural and engineering science components of such nanotechnology initiatives, and the purposeful amnesia that allowed those initiatives to portray nanotechnology as novel.[47] This book completes the trilogy by offering a prehistory of the ideology—responsible innovation—that has become the legacy of the social science and humanities components of those initiatives, and of the amnesia that allows responsible innovation to be portrayed as novel.

For, as figure 1.2 shows, there is indeed a longer history to responsible innovation; *pace* the editors of JRI, the concept "fully emerged" in the 1970s (or perhaps even earlier) rather than the 2000s. The term was not the most common buzzword for the kinds of initiatives portrayed in this book, but it was among those buzzwords; and the buzzwords it coexisted with—"socially responsible" and "societally relevant" in particular—clearly overlap with today's definition of RRI. It's therefore strange that the 1970s usage of "responsible innovation" has disappeared from memory so completely—especially because in some cases prominent individuals in RRI were themselves participants in and products of that earlier era. Beyond individuals, there are still-extant academic and policy communities that are closely involved in RRI today, such as Technology Assessment or Science and Technology Studies (STS), that arose in the late 1960s to early 1970s.[48] Those fields made the point then, as RRI makes it now, that scientists and engineers should be more aware of the social dimensions of their practice. In chapter 3, for instance, we will look at Stanford University, where one of the first STS departments (then known as the Values, Technology & Society program) was founded in 1971 by a coalition of engineers and humanists.

The long 1970s also gave rise to social and intellectual movements such as Appropriate Technology (or AT) that grew and attracted adherents before petering out in the 1980s—movements that trained a new generation of reformists who shaped the ELSI/ELSA initiatives of the 1990s and that continue to shape today's RRI. A few of the squares who appear in this book were involved in such movements—for example, the Stanford engineers in chapter 3. Most were not, but almost all would have been aware of things like the US Congress's Office of Technology Assessment (founded in 1972) or Ernst Schumacher's 1973 manifesto for AT, *Small Is Beautiful: A Study of Economics as if People Mattered.* All were in some sense responding to pressure—both top-down and bottom-up—for "societal actors and innovators [to] become mutually responsive to each other with a view to the (ethical) acceptability, sustainability, and societal desirability of the innovation process." Whether they achieved that will be open to question throughout what follows; my point here is simply that something *like* responsible innovation was at play in the 1970s, and sometimes that label was even used. Moreover, the extent to which square scientists and engineers did or didn't *achieve* responsible innovation then, and the reasons they did or didn't, can help us critique the aims and gauge the feasibility of today's RRI. I will leave those lessons more or less implicit through most of the chapters that follow but return to discuss them in the final chapter.

THE LONG 1970S AS PIVOT

So certain present-day terms such as *responsible innovation* and certain practices related to those terms were seemingly presaged in the 1970s. Yet only "seemingly": the milieu of the time was different, not least because responsible innovation did not have much of a history then but does now. The 1970s gave rise to a variety of concepts and practices that have modern analogues; yet those analogues meant something different then, if only because the context in which they were embedded was so different.

To understand *how* the 1970s were different from today, we will need to bridge two adjoining eras that have been the subject of more extensive literatures: the period from 1945 to 1970 and from 1980 to the present. With respect to the former, even before the Second World War had finished, academic and participant–historians were sketching the outlines of a Cold War mode of research that captured policy establishments in North America and Western Europe until the late 1960s.[49] Later histories of this Cold War research paradigm often end their narratives, sometimes abruptly, with the student protests of 1969–1970 and the activism of Schwartz and his allies.[50] Conversely, there is a much more interdisciplinary literature by science and technology studies scholars, sociologists, anthropologists, philosophers, and management scholars—with historians playing catch-up—on commercialized "technoscience" in the years after 1980.[51] The intervening decade from 1970 to 1980 is not exactly *terra incognita* but we still have relatively little understanding of how we got from the one era to the other.[52] That, then, is one aim of this study.

What pictures of American physical and engineering science pre-1970 and post-1980 am I trying to bridge? The consensus among historians is that the earlier era was typified by dramatic growth in the resources and number of personnel involved in research and development (R&D).[53] It saw much of the governance of research migrate from philanthropies such as the Rockefeller and Carnegie Foundations to the state (and especially the military)—though often philanthropic foundations still played an influential role in guiding state support.[54] Postwar funding for research and especially development was predominantly funneled through not just the military services but also through other national security–oriented arms of the state such as the National Security Agency, the AEC, and the National Aeronautics and Space Administration (NASA) (the subject of chapter 4). Much federal research funding now employed the new accounting vehicle of the contract with overhead; universities and some other research organizations saw the overhead on such contracts as a way to pay for growth in personnel and physical plant. These organizations therefore learned to compete for federal grants for research and graduate education.[55]

As a result, elite research organizations in the United States converged on a single model: the "university without students."[56] Universities such as MIT and Stanford put less effort than they had prewar into undergraduate education because it brought little overhead. Instead they turned their attention to graduate training and "sponsored research." Large high-tech companies abandoned their urban workshops and established isolated exurban laboratories designed to resemble the collegiate ideal.[57] There—in theory—PhD-trained scientists could think great thoughts without the distractions of a nearby manufacturing line. Government agencies such as the National Bureau of Standards built similar rural campuses, in part for the same reasons but also as a security measure: the sparsely populated regions around, say, Los Alamos, New Mexico, or Oak Ridge, Tennessee, made it easier to discourage espionage and contain toxic accidents near those sites' new National Laboratories. In chapter 2 we will visit another enactment of the university without students located far from the madding crowd, the beachside think tank; and in chapter 7 we will visit a working farm that doubled as a systems engineering consultancy and—literally—a testbed for the automation of agriculture.

The rurality of the postwar research system was underwritten by another feature of the postwar compact: a belief that research could be sorted along a spectrum from basic to applied, with pastoral scenery putatively more accommodating to basic research (i.e., research characterized as fundamental, curiosity-driven, or esoteric). Today, there is some debate among historians as to what people at the time thought basic and applied research meant and what they believed the relationship between them was.[58] There should be little doubt, however, that by the 1970s these categories were in widespread use among US physical and engineering scientists.

From the analysts' perspective, of course, basic and applied research are social constructions: we know that different groups drew the boundary between the two categories differently (or didn't draw it at all), and therefore the same piece of research could be basic for one person but applied for another. But the historical actors who invoked these concepts were fully aware of that point: they keenly felt *both* the interpretive flexibility and the reality of the distinction between basic and applied.

As Harvey Brooks, a semiconductor physicist at General Electric and then Harvard and an influential figure in policy circles, argued in a 1967 opinion piece for *Science*:

> In institutions whose missions include the application of research results to products or operations, the categorization of research into basic or applied is not too meaningful and has little operational value. . . . [A]ll research in a "mission-oriented" organization contributes or should contribute, however remotely in time, to the general objectives of the organization. On the other hand, there is clearly a spectrum of activities ranging from pure research on the one hand to technological development on the other, and to some extent one can locate research activities within this spectrum according to their "appliedness."[59]

The actors in this book were people who, like Brooks, had the will and ability to zig and zag across that spectrum of "appliedness." That is, they were middling and ambivalent not just in their politics but even in their technical work—the physical and engineering scientists I look at were neither engineers nor physicists full stop but rather could take on the trappings of engineering or basic physical science as needed.

Of course, that kind of flexibility is a privilege and a resource. By being able to move around on the spectrum of appliedness, physical and engineering scientists could emphasize the societal relevance of their work when doing so allowed them to win contracts or defuse protests; but they could also emphasize the fundamental character of their research when they needed to brandish "academic freedom" as a way to avoid oversight. In what follows, I will occasionally offer quotes from physical and engineering scientists or research policy administrators warning of threats to "basic research" such as budget cuts or societal litmus tests. Readers should keep in mind that the form of research in question was *ambivalently* or flexibly basic—these were warnings about threats to research that could be represented as either basic or applied, and threats to physical and engineering scientists' right to traverse the spectrum of appliedness at will.

Those who did routinely traverse that spectrum experienced the long 1970s differently from those who did not in part because of a postwar asymmetry in resource allocation that was more frequently contested from the late 1960s onward. In the early Cold War, national security agencies and corporations funneled resources into research that spanned the putative basic–applied spectrum. Even Charles Schwartz received military funding for basic research! In fact, at the 1970 APS meeting, Schwartz praised the military for supporting such work.[60] That is, the military could fund both basic and applied research; in contrast, the NSF was largely restricted to the basic end of the spectrum until the very end of the 1960s.[61] Similarly, academic physical and engineering scientists had license to conduct basic research on almost any topic or applied research that contributed to national security but didn't compete with industry; whereas academic applied research oriented to civil society was harder to find and fund.[62]

This asymmetry came under increasing pressure from all sides in the late 1960s. The military came to believe that basic research was not contributing to weapons systems as much as it had hoped.[63] Large firms came to believe that their basic researchers were not communicating enough with their manufacturing lines, and therefore that the beneficiaries of their basic research were smaller companies that could pirate the large firms' findings and commercialize them more quickly.[64] And reformers like Charles Schwartz argued that American researchers needed to focus more on topics of relevance to civil society—preferably assisted by funding for applied research from civilian government agencies such as the newly founded Environmental Protection Agency (EPA) or the Energy Research and Development Administration (successor of the AEC and forerunner of the Department of Energy).[65]

By the 1980s, civilian agencies were fully on board with funding applied physics and engineering science—provided that researchers in those fields could demonstrate their relevance to *economic* security.[66] Indeed, even military agencies now saw it as part of their mission to help civilian industries, such as semiconductor manufacturers, to compete with peers in other countries.[67] Some philosophers and historians have described this turn as symptomatic of an "epochal break"—sometimes dated to early 1980—in which technology and applied research gained the cultural prestige once accorded to basic science.[68] I contest the idea of an epochal break but agree that the research system post-1980 differed from that pre-1970 in a number of fundamental ways. For instance, corporate basic research began to decline after 1980, whereas academic research was becoming ever more oriented to (and funded by) industry. That shift was entailed by a new understanding of universities as sources of economic activity, rather than primarily as trainers of personnel.[69] The state started to conceive of itself as a parsimonious builder of networks connecting universities and industry, rather than as a generous market creator.[70] The fascination with networks stimulated decentralized models for organizing research, whereas hierarchical Big Science research organizations—typical of the early Cold War—lost favor.[71] "Network-ism" also accorded well with an increasing emphasis on interdisciplinarity. Finally, government, industry, academia, and a newly empowered cadre of venture capitalists became infatuated with biotechnology and other emerging technologies that promised to revive America's flagging economy.[72]

These same groups also became less patient, demanding faster returns on investment and greater accountability from scientists and engineers.[73]

No one has yet fully explained what happened between 1970 and 1980 to effect these transitions. I will not offer a comprehensive explanation either, but I will describe a number of mechanisms that would have to be part of a comprehensive explanation. In doing so, I build on the most perceptive scholars of the post-1980 paradigm, such as David Mowery and Paula Stephan, who argue that this sea change actually began around 1970 or a few years earlier.[74] Yet neither these scholars nor those who believe that the epochal break came in 1980 give much credit to the social movements and protests that roiled the American research system in the years around 1970. I will argue, in contrast, that we cannot understand the post-1980 paradigm without taking into account the messiness of the late 1960s—but we also can't understand it without tracking the *evolution* of macro- and microlevel adaptations to that messy environment over the course of the 1970s.

That is why the periodization of this book is the "long 1970s." I place the beginning of that era around 1966 with the peaking of military funding for basic research, the growing campus protests against that funding, and the military's public declaration of doubts about the efficacy of such funding. I mark its end with the Reagan administration's cancellation of alternative energy projects, the rapid fall in the price of oil, the break-up of AT&T, and other events of the early 1980s. Obviously, there were critics like Charles Schwartz before 1966, and Reaganomics didn't consume everything after 1983; other periodizations would have been just as valid. But by focusing on the long 1970s I hope to make central a period that has so far been at the margins of other histories. Most studies of postwar American science thus far have either *ended* with the "global 1968" or have *started* their story more than a decade later.[75] In contrast, I argue that it was not 1968 itself but rather the long stream of events *after* 1968 that prepared the ground for post-1980 technoscience.

The chapters that follow will present three main ways in which the stream of events in the decade-plus after 1968 mattered. First, responses to the protests and other varieties of pressure for reform of the late 1960s engendered counterresponses, frustration, and sometimes backlash. Second, the late 1960s pressure for reform elicited a wide variety of responses, which were then steadily winnowed down over the next decade. And third, events kept on happening, often with little connection to the cataclysms of the late 1960s: in particular, the oil crises of 1973 and 1979 reset the terms of debate. Obviously, these three mechanisms interacted: the experiments that arose post-1968 that remain today were the ones that could be adapted to the new crises of the 1970s and/or were the ones that were most palatable to scientists and engineers whose frustrated attempts at reform made them resent calls for societal relevance.

A LOT GOING ON

In other words, it was a complicated time—no monocausal story about the transition from pre-1970 to post-1980 will suffice. The intervening period was, as Philip Jenkins has put it, a "decade of nightmares" in which a series of perceived crises built on each other in nonlinear ways, often at a subconscious level of analogy and psychodrama.[76] The riots, assassinations, oil spills, and battlefield defeats of the late 1960s reinforced dire interpretations of the stagflation, oil crises, constitutional breakdown, and retreat from Vietnam of the 1970s.

At least early on, the war in Southeast Asia was probably the "nightmare" that personally affected square scientists and engineers the most. For a few, that was because of the protests and other reminders of their complicity in the war. Some responded to such criticism by becoming more hard-line, whereas others sickened of their defense research contracts and tried to cut their ties to the military. For most squares, though, the war mattered more because the winding down of US involvement in Vietnam (and the economic dislocations accompanying the war's denouement) led to a reduction in military spending and hence substantial cuts in R&D funding and the science and engineering workforce.[77] Suddenly, science and engineering students couldn't get jobs, established engineers were laid off by the thousands, and researchers couldn't afford the equipment needed for their experiments.

Funding cuts were especially deep for basic physical and engineering science research in universities and corporations. As the National Science Board (NSB) noted in 1976,

> Basic research expenditures per scientist and engineer in doctorate-granting institutions were almost 30 percent lower in constant dollars in 1974 than in 1968. The largest declines have been in the physical sciences. In private industry, inflation reduced real expenditures for basic research to the same level as in 1961.[78]

Those declines were shocking in part because they were so out of line with what had gone before. After World War II, and especially after the launch of Sputnik in 1957, American scientists and engineers had become used to rapidly rising public funding; for real funding to suddenly *fall* was galling for many.

Again, the NSB offered a representative view from the trenches: "Expectations by the science community regarding Federal support nurtured by long term trends in the '50s and '60s have been greatly upset since about 1966 or 1967. Confidence in the existence of a consistent *de facto* or implicit national science policy has disappeared."[79] That downturn ended up lasting fifteen years, as reflected both in budgetary numbers and in indirect figures such as the number of articles per year published in the APS's journals (figures 1.3 and 1.4). Because of the constriction of resources during the 1970s, ambitious researchers who had assumed perpetual access to new equipment and infrastructure suddenly had to make do with perpetual shoestring budgets.

Cuts in R&D funding also meant job losses, especially in the aerospace industry, the consequences of which will echo through this book's chapters. But money and jobs aren't the whole story. Science and engineering also faced a legitimacy crisis owing to those fields' perceived contributions to the war and other social problems. The legitimacy crisis was carried literally to scientists' doors by protests, sit-ins, takeovers of laboratories, and occasionally violent action such as arson and bombings.[80] On university

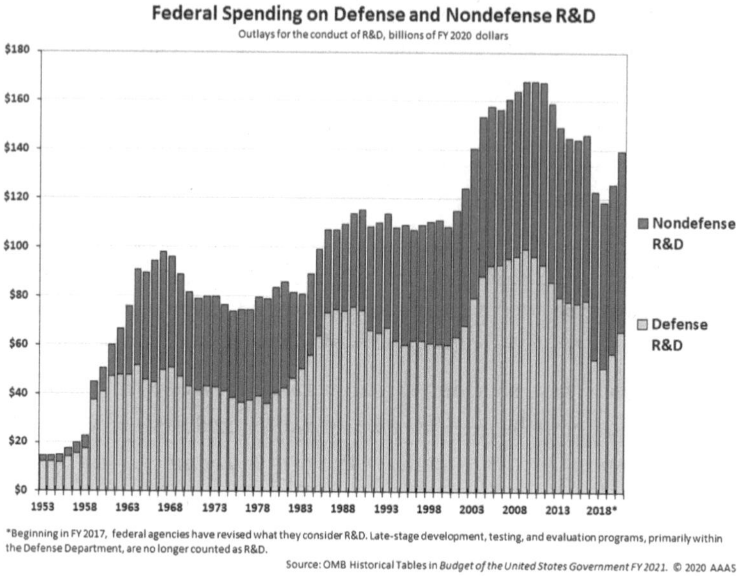

Figure 1.3

US federal budgets for research and development, 1953–2018. Note the drop-off after 1967, with return to 1967 levels only in 1985. *Source*: American Association for the Advancement of Science, "Historical Trends in Federal R&D." https://www.aaas.org/sites/default/files/2020-05/Function.png. Accessed October 12, 2021. Permission granted by the AAAS.

campuses, the funding and legitimacy crises also resulted in a rapid deterioration in science and engineering enrollments.[81]

Many square scientists and their allies saw protests and enrollment declines as a sign that young people were turning against science. But recent historiography has shown convincingly that in fact many young people—and their reform-minded elders—just wanted a *different* science, one more open to new ideas, less beholden to the military, more committed to solving civilian social problems.[82] We will see that many squares did embrace that vision of a new, more humane technoscience—but usually only temporarily. The aim of this book is to understand what encouraged them to move in that direction, and what drew them away from it.

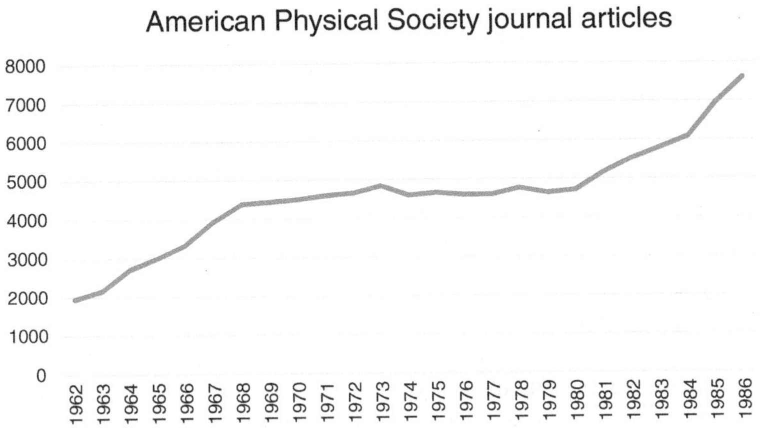

Figure 1.4
Numbers of articles in the APS's *Physical Review* family of journals. Note the steady rise before 1968 and after 1980 (a bit steeper after 1984), and the static level in between. *Source*: Calculated from the search engine on the APS journals website.

One important reason that many of these efforts were fleeting is that the crises of the 1970s cascaded, with each new crisis eclipsing the previous ones (figure 1.5). The long 1970s started with widespread discussion of an "urban crisis" that science and engineering could purportedly help solve; following the 1969 Santa Barbara oil spill, the public's attention moved toward the "environmental crisis," pulling effort away from urban problems. After the Organization of Arab Petroleum Exporting Countries (OAPEC) embargo of 1973, public discourse was dominated by an "energy crisis," which in part hindered action to resolve environmental problems unrelated to energy. Meanwhile, throughout this period the economy worsened and gradually became the center of political discourse; in figure 1.5 I've represented this by highlighting how the postembargo recession pushed attention to environmental and urban problems off the map. These swings frustrated many scientists and engineers. In a typical letter to the head of the NSB from 1975, one academic biochemist complained of "the lack of long-range planning and consistency with respect

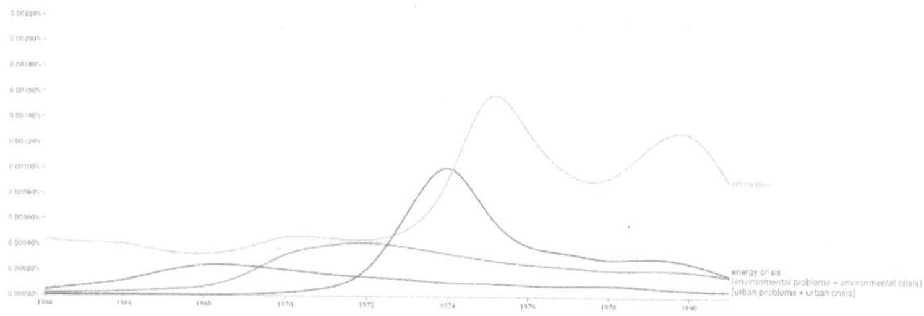

Figure 1.5
Cascading crises. I've chosen the search terms so that the four curves are at roughly the same scale; different search terms would change the scale but not the periodization of the different crises. The search terms also reflect discourse from the era: the "urban crisis" was used interchangeably with "urban problems," whereas references to the "energy crisis" were not interchangeable with the more rarely used construction "energy problems." *Source*: Google's Ngram service, American English, smoothing of 0.

to the allocation of funds and resources for basic research and education. Emphasis on 'environmental' work one day, emphasis on 'energy' the next . . . raises serious planning problems at the investigator and university levels with respect to 'basic research.'"[83]

The rapidly shifting winds of domestic politics meant that the decline in R&D funding was accompanied by administrative confusion and a growing regulatory apparatus, both of which made the remaining funding even more precarious. A few prior studies have, for instance, noted the upheaval wrought in the early 1970s by the Mansfield Amendment, which prohibited the military from funding research not directly related to its mission.[84] Little if any funding was actually cancelled because of the amendment, but fears that it would be led to much confusion. It also led to a handover of some programs from military to civilian funders. Universities such as Stanford that had become dependent on those programs dispatched streams of envoys to Washington to learn what the transition to civilian funding would portend. The news was not good: as the assistant director of Stanford's Center for Materials Research reported in 1970, "Many people in the DOD and Congress think that it is time both to allow the universities to 'cleanse themselves by fire'—i.e., suffer, and to cut off the academic fat cat professors who 'really don't produce anything of value for this country anyway.'"[85] Four years later things hadn't changed much: the NSF "dislike the 'prima donna attitude' of riding on past performance and an excellent faculty in order to justify a lack of specificity in our current proposals."[86] In other words, sticking to the early Cold War funding template wasn't going to work anymore.

Not that anyone—least of all the funding agencies—knew what would replace the old template. All that Tom Meloy, NSF's Engineering Division Director, could tell Stanford in 1975 was to "identify major sub-fields in terms of the technology's ripeness for development—identify these early and support them strongly" so that Stanford could "point to some coherent areas of results."[87] But when the university's envoys "probed a bit on what this would mean to the average faculty member making proposals, [Meloy] was more than a bit vague."[88] If we read between the lines we can see that Meloy's implied message was for Stanford to latch onto the crises depicted in figure 1.5. For example, in 1974 Stanford's envoys to Washington heard that the trendy directions in federal research funding were "'biological engineering' and 'scaling factors' in bringing laboratory and small-scale results into large-scale use [such as] microfabrication in electronics."[89] But they were also apprised that amid the debates about overpopulation and resource scarcity of the early 1970s (and especially after the oil shock of 1973), federal agencies' "priorities run to energy (naturally) and natural resources including food and minerals."[90] And amid the economic turmoil of the mid-1970s, they benefited from the Ford administration's view of "science and technology as a solution to our current inflationary problems [such that] scientific R&D will be . . . protected from the President's budget cuts."[91]

So Stanford's envoys to Washington brought back valuable, if somewhat ambiguous, intelligence. But they were also sent to DC to petition the school's funders to help its scientists cope with a growing administrative burden. As William Massy, Stanford's Vice Provost for Research, told the National Science Board in 1975,

> The general and administrative component of indirect costs . . . grew at a real rate of 5.9 percent per year [from 1967 to 1974]. We attribute much of this "negative productivity" to the adaptation to externally imposed requirements. . . . Examples of individually worthwhile but cumulatively burdensome requirements include: ever more extensive and complex human subjects reviews; animal care regulations; health, safety, and radiological hazards review and certification; affirmative action; and (potentially) increased requirements for property control and faculty time and effort reporting and documentation.[92]

Note Massy's ambivalence ("individually worthwhile but cumulatively burdensome"). Note also how these requirements squeezed Stanford from both ends of the political spectrum. The items at the beginning of Massy's list were broadly associated with social movements on the left. "Property control," however, is probably a reference to the "riot riders" that some conservative politicians attempted to insert into legislation so that universities where students had taken over laboratories would be ineligible for federal R&D funding. Similarly, "faculty time and effort reporting" reflects efforts to make academics more accountable in their spending of public funds—an initiative that by the 1980s and under the guise of protecting "taxpayers' money," came to be associated with the political right.

A distinct if related headache for square scientists and engineers was the era's economic tumult. The US economy faltered throughout the long 1970s but especially in the wake of the oil shocks of 1973 and 1979. At the same time, a number of important industries relocated abroad or were being increasingly outcompeted by their peers in other countries: textiles, steelmaking, consumer electronics, commodity chemicals, shipbuilding, car making. Other industries looked to be headed in the same direction, most notably semiconductor manufacturing.[93] We'll see that a rocky economy and increasing global competition offered opportunities for some square scientists and engineers; but for most it would have been one more reminder of lost privileges and growing insecurity. As Philip Jenkins argues, the national discourse of having won the Second World War and the postwar peace now rang hollow in light of headlines declaring military and economic defeat accompanied by disorder in the political system.

EXISTENTIAL SUCCESS

Jenkins aims to show that these defeats fed square Americans' desire for good news. In politics, that meant embracing the counterpunching optimism of Ronald Reagan. In science and technology, it meant an over-the-top celebration of new high-tech industries such as biotech and personal computing.[94] Jenkins's argument is useful, but it is also incomplete. It is important to remember that the "nightmares" he examines were not universally seen as such. Losses in US industrial competitiveness, for instance, have contributed to rising standards of living elsewhere, most notably in East Asia. Similarly, domestic cultural upheavals that many square scientists no doubt viewed with ambivalence or even distaste were necessary for other groups to make gains. From the mid-1960s through the long 1970s, the civil rights movement transformed into a "movement of movements" inspiring the antiwar and Free Speech movements, the environmental movement, second-wave feminism, the antipsychiatry movement, the gay rights movement, and movements for greater civil rights for prisoners, Latinx and Indigenous people, and others.[95] Notably, several of these movements made persuasive cases that scientists and engineers were complicit in perpetuating inequality and discrimination.[96]

This book is about people who generally did not join such movements and whose attitudes toward them ranged from positive to curious to icy (though rarely publicly hostile), but whose aims and actions nevertheless sometimes aligned with those movements' goals. We'll see that in science and engineering, as in American society more generally, these movements gave rise to more opportunities for women and people of color, and they encouraged attitudes toward minoritized views and lifestyles to soften, even among the squares. But by the same token this was the "me decade," or what Daniel Rodgers calls the "age of fracture," in which tolerance for individual expression merged seamlessly into neoliberalism's demand for individual initiative and disavowal of the kinds of collective expertise represented by the APS and other disciplinary organizations.[97]

The slippery slope toward cultural and economic neoliberalism presented a vexing long-term conundrum for science and engineering: how to be open and inclusive without losing some claim to a collective expert consensus on technical matters. Yet the people who appear in this book did not use the word neoliberalism, and most were probably unaware of that movement's leading thinkers. My argument is, in part, that many of the shifts that *later* came to be seen as part of a neoliberal agenda began in—were stimulated by—the era of "societal relevance" and initially had almost nothing to do with the views of Friedrich Hayek or Milton Friedman.[98]

That said, many of the people in this book were eventually embroiled in the growing influence of neoliberalism's interlocking cultural, political, and economic arms.[99] So let me define those terms before confusion sets in: by *cultural neoliberalism* I mean the trends described by Michel Foucault as the "entrepreneurial self"—the self as an individual project, uninhibited by solidarity with others.[100] By *political neoliberalism* I mean the campaign, associated with politicians like Ronald Reagan and Margaret Thatcher, to shrink the state and turn its responsibilities over to firms and individuals.[101] And *economic neoliberalism* here means denial of the possibility of a public good, thereby entailing a program to monetize all goods—including scientific knowledge—on behalf of private interests.[102]

Though it is largely beyond the scope of this book to prove it, I suspect that neoliberals were eventually able to coopt American science and engineering in part by amplifying the perception of defeat and crisis, and by taking credit for overcoming those perils even when the hard work of recovery was actually done by older forms of governance. In that way, neoliberals were able to label those older forms as responsible for defeat and incapable of victory. We will see some evidence of this in chapter

6, where Silicon Valley's late 1970s malaise was overcome through old-fashioned community and labor organizing and an activist state willing to curb pollution and discrimination—only for the Valley's rebirth to be credited to libertarian fantasies of heroic entrepreneurship.

I am on firmer ground in making a slightly different argument: namely, that the upheavals faced by square scientists and engineers were as much a product of their own *success* as of their nation's real or perceived defeat—and indeed that squares viewed the consequences of their success with a mix of dread and resentment. I term this condition *existential success*; it will be the guiding concept of chapter 4, but it overshadows the events of all the chapters. What I mean by it is that by 1970 the postwar research system had succeeded to the point of existential crisis. It reached that point because it had fulfilled the purpose for which it was erected: to provide weapons to defeat the Soviet Union on the literal battlefield and discoveries to fight the Soviets on the figurative battlefield of prestige. Obviously, postwar scientists and engineers did much more than just that, but that was the aim that made the postwar research system different from what came before and what came after. And by 1970 that aim had been met, at least to the satisfaction of much of the American political class and public. The question thus became, what now? With the existential justification for the research system gone, should it continue to exist? If it should, then what should be its new justification(s)?

In chapter 4, I illustrate existential success through the travails of the American space program in the 1970s; for now, let me illustrate it instead with a brief look at the evolution of the US nuclear weapons complex. Remember, there was no such thing as a nuclear weapon before 1945. America—certainly American science and engineering—had to be remade wholesale to bring one into being. After the war, the Manhattan Project could have been liquidated, like many wartime projects. But instead, it morphed into an archipelago of National Laboratories, weapons facilities, uranium mines, an AEC, and so forth. That complex was designed from the ground up to maintain the US lead over the Soviets—and anyone else—in nuclear weapons design and deployment.

By the early 1950s, the Soviets were pulling close to even in what was widely understood as an arms "race." Scientists representing a rather wide spectrum of political views—from Robert Oppenheimer on the left to Edward Teller on the right—pitched in to maintain the US lead.[103] Even so, with the launch of Sputnik, the Soviets seemed to have pulled "ahead," spurring a massive new round of investment in nuclear weapons development and in the US science and engineering enterprise more generally. That investment closed the "missile gap," and, by 1970, the superpowers were at a nuclear impasse.

That impasse presented the US nuclear community with two broad options. One group—associated with Teller—wanted to keep forging ahead to new technological vistas: anti-ballistic missile systems, neutron bombs, hyperaccurate counterforce missiles, nuclear earthmoving, and so forth.[104] The other group consisted of people like Hans Bethe and Herbert York: stalwarts of the nuclear establishment, not exactly doves, but over time finding themselves further and further to the left of Teller and his allies. That group called for something like an acceptance of existential success: lower the aims for development of new weapons systems, make sure any new weapons wouldn't create incentives for either side to attack, and advocate for shrinking both superpowers' nuclear arsenals.

Surprisingly, the latter group more or less carried the day, at least until the 1980s. Even though the Nixon administration's sympathies were closer to Teller and his allies, the administration recognized that scaling back weapons development would facilitate détente with the Soviets and save money under tight economic conditions. Thus, as Hugh Gusterson has shown, nuclear weapons development became much more boring than it had been in the cowboy days of the 1950s and 1960s: "polishing turds" as one of Gusterson's interviewees put it.[105] With this deprioritization of their existential justification, the National Laboratories were forced to find new missions. As a consequence, in the early 1970s they partly reoriented their research toward civilian priorities such as (as we'll see in chapter 4) improving public housing. The nuclear weapons complex had succeeded existentially, and out of that existential success came—perhaps temporarily, perhaps not—a new complex that put more emphasis on the problems of civil society.[106]

Crucially, that new complex retained much of the infrastructure of the old one. More generally, square scientists and engineers adapted to the long 1970s by taking features of the Cold War research paradigm—knowledge, personnel, materials, administrative procedures, experimental equipment, methods, funding streams, networks of trust, and so forth—and reconfiguring them for application to the urban, environmental, energy, or economic crises as well as other topics of societal relevance. That strategy is hard to argue with: square scientists and engineers were brought up in the Cold War paradigm, and its practices had been incredibly successful for them and (in their eyes) for their nation. As 1970 approached, many squares could see that the research system required some reform, yet still viewed it as a resource on which to draw. In fact, that was a widely held view. As we'll see in chapter 3, even some of the groups calling for radical change—for example, Students for a Democratic Society—agreed that the tools of Cold War science were so effective that they needed to be retained and adapted to research oriented to the problems of civil society.

We'll see that some of the squares who attempted to reform the Cold War research paradigm from the inside sincerely wanted to create a new paradigm. Others opportunistically adapted their research in the moment but made sure to leave a backdoor so they could return to their military patrons at a later date. Many who made attempts at societal relevance at the outset easily became frustrated and their commitment faded, though often without disappearing entirely. Overall, Cold War practices were adapted in a wide variety of ways in the long 1970s, corresponding to the variety of attitudes toward calls for reform. One of my primary arguments, though, is that by the 1980s most experiments with reform were weeded out, leaving behind only a few that were quickly abstracted away from their roots in the era of crisis. In particular, moves toward the market, and moves away from the collective disciplinary authority of professional science, were preferentially retained in the neoliberal 1980s, but without the aspirations for societal relevance that originally supported those moves. The experiments of the long 1970s created the conditions for neoliberal science to prosper, but few of those experiments were inspired by neoliberalism when they were first introduced.

OUTLINE OF CHAPTERS

Despite the rise of neoliberal individualism, though, scientists and engineers were—and still are—obligated to work collectively. The kinds of collectivities in which they work are quite varied. To survey the landscape of American physical and engineering science in the long 1970s, I have chosen to organize this book around some of the main varieties of collective technical effort. Chapter 2 starts with a very large-scale collective: the region. We will look at the region around Santa Barbara, California, as a microcosm for understanding how smaller-scale organizational types interacted in this era: public universities; think tanks associated with large, established firms; small, young companies; philanthropies; organizations associated with social movements; and so on. Most of the themes of the remainder of the book appear in some fashion in this chapter.

Next we move to a series of studies focusing on single organizations of different types. Chapter 3 looks at Stanford University: a private, wealthy, influential university that has been studied extensively as an exemplar of both pre-1970 Cold War science and post-1980 commercialized technoscience. This chapter asks what happened to Stanford between 1970 and 1980. In particular, I show that calls for interdisciplinarity arose as a means of solving Stanford's crises of funding and legitimacy but transformed over the 1970s as a way for Stanford to, in Matthew Jones's colorful phrase, "launder the Cold War for neoliberalism."[107]

Chapter 4 examines a government agency, NASA, as it adjusted to budget cuts and mission change in the wake of the 1969 moon landing. The problem of existential success was acute for NASA. It had staked its existence on beating the Soviets to the moon, and with that victory won NASA's scientists and engineers needed new missions to legitimate and fund their work. Crucially, NASA's success in getting to the moon demonstrated that those scientists and engineers were capable of solving problems. Thus, everyone with a social problem to be solved sought out NASA; and without a new mission the agency had little ability to resist their requests.

But scientists and engineers didn't answer such requests purely for instrumental reasons. Sometimes seeking out new missions also fulfilled squares' sense of patriotic duty or desire for interesting challenges. Chapter 5 looks at one such case involving the Nobel laureate Jack Kilby and a residential solar energy system he developed in partnership with his former employer, Texas Instruments (TI). In working toward a solar future, Kilby and TI borrowed heavily from the Cold War research paradigm while neglecting actors and methods located outside that paradigm. As a result, external support for TI's solar technology was shallow and evaporated quickly with the election of Ronald Reagan and the early 1980s' fall in oil prices, even though Kilby himself remained committed to the idea.

Kilby's confidence that TI could successfully adapt to the solar power market was based, in large part, on its earlier success in transitioning from oilfield services to semiconductor manufacturing, and in transitioning from making high-performance chips for military customers to making calculators and watches for civilian consumers. Chapter 6 therefore looks more closely at the US semiconductor industry's civilianization in a period of nascent globalization. The focus is Signetics, a leading-edge Silicon Valley firm of the 1960s brought low by the early 1970s recession. Signetics's owner, Corning Glass Works, sought a buyer and eventually settled on the Philips Corporation of the Netherlands. Curiously, few on either side of the Atlantic objected to the deal, even though Signetics was a primary producer of circuits for the US nuclear arsenal. Signetics's integration into Philips's multinational web of enterprises offers a way of understanding how American high-tech businesses changed over the longer term, on both domestic and international fronts.

Chapter 7 offers concluding remarks supplemented—like this introductory chapter—by a truncated case study. Here I look at Arthur D.

Hall, an influential early systems engineer who, like Kilby, became an independent inventor in the 1970s. Hall's relationship with his former employer, AT&T, was even more fraught than Kilby's: Hall was one of the expert witnesses who helped the Department of Justice break up AT&T in the early 1980s. With his earnings from consulting, patents, and book sales, Hall could afford to throw himself into a new and lonely obsession: the automation of agriculture. For most of the rest of his career he developed and tried to find support for an invention he called Autofarm, an integrated electronic system for growing crops. In the early 1980s he also tried desperately to gain a post in the Reagan administration. Hall and Kilby's abandonment of large organizations in favor of independent consultancies can be read as a harbinger of the new economy that Reaganism would accelerate. But both were also passionate advocates of environmentally friendly technologies that Reagan abandoned; both Hall and Kilby were bitterly disappointed that their green inventions were rejected by their erstwhile political allies. Their stories therefore remind us that the politics of the 1970s were profoundly different from today's, and that those politics momentarily had the power to move even rather conservative scientists and engineers toward the kinds of projects prized by today's RRI agenda.

The sites I've chosen are meant to be illuminating rather than representative. They are not randomly chosen, and most of them—for example, Stanford and NASA—were profoundly exceptional in some ways. I have tried to be systematic in selecting from the major *types* of organizations that square scientists and engineers at the time would've seen as their most likely employers: public and private universities, government agencies, big and small companies, and independent consultancies and think tanks. But in choosing cases from among those organizational types I was guided considerably by the accessibility of sources as well as insights and contacts acquired in my earlier projects.

The sources I draw on are a mix of published documents and unpublished materials housed in a variety of archival collections. Chapters 2 and 3 supplement documentary sources with oral history interviews. Chapters

2 and 3 also deal with people who figured in my earlier books on the history of microelectronics and probe microscopy. Those chapters, and to some extent Chapters 4 and 5, benefited from my ten-year association with the Center for Nanotechnology in Society at the University of California, Santa Barbara, which provided colleagues who gave feedback on drafts and funding to visit archives in California. Most of the sources for Chapters 4 and 6, meanwhile, were housed in personal and archival collections in Texas, where I was then working at Rice University.

That may sound like an unreliable way to choose cases, and perhaps it is. Yet if these cases are not representative, they are also *not unrepresentative*. The fact that I have run across square scientists doing socially relevant research in almost every archive that I've looked at makes it plausible that similar cases could be found almost anywhere. The test I would like readers to apply is whether they think US scientists and engineers in the 1970s would've recognized what was going on at these sites as continuous with what was going on at lots of other places. My sense is that they would. Where possible I will provide evidence that the actors I discuss believed that they were facing conditions similar to those of their colleagues across the country. Sometimes their responses to those conditions echoed moves made elsewhere, sometimes they were unique and idiosyncratic but still shaped by the national debate about what and how American scientists and engineers should research.

PROFESSIONAL SOCIETIES AND RELEVANCE TO SOCIETY

Readers will note that the overview in the previous section leaves out exactly the kind of organization that this chapter began with: professional societies like the APS. That's because the professional societies employed relatively few people and commanded relatively little attention from most square scientists and engineers on a daily basis. Little design or experimental work was done within the context of these societies, even if the societies provided the journals and conferences where such work could be discussed and vetted. Because my argument is partly that navigation of the 1970s was enacted through the design of experiments and other technologies, I choose to focus more on the organizations where those activities were more obviously carried out.

The professional societies did matter in other ways, however, so I will use this introductory chapter to peek inside them. They had roles in the accreditation of degree programs and licensing of practitioners; in communication, through journals and conferences; in political lobbying, though this had to be discreet and nonpartisan; and in the "boundary work" of claiming certain activities as their members' "jurisdiction."[108] The APS's activities in these areas have drawn a certain amount of attention from historians, some of whom acknowledge the turmoil of the long 1970s and the organization's attempts to navigate that turmoil through forays into societal relevance.[109] For instance, APS set up a Forum on Physics and Society in the early 1970s in addition to the committees on women and minorities mentioned earlier, and its Panel on Public Affairs increasingly addressed issues such as alternative energy and air pollution in that era.

For some square physicists, APS meetings were where they were most directly and disconcertingly confronted with pressure for reform. In a typical incident from 1973, a Los Alamos physicist

> presented an invited paper on Laser Fusion. Three men were allowed to stand in the front of the room during my entire talk, holding up placards, one of which read "Boycott the Los Alamos and Livermore Death Labs." This was done with the advance knowledge of the session chairman and apparently higher officials of the APS.[110]

Humans first created artificial fusion events by making and detonating thermonuclear weapons—hence the "death labs" label. But fusion energy was and is often presented as a form of clean energy superior to polluting energy sources such as nuclear plants and fossil fuels. The Los Alamos physicist could easily have framed his work as linked to environmental concerns rather than "death"—yet his letter to the APS studiously avoided doing so, and he apparently made no attempt to convince the protestors that he was on the side of "life." That is, both fusion researchers and their foes illustrate the excluded middle of American science and engineering in the 1970s.

Importantly, fusion researchers had other professional societies to turn to when the APS became less hospitable. The American Vacuum Society (AVS), for instance, was home to a significant number of people working on nuclear fusion. The AVS and a number of other societies were, like the APS, housed under the umbrella of the American Institute of Physics—yet they were much less committed to the "pure science ideal" that Joseph Martin paints as dominant within the APS.[111] Somewhat paradoxically, that meant that the APS was where debates about how to *apply* science to civilian concerns raged and polarized. Heated letters and direct-action protests of the sort that occupied the APS leadership are almost entirely absent from the records of more applied AIP organizations such as the American Geophysical Union (AGU), the American Association of Physicists in Medicine (AAPM), and the AVS. Instead, the leaderships of those societies struggled through the 1970s with maintaining their organizations' tax status, carving out distinctive subdisciplinary identities and deciding what stance to take on national adoption of the metric system.

Reformist calls for a new kind of science were not entirely absent from these squarer, more applied professional societies—they just generated less heat. Many of the AVS's members, for instance, were surface scientists who became more interested in catalytic phenomena in the 1970s as a result of public attention to (and funding for) pollution research. The AVS also sponsored conferences on solar energy. Tellingly, though, these conferences were attended mostly by employees of military–industrial firms trying to survive the 1970s recession by converting to civilian markets.[112] Similarly, the AGU responded to the civil rights, feminist, and environmental movements by forming both a Committee on Minorities and Women and a Committee on Environmental Quality during the 1970s.

Curiously, the AGU Committee on Environmental Quality was chaired by Fred Singer, then working for the EPA but later one of the "merchants of doubt" campaigning to undermine evidence for ozone depletion, the link between smoking and cancer, nuclear winter, and above all climate change.[113] Another future merchant of doubt, Fred Seitz, was similarly involved in the AAPM's attempts to combat legal limits on radiation exposure caused by public fear of radiation from X-rays. Seitz was a member of the AAPM Ad Hoc Committee on Radiation Dose to the General Population, and that committee's 1971 report features many of the tropes later associated with Seitz's climate denialism: for example, arguments that the benefits of exposure outweigh the risks, and that the risks from artificial hazards are swamped by those from natural sources.[114]

What this means, I think, is that while Singer and Seitz later became associated with bad-faith attempts to undermine reformist science, at this point in the 1970s they were constrained to contribute to square science's ambivalent adaptation to pressure for societal relevance. It was only with the lifting of that pressure in the 1980s, and the emergence of a countervailing pressure for market fundamentalism, that their second careers as merchants of doubt truly took off. There are some gaps in this story, though; I suggest that if we want to know more about the merchants of doubt and other conservative physicists, the smaller AIP societies would be a good place to look.

So far, unfortunately, these peer organizations of the APS are largely missing from the historiography of postwar American physical science. True, they were smaller and more specialized than the APS, and thus their archival sources are somewhat thinner and harder to interpret than those of the APS. But another reason historians have overlooked them is precisely because these societies were squarer, more applied, more oriented to industry and technology rather than prestigious discovery. That their squareness made them less visible was acutely felt by leaders of the smaller societies themselves. As a 1969 internal planning document for the AVS put it, "The AVS is small, new, and relatively unimportant by comparison with most technical societies. We have every opportunity to become an important society over the next few years, but we must guard against the 'snobbery' of pretending to be like the present 'big societies.'"[115] Yet for just that reason these smaller societies exemplify the way in which science and engineering were organized in the Cold War better than the APS. As Joseph Martin has shown, the APS was long dominated by an old guard that resisted the Cold War paradigm. The smaller physics societies adjusted more quickly, both to the early postwar situation and to the changed circumstances of the long 1970s.

WERE THE 1970S WHAT CAME BEFORE OR AFTER?

Still, the smaller societies were also capable of resisting the pull of their environment, both toward and away from the Cold War paradigm. As an illustration of their potential for resistance, I offer a final anecdote to bookend the long 1970s. Like Schwartz's peace symbol, this anecdote concerns a professional society meeting: in this case the November 1983 meeting of the AVS in Boston. In the immediate run-up to that meeting, the Federal Bureau of Investigation (FBI) had contacted the leadership of the AVS to request "registration badges, under assumed names, for agents to attend the symposium."[116] The AVS's then-president, John Arthur, "told them that anyone who desired to attend our meeting was welcome to do so. . . . The FBI . . . requested our cooperation in minimizing discussion of their presence at the meeting since this was part of a long and extensive investigation."[117]

That investigation came to a head when the FBI approached an East German physicist, Alfred Zehe, and arrested him at the conference hotel. Zehe was charged with espionage relating to a sting operation in which a civilian employee of the US Navy pretended to offer details about classified technologies to East German diplomats in the US and Mexico. The East German government had called in Zehe, who was living in Mexico at the time, to review the offered secrets and determine if they were real and valuable. Zehe's review of the "stolen" documents was the ostensible basis on which he was arrested. In hindsight, the government's case against Zehe looks flimsy. Despite prosecution claims, it seems unlikely that Zehe was a "'well-trained East German intelligence agent' and 'well-instructed in the techniques of espionage.'"[118] More likely he was simply a physicist interested in surface science who was called in to make a routine technological evaluation; his presence at the AVS thus meant no more than that he wished to attend the most prominent conference on those topics in the hemisphere in which he was living.

Why arrest Zehe so publicly and splashily, then? Why bother to register FBI agents for the AVS meeting instead of simply staking out Zehe's hotel? Two explanations seemed plausible, both of which were worrying to the AVS leadership. One was that the FBI thought that Zehe might have been there to meet US scientists who could be—or had been—convinced to provide classified secrets to the East German intelligence service. The other, less generous explanation, was that the FBI knew that Zehe was just an ordinary physicist but wanted to send a message to the East Germans (and their Soviet allies) and also to the American public, including American scientists and engineers: a message that the Cold War was back.

Superficially, the AVS might have seemed receptive to such a message, because the organization had deep roots in the Cold War paradigm for research. Its rapid growth in the 1960s was fueled by NASA's and the Pentagon's need for fast, compact circuits for rockets and spacecraft and for better information about the effects on materials of the vacuum of space. Yet despite the society's longstanding partnership with the national security state, the AVS leadership viewed the FBI's behavior in the Zehe case as a potentially existential threat. As Arthur put it in the AVS newsletter for November 1983,

> In the flurry of interviews with the press after the arrest, I was distressed to find many questions which implied that the AVS might have been lax in not "screening" the attendees at the meeting. I emphasized that the meeting was open to all, that the proceedings would be published in the open literature, and that it was not our practice to restrict attendance in any way. Last week, the FBI . . . requested the registration forms for the Symposium. Because . . . it is not the practice of the AVS to furnish an attendance list to anyone, I denied the request. I offered to allow the FBI access to our files in our office, but stated that we would release the files from our office only upon receipt of a court order, as suggested by our legal counsel.

This position was not favorably received by the FBI; however, as of today, a court order has not been presented to us.[119]

Let me quote Arthur at some length, because the position that he staked out in response to the Zehe affair offers some clues on how to situate the long 1970s within the longer arc of American physical and engineering science. As he explained to the AVS membership,

the most important issue [arising from the Zehe case] has to do with potential restrictions on information and/or participants at a symposium dealing with research which is clearly basic by all definitions currently used by federal agencies. The arrest of an alleged spy at our meeting suggests to the uninformed that he must have been seeking "high tech" information which should not have been discussed openly. While this is clearly absurd, we must make every effort to see that it is refuted whenever it is suggested. If the "spy" was not at our meeting because of its technical content, what then? He must have been there to make contact with a scientist with access to secret information, another spy, goes the logic. Possible, but I would prefer all of us to be regarded as innocent until proven guilty. The third possibility, that Mr. Zehe as a scientist simply chose to attend a highly regarded scientific meeting, should also be considered. I am concerned that the decision of the FBI to make the arrest at the meeting has in some sense compromised the good name of the AVS.[120]

Arthur framed the Zehe case in terms of scientific freedom and protecting the public good of basic research as well as (more parochially) the "good name of the AVS." Those are, in many ways, topoi typical of the Cold War paradigm. We will see, particularly in chapter 3, that some scientists and engineers who were reluctant to reorient their research to subjects of social relevance brandished their sacred devotion to basic research and scientific freedom as a way to ward off criticism. In that sense, the Zehe case is an example of professional scientists' ongoing resistance to the kind of threats to their professional authority that first arose in the late 1960s.

Yet one could equally frame the Zehe case as marking the decline of the Cold War as an organizing principle for American research. That might seem an odd framing given the timing of Zehe's arrest. After all, the Carter (1977–1981) and Reagan (1981–1989) administrations did much to revive the Cold War by boosting defense spending (including defense research funding); pursuing proxy wars with the Soviets in Latin America, Africa, and Asia; and adopting a more aggressive nuclear strategy. Reagan also applied his anti-Soviet rhetoric to domestic polices, for instance by characterizing 1970s-era regulations and agencies such as the EPA and the Solar Energy Research Institute as a threat to capitalism. Moreover, Reagan's election saw a return to the anticommunist suspiciousness of the McCarthy era.[121] Setting up an elaborate sting operation targeting East German intelligence and splashily arresting a hapless East German physicist were, thus, entirely in keeping with the resurgent Cold War of the 1980s.

Yet the AVS's response to Zehe's arrest shows that the Cold War's grip on scientific professional societies had also loosened considerably by that point. In the early Cold War, a professional society such as the AVS would likely have gone along with the FBI's demands in the Zehe case. Certainly, the APS acceded to McCarthyite campaigns against a number of its members in the 1950s.[122] It is true that some physicists and engineers had long worried that Cold War paranoia would restrict the free flow of scientific information and technological trade. But in the early Cold War, their ability and willingness to challenge such restrictions were limited. The public, vocal challenging of such restrictions—most notably protests against the Pentagon's ability to classify large swaths of university research—signaled the start of the era surveyed in this book. Thus, during the long 1970s, the policing arm of the national security state was pushed back, though it by no means disappeared.

When the Carter and Reagan administrations instituted new restrictions, therefore, square scientists and engineers faced a choice: go along, as many had in the 1950s, or resist chilling actions such as Zehe's arrest and fully embrace the freer atmosphere that had been gained in the late 1960s. I don't want to exaggerate—by no means did everyone choose the latter option. But it does seem that practitioners in the fields surveyed in this book were more ready to resist the national security state in the 1980s than they or their predecessors were in the original McCarthy era.

An incident from 1980 gives a sense of scientists' confidence in facing down a more muscular Pentagon:

> A NATO Conference scheduled for this summer by faculty at Berkeley and industrial members from Xerox and DEC was effectively shut off by DARPA's threat to withdraw their support from research programs at Berkeley. . . . The conference was to be held in Belgium in a NATO context, and was thus not particularly aimed at communist-bloc countries.[123]

This move provoked electrical engineering researchers at MIT to try to form a coalition to reverse the posture of the Defense Advanced Research Projects Agency (DARPA). As one MIT engineer wrote to another,

> I have conversed with Linvill at Stanford [John Linvill, one of the central figures in chapter 3], and [Edward] Wolf at Cornell (Nat'l Submicron Lab). Neither one is willing to restrict education and research activities, although both have been "careful" about communist-bloc access for some time now. . . . I agree with your suggestion that it is *now* time for leading academic institutions to make a concerted response to this problem. . . . I know that both Linvill and Wolf will want to cooperate, and I'm sure Berkeley must be upset. We too are very much involved, as we move toward our own expanded VLSI research (sponsored in part by DARPA!). I for one do not wish to become involved with a "classified" research program on campus, and I don't think the VLSI activity will succeed at MIT if it has to be off campus.[124]

Notably, DARPA's newly aggressive stance was motivated as much by economic as national security. Its leadership saw "potential competition from NATO member countries as severe as that from Japanese competitors, and is also concerned that technology could leak to the Soviet bloc countries through the French and Germans."[125] Protecting American industry was hardly a new motivation for national security restrictions on scientific information, although the barriers were raised higher in the late 1970s.[126] What's interesting here is the concerted and outraged move in the 1980s by very square engineers, themselves products of the Cold War, to attenuate those restrictions—where their predecessors (or in some cases their younger selves) had had no problem at all with "a 'classified' research program on campus."

Indeed, the Zehe case spurred widespread condemnation of the government's actions from the squarest bastions of American science and engineering. The AIP issued a public statement of "its intention to oppose vigorously all attempts to interfere with the free dissemination of unclassified scientific information at meetings of its member societies and in journals sponsored by itself or by its member societies."[127] Two weeks after Zehe's arrest, *Electronics* magazine editorialized against the Pentagon's "self-satirizing hodgepodge [of a plan] to cut off the flow of unclassified—that's right, *un*classified—papers."[128] *Science*—the journal of record of American science—hypothesized that "the Administration will use the Zehe incident to bolster its arguments for tighter restrictions."[129] And the editor of *Industrial Research and Development* warned that "Federal efforts to restrict the exchange of applied science knowledge . . . [jeopardize] what keeps the economic pot boiling in these United States."[130]

What does this tell us about the long 1970s? When historians first examined American science and engineering of the 1970s, their initial conclusion was that it was the "in-between decade," an interregnum in which radical changes were put forward only to be taken back in the 1980s.[131] One of the most salutary contributions of the literature on "groovy science" has been to show that something permanent *was* achieved in those years, even if the original motivations for those achievements faded over time. So far, the few studies of square scientists and engineers in the 1970s have noted some enduring changes but have pictured them as natural and inevitable steps in the evolution of research technology, encouraged somewhat by gradual changes in funding patterns.[132] Such studies don't confer much agency on the square scientists and engineers who contended with the decade's dizzying events. I will argue, however, that the squares were as experimental as their groovy peers in defining a research agenda attuned to the demands of the 1970s. Some of their experiments endured, meaning that the 1970s should not only be seen as the last gasp of the Cold War paradigm, nor as an interruption in that paradigm, but can also be understood as the beginning of something new and lasting.

The Zehe affair and the Berkeley–NATO conference mentioned earlier offer a Rashomon-like prism for understanding the long 1970s. On the one hand, they lend credence to the idea of the 1970s as an interregnum in the Cold War. Certainly, that is how many defense officials in the Carter and Reagan administration viewed the decade's developments. No doubt many New Left scientists and engineers such as Charles Schwartz also viewed the 1980s as undoing much of what they had gained. But the resistance of the AVS leadership and influential elite academic engineers (despite being funded by the military!) hints that we can also see the long 1970s as the start of something new. Square scientists who had begun the long 1970s ambivalent toward Schwartz and at ease "playing footsie" with the Pentagon were by the 1980s quite comfortable contesting the Pentagon's constraints on their agency. I'm doubtful that they had become any less ambivalent toward Schwartz and his allies, but they were less likely to see such scientific activists as an existential "problem" threatening their profession. Instead, they had moved toward a vision not dissimilar to that staked out by New Left: one in which American scientists and engineers were much less dependent on national security funders and much more oriented to the problems of civil society. That the squares had come to view the problems of civil society largely through the lens of markets rather than social movements is one of the (perhaps predictable) ironies of history.

But then, what is history if not a continuous flow of ironic and unintended consequences? The remaining chapters of this book will try to make some sense of those ironic twists and turns of the long 1970s, though our understanding of that decade will only ever be partial. It was an era in which much was alien and hard to comprehend; yet it was also the era in which many of today's commonplaces made their first experimental appearances. To travel to that time is to enter an uncanny valley between familiarity and strangeness, where today's rules apply—except when they don't.

2

TURN ON, TUNE IN, START UP: THE EXPERIMENTAL LIFE IN SANTA BARBARA

The Cold War can be pictured as a clash among nations and ideologies, but it was also the playing field for advancement and competition among subnational regions. Internationally, subnational conflicts sometimes became entangled with the interests of the superpowers.[1] Within the United States, new postwar industries and defense strategies benefited some regions much more than others. There is a robust literature on both economic geography and scientized regional planning in the United States and elsewhere from the early Cold War onward, some of it from the vantage of history of science and technology and some not.[2] Studies of the regions around Los Angeles, Boston, and San Francisco Bay/Silicon Valley, in particular, show that regional military–industrial agglomerations produced network effects that amplified economic activity and the capacity for research and innovation.[3] A number of smaller regional clusters have been examined as well, some because they failed to replicate what their leaders thought had happened in Silicon Valley and some because they succeeded despite lacking the supposed advantages of Silicon Valley.[4]

So the Cold War military–industrial complex clearly fed many regional agglomerations of economic and technological activity. But some canonical regional clusters—including Silicon Valley—were starting to form well before World War II or were only indirectly influenced by the postwar superpower confrontation. These include the Michigan–Ohio automotive cluster, the Philadelphia–New York chemical/electrical corridor, and the Dallas–Houston "energy capitals," all of which predated the war and spawned technosocial innovations that shaped the Cold War at least as much as the Cold War shaped them.[5]

More important for this study is that America's regional clusters had a life *after* the Cold War. Some, such as Detroit, went into decline because of globalization and other trends accelerated by the twilight of the superpower conflict. Others somehow reinvented themselves as clusters for new, less overtly militarized industries. The clusters that dominate the biotechnology and to a lesser extent the dot-com industries, in particular, seem to be built on the remains of military–industrial clusters around Boston, San Francisco, San Diego, and Washington, DC.[6] Just how these regions made that transition is unclear, though. Much more attention has been paid to how such regions arose in the first place (and why they didn't emerge elsewhere) than to how Cold War regions adapted to the post–Cold War era. When this issue has been broached (in both popular and academic histories), the counterculture has received considerable credit for some regions' transformations, especially the San Francisco Bay Area's.[7] As this chapter will show, the counterculture's exploration—and often commodification—of alternative tastes and lifestyles did indeed lend itself to both high- and low-tech entrepreneurship that facilitated shifts in regional economies.[8] Yet there were also regional clusters that developed in the late Cold War around technologies in which the counterculture was deeply interested—for example, personal computing and computer graphics—*even* in places that lacked much countercultural presence (such as the area around the University of Utah).[9]

Thus, square scientists and engineers with little interest in the counterculture must have played some role in shifting regional clusters to a post–Cold War economy. This chapter looks at a group of squares who helped transform their region's industrial base by knitting together networks spanning different kinds of organizations: universities, think tanks, established firms, philanthropies, civil society organizations, government funding and regulatory agencies, hospitals, start-ups, and so forth. They built those networks by experimenting with new ways of doing research and development—experiments stimulated by the crisis conditions of the long 1970s. Often, the same individuals were connected to a variety of different experiments, which to some extent made those experiments mutually reinforcing. Over time, though, the crises faded, and thus most of the experimental responses to crisis conditions petered out. Those that remained are the ones prized by regional boosters, venture capitalists, and university administrators today—that is, those that are seen to contribute to the formation of clusters of high-tech start-ups. But if promoters of regional clusters are looking for lessons from this chapter on how to "incubate" start-ups, they will be disappointed. Marginality, eccentricity, friction with management, and organizational upheaval were the main factors at play here—conditions few actors would or could wittingly reproduce.

THE THINKING INDUSTRY

This chapter will make these points with reference to a case study of the region around Santa Barbara, California. This is *not* a region that generally figures in discussions of economic geography, though to some extent it can be seen as an extension of the Los Angeles region and therefore is encompassed by the literature on that area's aerospace and other industrial clusters.[10] There is also a cluster of a couple dozen computing start-ups in Santa Barbara that arose in connection with the ARPANET (the military-funded kernel of what became the internet).[11] One early node of the ARPANET was based at the University of California, Santa Barbara (UCSB); its leader, Glen Culler, was a friend and collaborator of some of the people in this chapter. Today there is also a cluster of nanotechnology, materials, and instrumentation start-ups located near the Santa Barbara airport; my own acquaintance with Santa Barbara's high-tech economy came through that cluster, and this chapter can be read as one of its origin stories.

We begin, however, with a much stranger "industrial" cluster that exemplifies the economic geography of science in Cold War America: the Southern California think tank industry. Though the think tank is today a globally distributed organizational type, its origins lie in early postwar Southern California—the first and most famous, the RAND Corporation, having been founded in late 1945 in greater Los Angeles.[12] RAND was originally a subsidiary of Douglas Aircraft but formally separated in 1948, though it remained in the region to be close to Douglas and other growing aerospace firms. RAND's approach and personnel, however, were closely tied to bastions of the East Coast establishment such as MIT, and its analysts quickly became influential in Washington, DC.

RAND's outsize role in Cold War policy and in particular nuclear strategy is well known, not least because it employed a generation of defense intellectuals who came to dominate economics—and even, in the case of Daniel Ellsberg, national newspapers—from the 1950s through the 1980s.[13] Some of those who left RAND, most notably Herman Kahn, founded their own think tanks and spread that organizational model around the United States and around the world.[14] By the 1980s the think tank was a portable enough organizational form that it was appropriated by climate denialists as a means of undermining environmental regulation.[15] Yet we still do not have a very good sense of how this industry arose nor how its participants distinguished themselves from practitioners of seemingly related industries of a similar vintage, such as management consulting.[16]

At any rate, it did not take long after RAND's founding for Douglas Aircraft's military–industrial peers to decide that they needed their own units that could provide strategic advice, rooted in cutting-edge physics and the more physics-oriented social sciences, while also representing their parent firms' interests in policy circles in Washington and other seats of government. In the early Cold War it became popular to keep these units physically distant, and hence (it was thought) intellectually independent, from corporate headquarters. Indeed, some East Coast firms put an entire continent between headquarters and their new think thanks. Thus, many of RAND's peers were also founded in California: close to the original think tank and to the Los Angeles aerospace cluster, but also far enough away from headquarters to encourage independent thinking.

In that way, Southern California's think tank cluster was seeded by outposts of East Coast firms—just like Northern California's microelectronics cluster.[17] General Electric (GE) and its Santa Barbara–based think tank, TEMPO, exemplified this dynamic. As one of the leading manufacturers of fighter jet engines, nuclear reactors for submarines, and other military–industrial technologies, GE was eager for strategic advice that could better link it to the aerospace industry and policy circles in Washington, DC. So in 1956 it raided a few physicists from RAND and engineers from Douglas and set up its own California outpost, TEMPO. As a 1958 brochure put it,

> TEMPO is concerned with the world of 5 to 15 years from today. Social, economic, military, technological, and political trends which may affect a particular problem are projected to establish the environment in which the problem must be addressed. Alternative policies and systems reflecting this environment are then developed and tested.[18]

TEMPO was, therefore, an early facilitator in moving scenario planning techniques from the world of nuclear wargaming and into corporate strategy.[19] We will see in chapter 5 that scenario planning was one of many Cold War techniques that corporations applied to the new socially relevant technologies of the 1970s.

It's not entirely clear why GE chose to locate TEMPO in Santa Barbara. Proximity to the Southern California aerospace industry (and Camp Cooke—now known as Vandenberg Air Force Base—north of town) was surely one rationale. But the idyllic, even sleepy, milieu also played some role. As the 1958 pamphlet put it, "The Santa Barbara location was chosen to provide isolation from the day-to-day interchange with engineering and manufacturing functions of General Electric and to encourage independent and objective studies by the technical staff."[20] Santa Barbara was and is a relatively small city—there were roughly 169,000 residents in the whole county in 1960—squeezed between the beach and picturesque foothills. It is a beautiful place to live, removed from the sprawl that characterizes the Los Angeles–San Diego or San Francisco–San Jose corridors. That appeal was an important draw for TEMPO and later think tanks. For instance, EG&G (a think tank co-founded by MIT's Harold "Doc" Edgerton) established an office in Santa Barbara in 1959 because the "logic was, if I [EG&G executive Sandy Sigoloff] move to Santa Barbara, I can hire really great people because people really like to live in Santa Barbara."[21]

One member of this nascent industrial cluster was Defense Research Corporation, a Virginia-headquartered think tank working on topics such as counterinsurgency, computer-aided literature searching, and antiballistic missile technology. Like EG&G, and at about the same time, Defense Research formed a Santa Barbara outpost. In the early 1960s, Philip Wyatt, a young theoretical physicist who had recently obtained his PhD from Florida State University, joined that outpost's antiballistic missile group and started exploring ways to track missiles with lasers. In the spirit of "independent and objective studies," however, Wyatt began pondering other applications of lasers. By 1967 he had come up with a laser inverse-scattering method for particle detection and characterization—that is, a means of shining a laser through a fluid containing suspended particles and then detecting how the laser scattered from those particles to calculate the particles' size and shape. One could imagine a number of applications for such a system, but in the context of Defense Research Corporation, Wyatt's initial idea was to use it for detection of pathogens for biological warfare defense.[22] That was, after all, the kind of application that would attract interest from his employer's patrons. And in fact, Wyatt was soon able to find Army funding to develop the technology.

Because of deteriorating relations with his supervisors, however, Wyatt quit Defense Research and joined EG&G. As much of the literature on Cold War tech clusters notes, the ability to leave one firm and quickly find work with another local employer in the same industry is an important factor in any cluster's success. EG&G's contracts were mainly with the

AEC, for whom it recorded data from nuclear weapons tests; but they "liked [Wyatt's] ideas and considered the [biowarfare defense] contract a wonderful opportunity."[23] During his brief time there, EG&G gave Wyatt some management training and the opportunity to study with microbiologists at UCSB and University of California, Los Angeles (UCLA). Both proved essential when Wyatt then formed his own start-up company, Science Spectrum, in 1968 to commercialize laser inverse-scattering particle detection.[24]

Science Spectrum's founding came at the start of the retrenchment in defense research funding and growing public discontent with the Cold War research paradigm that set the stage for the events covered by this book. As Jennifer Light has shown, defense intellectuals at think tanks always had some interest in civilian matters, particularly urban and regional planning, but that interest was strongly connected to nuclear strategy early on. By the late 1960s, however, many of those defense intellectuals were retooling to focus on issues of civilian concern such as "citizen access television" (i.e., cable TV).[25] Whether because of the decline in military funding or growing opposition to the Vietnam war, think tanks and defense intellectuals had to show that they could tackle more than just military topics. That is, presumably, why Defense Research changed its name to the more generic General Research Corporation in this period.

While military funding was being cut, some civilian agencies were expanding or shifting their priorities in ways that made them attractive funders or partners for Science Spectrum. The NSF, for instance, began moving beyond its traditional role as a small funder of basic research by individual physical and life scientists by expanding its support for civilian applied research.[26] The EPA formed in 1970, partly as a result of the previous year's disastrous oil spill in Santa Barbara; it quickly became another promising funder of physics research.[27] Funding for the National Institutes of Health—and biomedical funding generally—also began a long upward trend, especially after Richard Nixon's declaration of the "war on cancer" in 1971; as we'll see, the conditions of the long 1970s stimulated some engineering and physical scientists to seek NIH money.[28]

Thus, although Science Spectrum's core technology was conceived in a military–industrial context, in the early 1970s the start-up focused more on civilian applications and agencies. For instance, Wyatt tried to capitalize on the environmental research boom bolstered by the 1969 Santa Barbara oil spill: as he put it in a public talk in 1970, "It seems that Santa Barbara is becoming a center for the study of ways to fight pollution, and this company is in an excellent position to help in that research" by using its technology "to study smog and haze particles in the atmosphere and find out what causes this pollution."[29] Indeed, Wyatt believed that Science Spectrum and other environmentally oriented start-ups offered a solution to the economic dislocations of the early 1970s: "the federal government should subsidize companies that can solve pollution problems, thus creating new jobs and making the effort to dispose of waste a stabilizing influence in the economy."[30]

Given the dynamic light scattering (DLS) technique's origins in antibiowarfare applications, it is perhaps unsurprising that Wyatt had more success with biomedical research customers. By 1972, Science Spectrum was working with Santa Barbara's Cottage Hospital to apply DLS to assay "the relative efficiency of a panel of antibiotics against patient bacterial specimens."[31] In 1975, the company received a grant from the Food and Drug Administration (FDA) to develop a means of detecting "veterinary drug residues in food producing animals" and animal products—research that Science Spectrum conducted in collaboration with the FDA and Department of Agriculture.[32] The company also worked with researchers at the Kettering-Meyer Laboratories and the Southern Research Institute on detection of chemotherapy compounds in cancer patients' blood.[33]

With its biomedical applications and customers, Science Spectrum might have come to be seen as an early player in the biotechnology industry had it been located in the San Francisco Bay Area instead of Santa Barbara. Like the first (recognized) biotech company, Cetus, its founder was a physicist looking to collaborate with life scientists.[34] Like some biotech entrepreneurs, Wyatt was a maverick and a bit of a showman: while getting his PhD he was written up in the *New York Times* for his unorthodox ideas about terrestrial antimatter impacts, and while at Defense Research he had been a finalist in NASA's search for the first class of scientist–astronauts.[35] And like the early biotech entrepreneurs Wyatt reached out to the budding venture capital industry. Tom Perkins—a legendary venture capitalist investor in both the microelectronics and biotech industries—was on one of Science Spectrum's early boards. Another venture capital outfit, the Value Line Development Capital Corporation, invested $500,000 in Science Spectrum in 1973 and handpicked the company's new president, Harry Brown.[36]

Science Spectrum never did attain anything like the success of Cetus, though. Yet early on it did show significant promise, which explains why venture capitalists like Perkins took a chance on it. For example, in 1972 its Differential II product won an Industrial Research 100 award (better known by its current name, the R&D 100). The Differential II benefited from another characteristic Wyatt shared with early biotech entrepreneurs: the ability to recruit academic researchers. In particular, Wyatt spent several months in 1969–1970 persuading David Phillips, an assistant professor of physics at UCSB, to consult, and then work, for Science Spectrum to improve the Differential II and other products.

Phillips proved especially skillful in adapting the Differential II for use by nonphysicists. As the firm discovered when it hired a microbiologist, that discipline's "inexperience with light scattering apparatus emphasized the need for a reliable, easy to use instrument instead of the maze of wire, tape, switches, and dials that had enveloped the original photometer" (figure 2.1).[37] A good measure of Phillips's importance is that *Industrial Research* gave him top billing, as "principal scientist," in its IR-100 Award citation.[38] Another indicator is that the main conference room at Wyatt Technology, Science Spectrum's successor company, is even today named after Phillips. Yet another is Wyatt Technology's own narrative: "[Philip] Wyatt began his investigations of the practical applications of the inverse scattering problem in 1967 with studies of means to differentiate bacterial species from one another. Together with his colleagues, most importantly

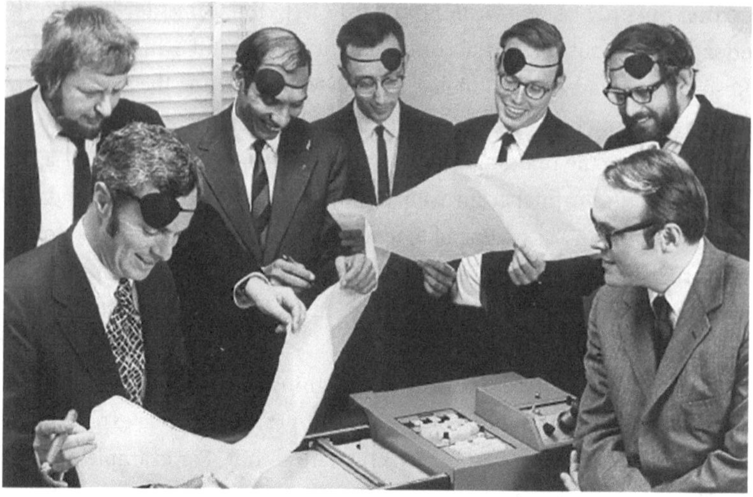

Figure 2.1
Members of the Science Spectrum staff inspecting one of the company's early products. Philip Wyatt is seated at left, with cigar; David Phillips is standing, second from right. *Source*: Courtesy of Wyatt Technology (http://www.wyatt.com).

Dr. David T. Phillips, he modified a traditional light scattering photometer to incorporate a laser light source."[39]

THE PARAPSYCHOLOGY INDUSTRY

Thus, there is good reason to believe that Phillips was a competent and serious scientist who made important contributions to DLS instrumentation. Yet we can infer that at least some of his UCSB colleagues doubted Phillips's competence, or at least his fitness for their profession, because his departure from the university was not by choice. The UCSB physics department was at the time in the middle of a long-term strategic reorganization interrupted by an acute budgetary crisis. The confluence of rising departmental standards and falling departmental budgets narrowed the tenure prospects for all of the department's assistant professors—but especially for Phillips, whose forays to the outer boundaries of squareness led some of his more obviously square colleagues to view him with suspicion.

To understand Phillips's precarious position, we have to shift our focus from the think tank industry and its spin-offs to the UCSB physics department. Again, the literature on Cold War high-tech clusters says that they almost always had at least one university with a global science and engineering profile nearby.[40] By that standard, Santa Barbara was an unlikely place for a tech cluster to emerge in the 1970s. Today, the university—and in particular the physics department—has a global reputation bolstered by a crew of current or former faculty members who have won Nobel Prizes as well as less well-known but influential entrepreneurs, science administrators, and of course researchers. But when UCSB's physics department was created in 1960, the school's reputation was for surfing, not scholarship. The department's early leaders therefore decided that the best way to change their fortunes was to follow the model forged by Stanford, MIT, Cornell, and other successful postwar research universities (see chapter 3). That meant deemphasizing undergraduate teaching in favor of PhD training, research, and securing federal grants (particularly from national security oriented agencies).[41]

Phillips had been an early hire in that campaign, but by 1970 he had supervised only one PhD dissertation, and his publication record was undistinguished.[42] In fact, he published much more *after* he moved to Science Spectrum than before. Moreover, the articles that Phillips published while he was at UCSB were a poor fit with the department's strategy. He was probably proudest of an article (for a journal focused on undergraduate pedagogy) describing how to build a cheap "poor man's nitrogen laser"—an insignificant, or even counterproductive, topic in the eyes of colleagues who were trying to move their department away from its collegiate past by securing lavish federal research grants.[43]

Phillips's lack of enthusiasm for the role that his department expected of him became especially problematic when Vincent Jaccarino became department chair in 1969. According to a departmental history, "Largely as a result of Vince's leadership, there was a significant change in the department's attitude regarding appointments and promotions—particularly for assistant professors. The department realized it had attained the position that allowed it to attract and promote only those physicists that met the highest standards. This meant making some difficult decisions but the department has been the stronger for it."[44]

But Phillips faced an additional problem beyond his inability to meet "the highest standards" in terms of publications, graduate students, and grants: namely, his deep interest in parapsychology and mystical experience was a cause for alarm among some of his senior colleagues, including Jaccarino. Phillips's involvement with parapsychology probably took off only after he moved to Santa Barbara in 1966; by the early 1970s, though, he was sufficiently ensconced in the area's parapsychology community to become research director of the Southern California Society for Psychical Research (SCSPR). He was also acquainted with the West Coast's leading lights of early 1970s parapsychology, including Charles Tart at the University of California, Davis; SRI International psychokinesis investigator Hal Puthoff; artist and remote viewer Ingo Swann; and Elizabeth Rauscher, a leading member of Berkeley's Fundamental Fysiks group.[45] Phillips even spoke at a workshop on parapsychology at Esalen, the Central California institute promoting expansion of human potential through a variety of mind-altering means.[46] Also on the program at that event was another Fundamental Fysiks researcher, Nick Herbert; headlining the workshop was the Israeli showman and spoonbender, Uri Geller.[47]

At this point, some readers may ask why I would categorize Phillips as a square scientist. Wasn't parapsychology a countercultural pursuit? Well, yes and no. Some who embraced parapsychology did so in parallel with a wider-ranging rejection of the establishment and conventional lifestyles. But parapsychology long predates the 1960s counterculture, and it has always been embraced by elements of both the establishment and the antiestablishment of science.[48] Even relative to that long history, though, the long 1970s saw exceptionally widespread interest in parapsychology, not least among otherwise quite square scientists. Harry Collins's more than forty-year ethnography of gravitational radiation detection physics, for instance, shows that in the 1970s it was acceptable to bring purported psychics into laboratories to try to influence experimental equipment with their minds—something the gravitational radiation physics community would've regarded as preposterous both before and after the long 1970s.[49]

Phillips, with his daily costume of black suit, white short-sleeved shirt, and skinny black tie, seems to have identified with the squarest wing of the parapsychology community. Here, again, I should emphasize that square and groovy were relational attributes: Phillips was much squarer than most of the Fundamental Fysiks crew, yet at least some of his departmental colleagues saw him as too groovy to be taken seriously. As he put it in a letter to a fellow parapsychology researcher, his department contained "several distinguished physicists [Phillips mentioned Vincent Jaccarino by name in this regard] who are not sympathetic to psychical research. Involvement has psychological risks, these are strong willed people."[50] Nevertheless, Phillips persisted out of a strong personal desire for "spiritual development" and greater "intuitive awareness."[51] But Phillips also believed—as a *physicist*—that psi phenomena merited scientific investigation, and that some mystical traditions' views converged on quantum mechanics and scientific cosmology.[52]

In other words, Phillips adopted what he believed to be a scientific approach to psi phenomena that cut between credulity and skepticism. His writings show that he clearly believed in the existence of some phenomena that would usually be labeled paranormal. But at the same time, he was disappointed with "the everyday occult mish-mash" and "foolish trickery that abounds in southern California" concerning paranormal phenomena.[53] He summarized his philosophy (and his precarious career prospects) thus in resigning from SCSPR's board of directors in 1974,

> My employment situation has remained unsettled and several psychic interests have developed here in Santa Barbara and I find that I just don't have time for real involvement in the societies [*sic*] activities in Los Angeles. . . . Uncritical belief in any and all claims can be reserved for religious groups, systematic examination of minute details of particular phenomena may be left to the scientific journals, but a good down to earth examination of the local psychic folk can do a lot to put the whole field in perspective for society members and the man on the street.[54]

Despite the shadow that it cast over his career, Phillips's interest in parapsychology lasted the rest of his life. As his son Glen (frontman for the band Toad the Wet Sprocket) put it in a song memorializing his father, "Dad's in the garage/place of legend and fable. . . . 43 boxes of textbooks on physics/20-odd more on parapsychology/Well there's DTP's papers, DTP's projects."[55]

With his employment "unsettled," Phillips looked to the projects that he cooked up in that garage to bring in extra income. Notably, many of those semicommercial projects combined the "textbooks on physics" with those on parapsychology. At about the same time that he left UCSB's tenure track and joined Science Spectrum, Phillips founded his own start-up, Glendan Company. Named after Phillips's sons, Glendan was what management scholars call a "lifestyle company"—that is, the line between its products and Phillips's hobbies was blurry.[56] Accordingly, some of those products were sold for income, whereas others were given away—with the local parapsychology community a frequent beneficiary of Phillips's gratis tinkering.

One of the projects that Phillips was most keenly interested in, for instance, was an amplifier that he built for Bill Welch, a devotee of the "voice phenomena" (i.e., sounds found in tape recorder hiss and other electronic noise that Welch believed were of paranormal origin). Phillips's correspondence with and about Welch speaks of his desire to know what the voice phenomena are and how to make them clearer; Phillips doesn't seem to have been interested in turning his amplifier into a commercial product or attempting to derive any revenue from it.[57] Yet he did harbor larger ambitions for Glendan. In the mid-1970s Phillips solicited at least two angel investors to put money into the company: Winifred Babcock (a New Age author) and Irving Laucks, a prominent chemist, industrialist, and local peace activist.[58] His approach to Laucks even resulted in a formal "Agreement Regarding ESP Teaching Devices" in which Phillips agreed to pay Laucks a royalty on Glendan's sale of an extrasensory perception (ESP) testing device.

In other words, Phillips and other Southern California parapsychology enthusiasts were doing the kinds of things that lead to the formation of regional industrial clusters: forming "homebrew" clubs, trading equipment and knowledge for free, while also starting companies in their garages. That there was a regional aspect to interest in parapsychology is unsurprising. California was widely recognized as a leading site for a variety of related social movements that intersected with or spilled over into parapsychology. It's important not to exaggerate—parapsychology enthusiasts and even researchers could be found almost everywhere in this era, and Phillips certainly kept his eye on developments around the world. But there's little question that it was relatively easy for Phillips to find interested audiences and fellow parapsychology researchers in his own region.

Despite these similarities to other innovation clusters, historians of science and technology have been reluctant to frame parapsychology and related social movements as forming the basis for an *industry* with the same kinds of regional network effects as the microelectronics or personal computer industries in Silicon Valley. Yet if you added up sales of New Age books, crystals, aromatherapy oils, and so forth, you would have a modest but profitable industry; if you include related sectors such as the yoga industry, you get something quite impressive. (Today more than 10% of Americans practice yoga.) Historians of groovy science, however, have been more interested in the epistemic than the commercial implications of parapsychology research; whereas historians of commercialized technoscience have ignored parapsychology in favor of industries like biotech that had more prominent political and scientific backers. If we drop some of our prejudices about what constitutes technology and innovation, though, we might learn a great deal from the similarities between parapsychology and canonically high-tech industries such as biotech.

Indeed, Phillips behaved much like any budding tech entrepreneur, just one who happened to operate in the parapsychology and New Age spirituality market. For instance, when he saw articles in the *Los Angeles Times* and *Psychic* magazine on Soviet experiments with acupuncture and detection of "bioplasm," he developed and then sold a gadget based on the Soviet research.[59] He advertised this handheld transistorized pen, the

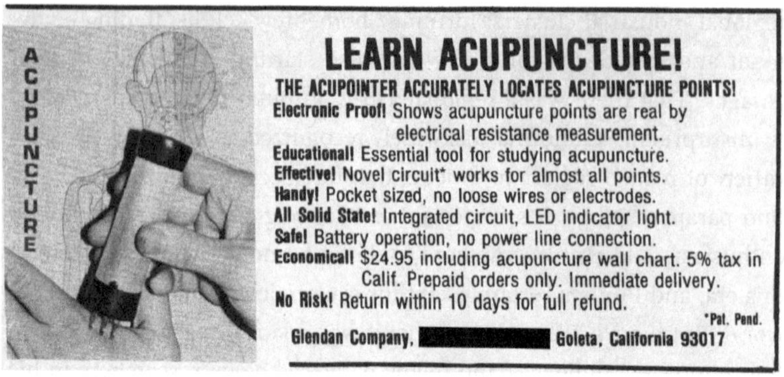

Figure 2.2

Ad for a transistorized acupuncture-point locator from the Glendan Company. The address of Glendan Company has been removed; note that Glendan Company no longer exists and this product is not available. Permission to republish granted by Linda Phillips. *Source*: *Today's Chiropractic* 2.5 (October/November 1973): 32.

Acupointer (figure 2.2), in several different New Age and chiropractic journals in 1973, and even wrote (on Glendan stationery) to the editors of the *Journal of the American Medical Association* chiding them for dismissing acupuncture and announcing, "To meet the need for an effective, economical device to locate these [acupuncture] points, Glendan scientists have developed the ACUPOINTER."[60]

Phillips also sought to create network effects by taking advantage of the regional influence of UCSB. Although he had left the tenure track in the physics department, Phillips still supplemented his income by teaching at the university. In part he did so through the Tutorial Program, UCSB's testbed for alternative pedagogical philosophies. Phillips's entrée to the tutorial program was his friendship with Robert Morris, a well-known parapsychologist who was employed there from 1974 to 1978. (In 1985 he took the world's first endowed chair in parapsychology, at the University of Edinburgh.) For several years, Morris and Phillips cotaught a course on parapsychology in the tutorial program.[61] They also collaborated on research, to the point of writing a proposal for NSF funding for investigations into "noncausal correlation" (i.e., psychokinesis or remote viewing) and even submitting an article on "Observer Influence upon Measurements of an Electrical System" to *Science*.[62]

In presenting his parapsychology work to various audiences, Phillips tended to list his affiliation with UCSB rather than with Science Spectrum. For instance, when he hosted an "audience participation clairvoyance test" in Los Angeles in 1973, he told reporters that he was a "physicist and research associate at UCSB."[63] Likewise, when prominent parapsychologists came to town, Phillips made the most of his campus connections. Again in 1973, Phillips brought the Dutch psychic detective, Gerard Croiset, into the laboratory of a new hire in physics, Paul Hansma, to try to mentally influence Hansma's experimental equipment. Afterward, Phillips wrote a friend that

> My experiments with Croiset turned out rather poorly. The change in the apparatus we observed when he was in the lab turned out to be a familiar consequence of drift in the bias current in the detector to an unstable point. The designer of the experiment, Prof. Hansma, said that it was nothing unusual. Unfortunately, no one was present at the time who understood the apparatus thouroughly [*sic*].[64]

Phillips also tried to get one of his angel investors, Winifred Babcock, to make a donation to UCSB so that Phillips could use the university's facilities to pursue parapsychology research. And, as we will see, he worked with a few UCSB students to build parapsychology experimental apparatus.

A PEDAGOGICAL EXPERIMENT

In addition to his work with Morris, Phillips had another campus connection that fed into his off-campus entrepreneurship. At the same time that he founded Glendan and joined Science Spectrum, Phillips became involved in an unusual pedagogical experiment centered on Virgil Elings, one of his former colleagues in the UCSB physics department. In 1971, Elings, at the time an assistant professor, took charge of a new degree program offered by that department: a Master of Scientific Instrumentation (MSI). Others had had a hand in starting the program, including the department chair, Vincent Jaccarino, whose harder line on tenure and parapsychology caused Phillips to move to Science Spectrum. Phillips himself was the program's coinstructor for the first few years. Yet whatever the contributions of others, Elings was always the program's most visible figure. As an external review committee put it in 1974, the MSI program's

> major strength is clearly the interest and dedication of its faculty director, Dr. Virgil Elings. . . . Herein, however, lies the major weakness of the program. It is entirely dependent upon the continued interest and activity of Dr. Elings. If he should become disabled, find another position, or take a sabbatical leave there is no one to continue the necessary leadership.[65]

According to the review committee, other tenure-track faculty took little interest in the program, even though the whole department benefited from it: "The other 19 ladder [tenure-track] faculty members [besides Elings] supervise 26 graduate students. The Physics Department is being credited for 20 graduate students (MSI) who are being handled by one ladder faculty member [Elings], thereby giving the other faculty members an exceptionally light load."[66]

Those twenty MSI students were much needed, because without them the department's graduate enrollments would have stagnated for almost a decade at about two thirds of their 1969 peak. Indeed, Elings later claimed that the MSI program was created "because we thought the PhD program was going to die."[67] Yet teaching a master's program on instrumentation was not what Elings was hired for. Trained as a theoretical high-energy physicist, Elings (along with three others) had been hired to form a specialization in high-energy research that would bring prestigious and lucrative AEC funding to UCSB.

By abandoning high-energy physics (and forsaking PhD students and AEC grants) Elings risked lowering himself in his colleagues' eyes—particularly because his new career path centered on low-status terminal master's students. Thus, the 1974 external review expressed concern about "the serious threat to [Elings's] career in academic physics if he stays with the MSI program for more than just a few years . . . [rather than] go back to the mainstream of physics."[68] In fact, Elings spent eight years moving from assistant professor to associate professor, and another eight moving from associate professor to full professor. During that time, junior colleagues who worked with PhD students were promoted much more quickly. For instance, Paul Hansma, the electron tunneling spectroscopist mentioned previously who was hired as a lecturer in 1973 and assistant professor in 1974 (and whose fortunes would later be tied to Elings's), was promoted to associate professor after just three years on the tenure track and promoted to full professor after just three more.[69]

Elings's new direction endangered his career because he was no longer seen as contributing to the strategy that his department had articulated in the early 1960s for achieving "a sudden rise to international distinction in a period of only a few years."[70] Prestige subfields such as high-energy physics that could attract the best PhD students and generate large grants from the AEC and other national security funders were at the core of that strategy. By the late 1960s, however, several trends combined to undermine that strategy and open a space for experiments such as the MSI program. First, federal R&D funding reached a peak of 3 percent of Gross Domestic Product (GDP) in 1964 and declined steadily until the 1980s; in terms of inflation-adjusted dollars the peak came in 1967, but certainly by 1970 it was clear that federal support for R&D was waning.[71] Beyond funding, there were also new constraints on federal R&D funding. In particular, the Mansfield Amendment curtailed the ability of military funding agencies to sponsor basic research (such as most academic high-energy physics) that was not directly applicable to battlefield technologies.

At the same time, federal civilian funding agencies were becoming more friendly to applied, interdisciplinary research than they had been in the early Cold War. In particular, the NSF was transitioning away from its traditional role as a funder of individual basic researchers and toward a new role in funding collaborative, interdisciplinary, and applied science and engineering. Emblematic of that shift was the NSF's Interdisciplinary Research Relevant to Problems of Society (IRRPOS) program, founded as a result of Congressional pressure in 1968. In 1971, IRRPOS rebooted as the more successful Research Applied to National Needs program (RANN), which eventually grew to about 10 percent of NSF's research portfolio. RANN supported research on *very* applied topics, from new mining and excavation techniques to new ways for New York City to pick up its garbage.[72]

As a UCSB Office of Research report from 1972 indicates, the university believed that RANN and initiatives like it portended a rapidly changing environment for physics and other science and engineering departments.

> Shortly after World War II, basic research flourished, however, in recent years a new emphasis has evolved toward applied research. In fact, there is a growing concern that even the National Science Foundation may be failing in its original mission and may in fact assume the role of directing instead of supporting research. This change in trend for NSF is evident in its RANN program (Research Applied to National Needs) and in other new approaches that are in their proposed budget. On this campus over 90% of the 313 active projects [across all departments] continue to be directed toward the study of basic research. The increased competition for the shrinking basic research funds has undoubtedly contributed to the 22% decrease in the value of awards and an 8% decrease in the number of awards recorded for this reporting period.[73]

As we'll see in chapter 3, some square scientists and engineers initially embraced RANN and similar programs, only to become fierce critics later. After about a decade, RANN folded owing to that opposition. But in the early 1970s UCSB physicists and administrators had every reason to think that RANN was the way of the future (as, in some ways, it was).

The second, related trend affecting physics departments was the souring job market. In 1970 alone, "sixteen-thousand scientists and engineers, each holding advanced degrees, lost their aerospace-industry jobs. . . . The number of academic [physics] jobs on offer at the [April American Physical Society] meeting dropped . . . [by] a factor of four" from 1968 to 1971.[74] With fewer grants to support PhD students, and dimmer job prospects for graduates, enrollment in physics PhD programs waned. As David Kaiser has shown, by 1974 the number of physics doctorates awarded annually by American universities dropped by a quarter from its 1971 high; the figure had dropped by more than a third by 1978.[75] Likewise, at UCSB, the number of physics PhD degrees awarded per year dropped precipitously from 1971 to 1972, and did not exceed the 1971 peak until 1984 (figure 2.3).

There was a third trend affecting academic physics departments in the Vietnam era: disaffection arising from the war and other cultural changes. Undergraduates who a few years earlier had seen a physics degree as a path to patriotic public service, stable employment, and deep understanding of nature now stayed away. At UCSB, undergraduate enrollments in physics, according to former department chair Paul Barrett, "peaked at 2.35% [of total UCSB undergraduate population] in 1959 (the post-Sputnik bulge), fell to a low of 0.64% in 1976 (post-Vietnam reaction), and rose to only 0.74% in 1981." According to Barrett, even the recovery after 1981 was more an artifact of UCSB's engineering departments' raising their standards than of physics's overcoming its legitimacy deficit.

As Stuart W. Leslie, Matthew Wisnioski, and others have documented, student disaffection stemmed from the perception that physical and engineering science departments were too narrowly focused on research funded by national security agencies, and were too technocratic and elitist and thereby too insulated from civil society's concerns. Student activists at these schools opposed not only applied, defense-oriented research but also *basic* research on the grounds that both did little to solve civilian problems such as pollution. The same activists also criticized the power of the traditional academic disciplines (and research conducted within them). We will see in chapter 3 that they believed that applied, defense-related research was effective in meeting its goals because it was more interdisciplinary than basic research; they therefore argued that research applied to civilian issues should also be interdisciplinary.

Members of the UCSB physics faculty were keenly aware that students wanted them to move toward civilian, applied topics and adopt a more interdisciplinary, humanistic outlook. Department brochures from the 1970s tried to appeal to that desire, though in a somewhat plaintive tone:

> There are serious misconceptions about science. Science is not, as many believe, a mere collection of facts. Scientists are not neutral and colorless professionals.

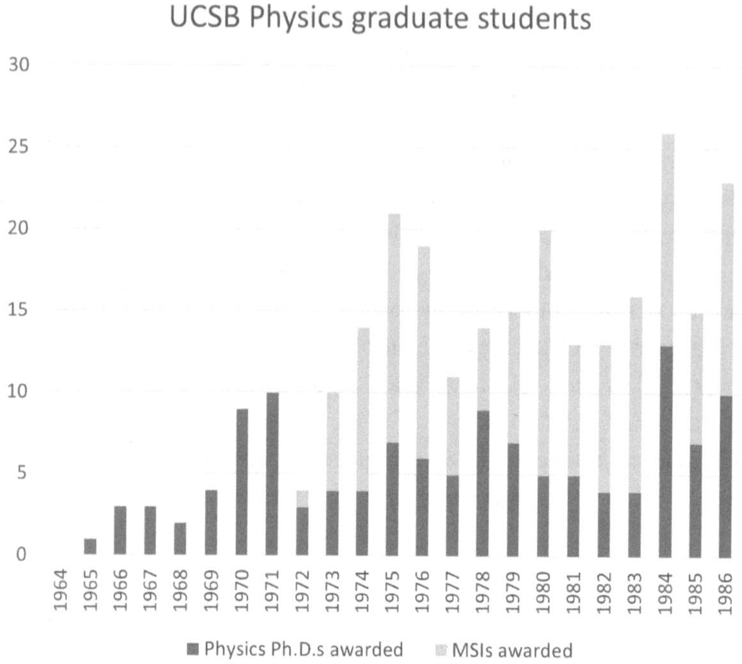

Figure 2.3

Number of graduate (both PhD and master) degrees awarded in the UCSB Physics Department. Note the fall from a peak in 1971, followed by a recovery from 1973 onward that was made possible by an influx of enrollees in the Master of Scientific Instrumentation program. Without them, the number of graduate degrees would've hovered around 5 or 6 per year from 1973 to 1984; instead, the number was closer to 15 per year. *Source*: Graph compiled from data presented in Paul Barrett's departmental history, 1987.

Science is not a set of special tricks for elite minds. . . . [P]hysics is also practical. It serves as a fundamental science in preparing the base for all technological advances. In the same sense, it serves as one of the most important guides in making difficult technological decisions.

Physics is far more than a technical study. It is part of the totality of our human experience. The influence and interconnections physics has had on our history, on our religions, on our philosophies, on our languages and literatures are deep and broad. To study physics is to begin to see these interconnections and, eventually, to give ourselves a more complete picture of our place in our intellectual heritage.[76]

Of course, change is difficult. The square scientists who faced calls for change in the 1970s had been rewarded during their whole careers up to that point for being "neutral and colorless professionals." Most of the UCSB physics department therefore remained focused on basic research of the sort valued by their disciplinary colleagues rather than society at large.

Still, some members of the department spent the 1970s experimenting with a new kind of physics shaped by the criticisms voiced in the brochure. For instance, Melvyn Manalis (a 1970 UCSB PhD recipient and then a research physicist and lecturer) taught a course on environmental physics intended to show that "physics is very relevant to modern environmental topics."[77] Similarly, Herb Broida—for whom the current UCSB physics building is named—established a Quantum Institute in 1969 to serve as "a problem oriented, interdisciplinary research unit . . . designed to respond rapidly to requests for assistance in the solution of problems important to governmental agencies, private industry, and research foundations."[78]

The department also created a Physics Learning Center (PLC) to make experimental equipment available to "students from all disciplines. With hundreds of demonstrations maintained in an atmosphere of free investigation, the PLC has established itself . . . in the service to local schools and the community."[79] This was possibly a response to a demand made on many campuses for science departments to become more involved in their local communities. Notably, though, the PLC does not seem to have been a top-down initiative. Instead, it was the brainchild of Tony Korda, one of the technicians in the department's workshop who was, like Phillips, deeply interested in parapsychology and New Age spiritualism. Indeed, he told Phillips that he possessed a "powerful 'energy' or Kundalini experiences . . . including the flame and many others" and that the PLC was inspired by a Kundalini-esque "experience . . . of poetic and artistic inspiration, together with strong feelings of love for others."[80]

The PLC and Manalis's environmental physics course offer clues for putting Elings's Master of Scientific Instrumentation program in context. These new departmental initiatives all offered some relief from the department's budgetary, demographic, and cultural legitimacy crises. Each was, as an external review described the MSI program, "an imaginative response to what seems to be a real need."[81] Yet in each case, the response to crisis was placed in the hands of someone at the margins of the department (Manalis, Korda, Phillips) or someone moving toward the margins because of his involvement in the initiative (Elings). The rest of the department, meanwhile, continued along much the same trajectory it had before, if under more straitened circumstances.

THE ESOTERIC AND THE TANGIBLE

Thus, while his department was—in its official outreach—complaining about "serious misconceptions about science" prevalent in wider culture, Elings shaped the MSI program to appeal to young people's desire for a new kind of science influenced by those "misconceptions:" that is, a reformist, publicly engaged science oriented to solving civilian social problems in a manner liberated from traditional disciplinary constraints. As the original program proposal to the NSF put it, "The program will satisfy a long felt need for interdisciplinary training."[82] Ads told potential applicants, "The UCSB Master's Program in Scientific Instrumentation is looking for Creative, Hardworking Bachelor's Degree Scientists who want to Solve Real Problems. . . . The one-year program, supported by the National Science Foundation, provides opportunity and experience in interdisciplinary problem solving . . . on campus and in nearby hospitals and industrial laboratories."[83] Or, as Elings told a local reporter, the MSI program was to be "a leader in the growing movement to involve students directly in interdisciplinary research by tackling problems in areas outside their formal training."[84]

The projects that early MSI students worked on reflected that desire for interdisciplinary, civilian, problem-oriented research. These included "an image stabilizer to aid in early detection of cancer in internal organs; a color television speech display system to teach the deaf to speak; a digital heart rate monitor for use by physiologists";[85] a tactile sound mixer built for a blind audiophile undergraduate;[86] ocean monitoring instrumentation packages;[87] and "a portable device to detect and measure the quantity of lead in gasoline."[88] The origins of the speech display system (known as the Chromophone) were typical of the program's interdisciplinary reach:

> The need for such a device was voiced by the staff of the Speech Communication Research Laboratory of Santa Barbara in a conversation with Prof. Glen J. Culler of the department of electrical engineering. . . . One evening he mentioned the project to physicist David Phillips, and the next day Dr. Phillips and Donald Holly, a graduate student in the scientific instrumentation program, were at work with some discarded color TV sets.[89]

Students also acquired interdisciplinary skills and benefited the university by building free experimental apparatus for researchers in many UCSB science departments, especially physics. In lean budgetary times, free equipment must have been desirable; during the first year of the program "the number of cooperating labs on campus increased and soon there were more labs requesting students than there were students."[90]

Many students also used their master's projects to build bridges between the university and community organizations both around Santa Barbara and in other parts of the country. Organizations such as the Santa Barbara Cancer Foundation, the San Francisco VA Medical Center, St. Christopher's Hospital for Children in Philadelphia, Cedars-Sinai Hospital and the John Tracy Clinic in Los Angeles, Goleta Valley Cottage Hospital, and the Santa Barbara Heart and Lung Institute hosted MSI students or field-tested their projects. Elings and Phillips were aware that working with MSI students placed some burdens on these organizations, but

> [t]he experience also provides the companies an opportunity to get acquainted with possible employees. . . . Such experience will benefit the student by letting him [*sic*] see what is going on "outside." Also, we feel that the ties which are built between the university and local industry in this way will provide valuable research ideas for both the campus and industry and future employment opportunities for students.[91]

In fact, the MSI program seems to have done quite well in cultivating "future employment opportunities" for graduates. As the UCSB press office proudly noted in 1972, "In contrast to the experiences of students in some other scientific fields, every one of the program's continuing students found employment in instrumentation last year."[92] The program even drew national attention: *Physics Today* lauded it for "preparing students for physics-related jobs."[93]

Thus, for students, the program offered both an outlet for baby boomer idealism *and* a pragmatic solution to the problem of a poor science and engineering job market. According to Mike Buchin, a mid-1970s MSI student whose projects during his time in the program included the SPOT (speech-optical translator, a variant of the Chromophone—figure 2.4) and a thermodilution method for measuring blood flow,

> I got involved in this project mainly out of my own interest in medically-oriented projects. Also, I like getting involved with projects that will help people—applied *project work*. In the long run, my work has paid off tangibly as well as esoterically (?). I am going to be working as a Research Physicist for Searle Analytical and Radiographics, a firm in Des Plaines, Illinois deeply involved in nuclear medicine.[94]

Despite such successes, the MSI program remained rather eccentric and hard to explain to the outside world. The website for the small photonics company that Buchin later founded, for instance, says he has a "Master's degree in Electronics" not scientific instrumentation. Yet it's striking that despite its oddity, in some respects the MSI program closely resembled initiatives by square physicists and engineers elsewhere. In particular, the topics of those early student projects resonate with what we'll see in chapters 3, 4, and 5: new technologies intended to help people with disabilities (particularly people with impaired vision or hearing),

Figure 2.4
A Master of Scientific Instrumentation student, Mike Buchin, demonstrating the SPOT (SPeech Optical Translator), a device for helping users pronounce vowels (and some consonants) correctly by modulating their voice to move a cursor to the intended part of the screen (here corresponding to the short "e" as in "left"). *Source*: Courtesy of the Department of Special Collections, Davidson Library, UCSB.

to extend life or make health care cheaper, to monitor pollution, or to foster alternative energy, all developed by physicists and engineers, rather than specialists in biology, medicine, atmospheric chemistry, or electrochemistry.

COUNTERCULTURAL CURRICULUM

Over time, MSI student projects became somewhat more drily technical and less idealistic. Of more lasting influence than the projects themselves, though, was the mildly countercultural pedagogy that informed those projects. To some extent, David Phillips was a conduit for certain countercultural preoccupations to enter the MSI curriculum. Throughout the degree program's run, Elings relied on a junior partner to give lectures and advise on projects—and for the first five years or so, Phillips was that partner. Unsurprisingly, he brought his parapsychology interests along. As he told the education director of the American Society for Psychical Research in 1972, "I wish to include some ESP and OOB [out of body] experiments in the curriculum of the Scintific [*sic*] Instrumentation Program at UCSB."[95] At least one MSI student, John Placer, built parapsychology instrumentation for his master's project (and published with Phillips and Robert Morris).[96] At least one other, George Jahn, helped Phillips and an undergraduate, Agi Ban, build apparatus for telepathy experiments.[97]

The more significant vectors of mild countercultural influence, though, were the MSI students themselves. In the 1970s, many young people questioned the traditional social order of the classroom and pushed for alternatives. For instance, a prominent demand in many of the campus protests of the late 1960s was for student-run courses. Unsurprisingly, then, MSI students rejected the traditional pedagogical format with which Elings and Phillips started the program: "The too familiar pattern of factual presentations of material chosen by the instructor was more tolerated than appreciated. . . . One year's experience indicates that experimental science students learn more from doing their 'own problem' than from passive reading and exercises."[98] As a result, Elings became a lifelong advocate of learning by doing and a harsh critic of conventional pedagogy. As he described it later, the MSI program was so unconventional in insisting that students learn things for themselves that for some students it made for "a rude awakening from the spoon-feeding of most undergraduate experiences."[99]

Over time, Elings's experience with MSI students inspired, or perhaps amplified, his own populist distrust of authority and pedigree. That, in turn, fed back into the program's curriculum and the *technological* approaches that students absorbed. Here, again, the national economic and cultural dislocations of the Vietnam era played an indirect role. As Christophe Lécuyer and others have shown, Robert McNamara's attempts to trim military budgeting in the mid-1960s led to a flattening of semiconductor firms' profits known colloquially as the "McNamara Depression."[100] In response, some firms began pursuing consumer markets more aggressively. Doing so, however, required manufacturers to significantly lower the cost per chip. By the early 1970s, Intel and a few other firms had succeeded to the point that electronics hobbyists could easily buy chips that a few years earlier would've been beyond their means.[101] From there, it was only a small step for hobbyists to begin building "personal computers" of the kind that counterculture icons such as Ted Holm had been advocating for years—a story now entrenched in the supposedly subversive mythography of Apple, the Whole Earth 'Lectronic Link, and other Silicon Valley icons.[102]

Less noticed in both popular memory and academic history, however, is that the sudden availability of cheap digital components also made certain kinds of scientific instrumentation easier to build.[103] Elings and Phillips were quick to embrace cheap digital circuits for their own projects; note, for instance, in figure 2.2, Phillips advertised the Acupointer as "All Solid State! Integrated circuit, LED indicator light." Phillips and Elings also soon concluded that cheap integrated circuits made possible a new approach to teaching master's students how to build instrumentation. Because the new, cheap microprocessors could be reprogrammed for a variety of tasks, students could be taught the basics of microprocessor programming as a universal solution to any instrumentation problem. As Elings put it a few years later, "One exciting aspect of microelectronics is that the resulting 'intelligent' instruments may be easily redesigned to meet altered specifications, often by simple software changes."[104]

Learning to design digital circuits therefore gave MSI students skills that they could adapt to almost any employer's needs—because microprocessor technology provided a flexible tool for problem solving not just in instrumentation but in other industries as well. Thus, "students now find that a little background in applying electronics to measurement problems opens doors to many jobs that they would otherwise not have access to."[105] As Elings began exposing MSI students to the advantages of digital circuits, however, he came to see that such circuits cast the necessity of a university education in doubt. Anyone, it seemed, could learn to design digital circuits: "one no longer needs a degree in electrical engineering to design state-of-the-art electronic instruments. . . . It isn't true that one needs a course in differential equations or Fourier analysis in order to design top-notch instruments."[106] If that was the case, then was a degree in electrical engineering worth it? Elings started to think not. As he put it in 1995, "The areas that [MSI] students had done undergraduate work in made little difference in their ability to design instruments. . . . All those esoteric courses made little difference."[107] Some of Elings's best students came in with very unlikely backgrounds: for instance, one student whom Elings still refers to often and who contributed greatly to his later endeavors entered the MSI program having done his undergraduate work at UCSB in psychology and religious studies.

Eling's growing disregard for academic credentialing, grounded in his experiences with the MSI program, aligned him with antiestablishment science groups with a more obvious New Left or countercultural bent such as Science for the People and the New Alchemists.[108] Yet Elings himself remained quite square. Pictures of him from this era show his habit shifting from black suits and skinny black ties to jeans and cowboy boots, but nothing wilder; and although he was willing to sign mild antiwar and proenvironment petitions, he wasn't an activist.[109] As we'll see, he occasionally collaborated in David Phillips's parapsychology pursuits, but seemingly more out of commercial than spiritual interest. Both Elings and Phillips occupied the excluded middle that is invisible in the historiography and had only a precarious existence at the time. They weren't part of the counterculture, but they were more friendly to it than to the countercountercultural backlash; their research aligned with some countercultural and New Left aims, but lightly enough that they were easily distracted in new directions as the counterculture faded.

INCHING TOWARD THE MARKETPLACE

In fact, Elings and Phillips had much else in common: both grew up in Iowa, graduated from Iowa State in the early 1960s, went off to big-name physics PhD programs (Elings MIT, Phillips Berkeley), and were hired at UCSB in 1966.[110] Prior to joining UCSB they had, like many squares, worked in the trenches of the military–industrial research complex: Phillips at the Naval Research Laboratory and Elings at Union Carbide Nuclear Division (where he helped "in the design of a nuclear reactor") and Autonetics Division (where he performed "an analysis on the gyroscope used in the Minuteman missile").[111]

Also, for both Phillips and Elings science and business were a family affair. Phillips's garage tinkering was often conducted in collaboration with his sons (who, along with their friends, were occasional subjects of Phillips's parapsychology experiments). His company, Glendan, comarketed with a similarly small outfit, Information Connection, run by his wife Linda; Information Connection offered a bibliographic search service and did contract research and editing. Likewise, both of Virgil Elings's sons, Mike and Jeff, were involved with his later entrepreneurial ventures, and Virgil sometimes went into business ventures with his wife Betty, a successful realtor.

When they got to UCSB, neither contributed to their department's PhD program to quite the extent that their colleagues had hoped; Phillips supervised one dissertation, Elings none. Perhaps they were distracted by outside pursuits: Phillips had his parapsychology and Glendan Company; Elings also had a garage start-up, Santa Barbara Technology (SBT). Indeed, there must have been some cooperation between the two endeavors. For instance, in 1973 Elings advertised a product virtually identical to Glendan's Acupointer under the brand name Tobiscope.[112] Similarly, speech therapists, schools for the deaf, and language instructors could order SPOT units from SBT as early as 1973, and from Glendan as late as 1981.[113]

Like Glendan, SBT's earliest products were not associated with the MSI program. In addition to the Tobiscope, SBT was perhaps best known for a toy consisting of two parabolic mirrors stacked with their concave sides facing each other, such that an object placed on the bottom mirror appears to be floating above the top one. According to legend, this effect was discovered by accident by a UCSB student (in some versions, a janitor or maintenance man) doing some housekeeping in a lab.[114] More certain is that Elings and Caliste John Landry, a PhD student in electrical engineering, patented the device together.[115] Eventually, Elings licensed the patent to another start-up, Optigone, which makes most of the versions of this product that are seen in science museum gift shops today.

Both Glendan and SBT, however, later marketed products based on MSI projects, such as the SPOT and Mike Buchin's thermodilution cardiac computer. Notably, Elings and Phillips (but not Buchin) took out a patent for the thermodilution cardiac output computer *after* it had been the subject of Buchin's MSI project.[116] That places the patent in an ethical and legal gray area, certainly by today's standards for academic patenting. However, given the ad hoc nature of both the early MSI program and university intellectual property practices at the time, Elings and Phillips were the only people in a position to make MSI students' projects available to organizations such as hospitals and schools for the deaf: no one else, certainly not Buchin, was going to make a commercial version of his thesis project. Glendan and SBT may also have served as career models for graduates of the program, because a fair number of graduates became freelance technicians who did small contract projects for local high-tech companies, either full time or to supplement their income—work that resembled both what they had done in the MSI program and what Elings and Phillips were doing with their garage companies.[117]

In any case, SBT sold cardiac computers until the early 1980s, and biomedical researchers reported using them in journal articles published well into the 1990s. SBT's other major product in the late 1970s was a "dermofluorometer" used to tell whether injured skin flaps were healing properly. The cardiac computer and dermofluorometer marked a new phase in Elings's career in which the MSI program, his own on-campus research, and his off-campus entrepreneurship converged. As

Elings's departmental biography noted, once he began directing the MSI program, "He shifted his research interest to medical instrumentation. Many MSI students gained valuable experience working on some of these projects. Among them was a lung water measurement system, and a light-scattering immunoassay instrument."[118] Today, Elings ascribes this shift to his dissatisfaction with the anonymous teamwork typical of high-energy physics—a reason given by other high-energy physicists who moved toward biomedical entrepreneurship in the 1970s, such as Cetus cofounder Donald Glaser.[119]

Surely another reason, though, was that turning toward biomedical research opened up new funding possibilities. Unlike most of his departmental colleagues, Elings could now solicit funding from places like the American Heart Association. That was the case for a project that he struck up with George Jahn, an MSI student who left UCSB to become a cardiologist, and John Vogel, a doctor at the Santa Barbara Heart and Lung Institute who had previously mentored MSI students such as Mike Buchin. That project resulted in the dermofluorometer line that SBT began selling.[120] About the same time, Elings also began collaborating on biomedical instrumentation with a new UCSB physics assistant professor, David Nicoli. On joining the physics department, probably in the 1976–1977 academic year, Nicoli seems to have taken over David Phillips's role as junior partner in the MSI program. In 1978, Elings and Nicoli formed another start-up company, Nicomp, to sell an instrument that they had developed on-campus for determining the size and morphology of particles suspended in solution. That is, over the course of a decade Elings's maverick pursuits had turned him into a serial—and possibly somewhat accidental—entrepreneur.

AFTER THE LONG 1970S

The MSI program, meanwhile, continued on, but with less passion and publicity. In the program's earliest days, the best students had been treated as Elings's near-peers intellectually and socially. He coauthored with at least four of the early graduates, boosting his own publication productivity while giving them a useful credential. Even as early as 1974, though, the department's external reviewers observed that the MSI program was no longer attracting as high a caliber of student, and warned against "the danger of attempting to maintain constant size, rather than setting a floor on incoming qualifications and letting the class size be determined by the number of qualified applicants."[121] After that the program became a more routine, less urgent, and less idealistic fixture of the UCSB landscape. Although Elings remained committed to it, his attention was increasingly diverted by his entrepreneurial pursuits.

Like the MSI program, Science Spectrum also drifted in the late 1970s, as cash flow problems forced cutbacks even among its senior staff. David Phillips fell back on consulting jobs and, for a few stretches, employment at other local firms such as Renco, a small electronics manufacturer.[122] By late 1980, Philip Wyatt had decided to close down the company. The late afternoon celebration of Science Spectrum's demise turned into an open-door wine tasting, and then into a laboratory experiment when someone decided to test the wines with the company's light-scattering instruments.[123] As a last hurrah, Wyatt sent a paper to *Science* comparing the ratings of the "nonprofessional consumer panel" consisting of a "physicist, mathematician, electrical engineer, office manager, x-ray technician, coin dealer, electromechanical assembler, and attorney" with the output of the light-scattering measurements.[124]

Then, to hear Wyatt tell it, a miracle happened. His letter to *Science* was picked up in various outlets, resulting in overtures from soft drink companies to perform analyses of their products with his light-scattering instruments. Those requests then led to further articles and publicity, snowballing into interest from the military and other customers.[125] In 1982, Wyatt reincorporated under the name Wyatt Technology. Before long, David Phillips was consulting for Wyatt again and eventually joined the new firm. Interestingly, Phillips's notebooks show that he was still dabbling in parapsychology, but with a 1980s twist. There, right next to computer code for Wyatt instruments and notes on meetings between

Wyatt engineers and representatives from the Army, are data Phillips collected from possible ESP adepts trying to predict the value of the next day's Dow Jones index or the price of silver![126]

For Elings, the early 1980s were less eventful, but his career took an abrupt turn in the middle of the decade. This was a two-step process. First, at a department talk by a Spanish physicist, Nico García, Elings learned the details of the scanning tunneling microscope (STM), a technique that García and Elings's departmental colleague, Paul Hansma, had each been working on since 1983. During the talk, Elings came to the conclusion that his expertise with digital circuits would allow him to build and sell STMs that were as good as the home-made instruments Hansma and García used. After attending one of the first international STM conferences, he realized that STM was hot enough—its inventors having just received the Nobel Prize—that there was a sizeable market for a commercial instrument. Within a few months, he had sold his share of Nicomp and invested the funds in cofounding an STM company, Digital Instruments (DI), with John "Gus" Gurley, one of his best MSI students.

Elings's second shove in a new direction was the—apparently quite sudden—cancellation of the MSI program. With the mid-1980s rebound in PhD and undergraduate enrollments and federal research funding, the exigencies that had led to the program's creation disappeared. Longstanding qualms about how a terminal master's program affected the department's reputation now became more urgent. With Nicoli having left to run Nicomp full-time, Elings was essentially alone in running and championing the MSI program. So when a new department chair took over in 1986 and quickly cancelled the program, Elings had no allies to help him resist.

I've written extensively elsewhere about what happened next.[127] Digital Instruments turned a profit in every year of its existence by supplying approximately half the world's scanning probe microscopes at the time. Its success was due in no small part to the MSI program, which gave DI several key personnel, a successful technological and managerial philosophy, and—through SBT and Nicomp—much-needed seed funding.

Elings and Paul Hansma, despite some ups and downs, established an extraordinarily lucrative (for all parties) technology transfer procedure. Yet in commercializing a fellow professor's inventions, Elings strained relations with UCSB in ways that he had not when commercializing MSI students' projects. Objections from departmental colleagues led Elings to quit his faculty position in 1990. Over the next decade, Elings turned his firm's attention toward industrial customers, which paved the way for DI's merger with a semiconductor process equipment company, Veeco, in 1998 for approximately $150 million. Since then, Santa Barbara has become home to a cluster of nanoscience instrumentation companies founded by veterans of DI, the Hansma group, and other UCSB research labs, as well as older companies like Wyatt Technology. In fact, for a time, Wyatt's company even shared a building (and Wyatt his entrepreneurial experience) with one of DI's spin-offs (the wryly named Asylum Research).

That collocation represents the kind of advantage that economic geographers cite as an important factor in the success of regional high-tech clusters: knowledge transfer between generations made possible by the density of start-ups in a small area (in this case just north of the Santa Barbara airport). But that literature hasn't plumbed quite all the regional effects of high-tech clustering—effects which, in the Santa Barbara case, bear the imprint of the long 1970s. Particularly in Virgil Elings's case, entrepreneurship has led to wealth, and wealth has led to generous philanthropy. Outside Santa Barbara, Elings has given money for, among other things, buildings at his alma maters, Iowa State and MIT. Locally, he and his ex-wife Betty Wells, a successful realtor in an area with fantastically expensive real estate, donated funds for Elings Hall, the home of UCSB's California NanoSystems Institute. Virgil and Betty have also each independently supported a number of museums and public institutions at UCSB and around Santa Barbara, and the whole family (Virgil, Betty, and their sons Jeff and Mike) gave the money that made possible Elings Park, a 230-acre public space that dominates the mesa southwest of Santa Barbara's downtown.

Campus buildings, parks, museums—these fairly conventional forms of philanthropy are among the less-acknowledged outcomes of the formation of high-tech clusters. Some of Virgil's donations, however, are less conventional because they are informed by lessons from the long 1970s. In particular, Elings often refers publicly to the MSI program as a model for the future of science and engineering education, and to back up that claim he has donated a million dollars to Dos Pueblos High School's Engineering Academy to support a problem-based, learn-by-doing approach that closely resembles the pedagogical philosophy of the MSI program. The Academy's founder, Amir Abo-Shaeer, takes much the same populist stance regarding conventional teaching as Elings—though it is perhaps an indication of the extent to which "countercultural" ideas about education now suffuse elite opinions on education "reform" that Abo-Shaeer has won a MacArthur Fellowship and is the subject of a best-selling nonfiction book entitled *The New Cool.*

CONCLUSION

Cool or not, the type of science practiced by Phillips, Wyatt, and Elings did become the new normal. High-tech start-ups—especially those founded by or closely connected to university faculty—were controversial enough to force Elings out of UCSB in the early 1990s, but within a few short years they were being widely hailed as icons of economic gain and social liberalization.[128] Similarly, with the end of the Cold War, the reorientation of the US research enterprise toward more civilian, applied research came to be seen as a critical part of the nation's "peace dividend," and thus the kind of research that Science Spectrum and the students in the MSI program had been working toward in the early 1970s took on a new life.[129] At the same time, the 1990s saw a dramatic increase in federal funding for biomedical research, causing academic physicists to form collaborations with clinical researchers and apply for funding from the NIH. Thus, the biomedical instrumentation projects and start-ups that had made Virgil Elings so different from his colleagues in the early 1970s came to be a routine, even mandatory, part of many university physics departments' research portfolios.

Not everything from the 1970s was reborn in the 1990s, of course. The overt parapsychological research championed by David Phillips, for instance, receded in the 1980s. Yet, as David Kaiser has shown, that was in part because the mainstream of physics had internalized topics such as Bell's theorem that were originally associated with parapsychology.

When historians, sociologists, and economists of science have looked at these changes in the American research enterprise, the role of the counterculture and the budgetary and cultural upheavals of the Vietnam era have largely been absent. On the one side, those who lament these changes, such as Philip Mirowski and Paul Forman, have generally traced their origins to the success of a conspiracy of neoliberal economists around 1980.[130] On the other side, those who take a more positive or neutral stance have variously chalked up these changes to the effects of the Bayh-Dole Act, the biotech industry, professionalization of university technology transfer officers, and so forth.[131] The fact that these events were all taking place in the shadow of student protest, countercultural ferment, the environmental and civil rights movements, and a hot war in Southeast Asia is treated, if at all, as coincidence or at most as backdrop.

What I've tried to show in this chapter is that for at least some square scientists in the long 1970s, that era's social dislocations did matter. They prompted the squares to diversify their interdisciplinary collaborations; to follow the shifting winds of civilian priorities into new lines of research, especially related to biomedical, environmental, and disability technologies; to experiment with new ways of reaching young people; and to forge new ties among industrial, academic, government, and civil society organizations. One vehicle for all of these strategies was to found their own start-up companies—hardly unprecedented in the history of American science, but also not common or valorized in the early Cold War, especially among university faculty.

Since the early 1990s, the start-up company has become a common sense, much-lauded vehicle for translating a more interdisciplinary, civilian-minded kind of science out of the laboratory (especially the academic laboratory) and into contact with both the public and various organizational partners. There is plenty of room for debate as to whether that is (or can be made to be) a beneficial state of affairs. But if we want to understand where some of those lofty, transformative visions for high-tech start-ups originated, we should look to the Vietnam era's disruptions and the unorthodox responses fashioned by the likes of Wyatt, Phillips, and Elings. My argument, though, is that at least as important as understanding where start-up culture came from is recognizing that other experiments accompanied start-ups at the beginning but were winnowed out in the 1980s. The 1970s were a time of inventive experimentation; founding start-ups was one—but *only one*—symptom of that. The forces that have given rise to our present model of academic capitalism are not the ones that stimulated academic start-ups in the first place; rather, our present world was created by the forces that allowed start-up culture to transcend its origins while the other experiments of the long 1970s did not.

3

A FEDERATION OF BULL SESSIONS: INTERDISCIPLINARITY AT STANFORD

Few words are uttered more often or with less irony on university campuses in the United States today (or, indeed, at universities anywhere) than *interdisciplinary*.[1] In the mouths of campus administrators, funding agency grant officers, journalists, alumni donors, and corporate and civil society partners, interdisciplinary has become a synonym for "good" when speaking of academic teaching and research. This book is a product of that worldview, because its author was trained in an interdisciplinary department and works in an interdisciplinary faculty. Yet despite the advantages that interdisciplinarity can offer, and especially despite its champions' representation of it as a timeless and universal good, interdisciplinarity has a politics and therefore benefits some more than others. In recent years, critics located in the traditional disciplines have cogently pushed back against interdisciplinarity discourse—as indeed have scholars from inherently interdisciplinary fields such as science and technology studies.[2] Interdisciplinarity has become the kind of commonplace that can and should be subjected to academic critique yet is so much a part of the air we breathe that critique is almost self-defeating.

This chapter is not so much a critique of interdisciplinarity as a genealogy—though it is the kind of genealogy that interdisciplinarity's

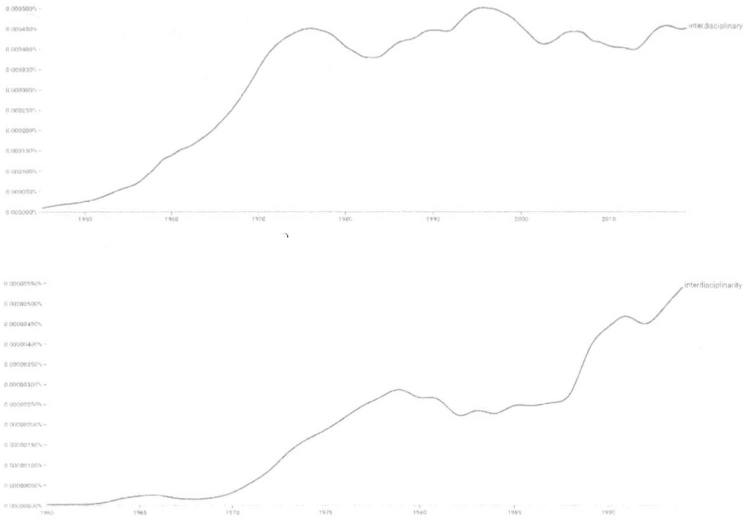

Figure 3.1
Google Ngrams for "interdisciplinary" and "interdisciplinarity." The former rose to popularity in the 1960s; its frequency is more or less level from 1970 to 2019. "Interdisciplinarity" followed a similar curve until the end of the Cold War but has become steadily more popular since 1991; I exclude most of the post–Cold War period to highlight its rise from 1968 to 1978. American English, smoothing of 1.

critics might take some comfort in. As figure 3.1 indicates, interdisciplinarity has a history in which it has been more popular at some points than others. In particular, its popularity (at least in the anglophone world) rose rapidly right at the onset of the long 1970s, during the period of student protest and severe research funding shortfalls. Like academic entrepreneurship in chapter 2, interdisciplinarity is taken for granted today, yet originally it was one of many inventive, oft-maligned, and hence precarious experiments in the face of crisis. Too often, the experiments that it accompanied and complemented at the time have since been forgotten.

Many of the justifications for interdisciplinarity in the 1970s have also been forgotten. At the time, interdisciplinarity was offered as a solution to a host of problems, many of which no longer exist. So why do we still prize it? And what did the days of crisis force us to give up that we no longer remember we gave up?

STANFORD BEFORE 1970

To unpack interdisciplinarity in the long 1970s, this chapter takes Stanford University as its case. If *interdisciplinarity* is a ubiquitous keyword of American academia today, so is *Stanford.*[3] Often they are uttered in the same breath. Stanford has become a much-envied (though often inappropriate) model for many American universities—primarily for its links to Silicon Valley and its start-up culture, but also (and relatedly) for its tradition of interdisciplinarity.[4] Few universities have been as successful as Stanford in projecting an image of both commercial and institutional entrepreneurialism—including the entrepreneurialism associated with "overcoming" the purported strictures of the purportedly traditional disciplines.

Stanford is not, therefore, a representative example of late twentieth and early twenty-first century American academia. But it still makes for a good case study precisely because so many universities want to be like Stanford, and because there are so many persistent myths telling them *how* to be like Stanford. There are also quite a few excellent histories that feature Stanford prominently, though their lessons do not seem to have checked the spread of Stanford's mythology as much as they should have. What both the academic histories and the myths have in common, though, is that they have little to say about the long 1970s. There are histories and myths rooted in the prewar and early Cold War period up to 1970, featuring Silicon Valley icons such as the Varian brothers and Hewlett-Packard.[5] And there are histories and myths drawn from the post-1980 era and the world of Google and Genentech.[6] The latter stories do sometimes start in the early 1970s, but without emphasizing that Stanford was so riven by protest in those years that many (both on- and off-campus) thought it had become ungovernable and might not survive. Historians of Stanford in the early Cold War do acknowledge that, but at the *end* of their narratives. That is, few studies show how Stanford not only regained its footing but also became the confident, globe-bestriding model of entrepreneurial interdisciplinarity it is today.[7]

Perhaps one reason no one wants to talk about Stanford's experience of the long 1970s is that this was an excruciatingly difficult period for the institution. In 1969 to 1970 alone, protestors occupied Stanford's Applied Electronics Laboratory for a week, the university was forced to divest its flagship Stanford Research Institute, classified research was banned from campus (with a consequent $2 million budget shortfall), and President Kenneth Pitzer was forced out. Though protests quieted after 1971, strained funding left labs scrambling for another decade. Throughout, interdisciplinarity was a ubiquitous tool in Stanford's response to campus protest and funding shortfall.

Stuart W. Leslie and Rebecca Lowen argue that the events of 1969 to 1970 were the "reckoning" that Stanford paid in return for its rapid advancement in the early Cold War—that is, the protests were a moment of what I am calling "existential success." Stanford started the Cold War as a sleepy, conservative backwater catering mostly to undergraduates descended from California's elite. Through planning, determination, and a great deal of luck, Stanford figured out the Cold War research paradigm earlier than almost any other institution: win federal grants with high overhead rates; carve a niche mediating between the military services and the aerospace and electronics industries; elevate physics as the model for all other fields; and emphasize graduate research at the expense of undergraduate instruction. Other institutions moved in the same direction at the same time, but most of them—for example, Johns Hopkins, MIT, and Cornell—were already influential before the war.[8] Stanford therefore came farther faster in the early postwar period than almost any peer. But riding that wave led to an almost inevitable crash. When military funding retrenched, Stanford's researchers ran short of money, its graduates struggled to find jobs, and its neglected undergraduates demanded to be heard. If Stanford's rise was faster than its peers, its collision with the reality of 1969–1970 was that much harder.

This chapter will show that that collision led to more, and more far-reaching, talk about interdisciplinarity. There was also some uptick in the practice of interdisciplinarity, though words probably outweighed deeds. Now, that's not to say that there was no interdisciplinarity at Stanford before 1970. Readers should take this chapter as just one of several possible genealogies of modern forms of academic interdisciplinarity; other genealogies extend well before 1970, at Stanford and elsewhere. But the forms of interdisciplinarity seen at Stanford before 1970 were rather narrow, in both their enactment and their aims, compared to interdisciplinarity after 1970. Before the late 1960s, the typical interdisciplinary interactions that Stanford administrators fostered were collaborations among electrical engineers, metallurgists, and applied physicists.[9] By contrast, after 1970 the same departments sought partners in music, medicine, philosophy, religious studies, and other areas.

Moreover, the interdisciplinarity of the earlier period was more of a one-way street, whereas in the long 1970s multidirectional interdisciplinarity came into vogue. In the 1950s, Stanford's legendary postwar provost, Fred Terman, saw interdisciplinarity largely in terms of reorganizing the life, social, and engineering sciences around tools borrowed *from* the basic natural sciences, particularly physics. Even though he was an engineer, Terman returned from World War II believing that physicists had contributed more to the war effort than engineers and that his disciplinary colleagues therefore needed to look to physics.[10] Thus, he pressured the electrical engineering department to abandon power transmission research and take up physics-based fields such as microwave communications, and he remolded old-fashioned "metallurgy" into "materials science." (The Stanford Metallurgy Department was second in the country to change its name to Materials Science.)

Nor did Terman limit physics' disciplinary imperialism to forays into engineering. For instance, he pushed political science to hire behavioralist faculty familiar with quantitative techniques borrowed from physics.[11] In biology, he sidelined older faculty who were not reductionist, quantitative, and successful in winning federal grants—characteristics that Terman thought biologists should adopt because they were fundamental to physicists' earlier success.[12] Thus, in the early Cold War, Terman and other Stanford administrators viewed interdisciplinarity as a means for enhancing the school's profile in basic research by reorganizing applied, holistic fields around more abstract, reductionist topics that national security funders and military–industrial partners wanted universities to specialize in. When national security funding waned, and Americans began clamoring for holistic, applied civilian research, that model of interdisciplinarity couldn't keep up.[13]

WEAPONIZING INTERDISCIPLINARITY

The long 1970s opened the door to more, and more diverse, interdisciplinarity because the public (and many of the politicians who represented the public) wanted American academic researchers to start contributing more to solving civilian social problems other than the "problem" of global communism. That aim was communicated to scientists and engineers through a variety of different means: budgetary changes, newspaper op-eds, direct protest, and so forth. An extended anecdote from a Stanford faculty member gives some sense of how direct such communications could be:

> in the late 1960s, early 1970s, I had a contract with the military entitled "Electronics for the Battlefield." Well, when they saw that title the demonstrators thought, "This is really contributing to the war effort in Vietnam and we're going to stop this." So, several dozen of them came to the building where I had my office. They had a megaphone and they said, "We don't want any research done here on electronics for the battlefield. We want to picket the operation." I was sitting there and I looked at them and I listened, and so I went outside. This is an incident that's unforgettable for me. They told me what they didn't want. They didn't want

electronics for the battlefield that was going to have an influence on what was happening in Vietnam. After they were finished they handed me the megaphone and I said, "Well, the work I'm doing for the military is really looking at very basic questions on . . . devices and circuits. That information allows me to build devices for use in research with medical school faculty on how to prevent heart rejection and organ rejection when you do transplants. I couldn't do this research for the Medical School if I didn't have this basic research support from the military." I handed him the megaphone and they walked away.[14]

As the anecdote shows, appeals to interdisciplinarity (and to the ambivalence of the basic–applied spectrum) were helpful in turning aside accusations of complicity in the war.

The anecdote also shows that despite the fractious nature of American society in the long 1970s, American scientists and engineers shared a common understanding of what was being asked of them: the engineering scientist quoted here knew that his gesture to biomedical applications would undermine the protestors' objections. Biomedical research was one of several social problems that were consistently identified as constituting the public's priorities for American scientists and engineers. For instance, Rudi Kompfner, a Bell Labs engineer who retired into a Stanford professorship, compiled a representative list in 1971, asking "What are the most urgent problems facing the nation today?" that physical and engineering scientists like himself could help to solve.

- Decay of the big cities
- Mass transportation
- Over-Population
- Pollution of the environment
- Availability of health care
- Natural resources and energy planning
- Un- and under-employment
- Integration of races, etc.[15]

Compare that with another list from 1971 compiled by John Baldeschwieler, a Caltech chemist and Richard Nixon's deputy science adviser: "Urban decay—crime; Drug abuse; Environmental degredation [*sic*]; Health care; Unemployment; Energy; Education in deprived areas; Productivity and foreign trade; Disease care and prevention; Transportation; Natural disasters."[16] Compare those topics, as well, to the MSI students' theses in chapter 2 and the NASA projects covered in chapter 4.

So in an era when there was little consensus on anything, there was near unanimity on the civilian topics that physical and engineering scientists should adopt. There was also considerable agreement that those problems were too complex for monodisciplinary investigation. As Stephen Kline, a Stanford mechanical engineer, put it: "The kinds of questions that do and should concern the students are: Do you build the SST [supersonic transport], and what is being done about smog? Questions of this sort cannot be seen clearly through the viewpoint of any single discipline."[17] Notably, Kline was a cofounder of Stanford's Values, Technology, and Society program—today known as Science, Technology and Society (STS). This was an era in which the first STS programs formed, largely as a reaction to the growing public sense that Cold War science had contributed to the social problems listed by Kompfner earlier. Invariably, STS programs offered their interdisciplinarity as a distinguishing feature. Values, Technology, and Society courses, for instance, were "developed and taught by various combinations of scientists, engineers, philosophers, historians, anthropologists, psychologists, psychiatrists, sociologists, ethicists, and theologians—all working very closely together."[18]

Even academic administrators who were skeptical of civilian "relevance" emphasized the interdisciplinarity required to address the public's favored problems. For instance, the Stanford *Campus Report* quoted Vice Provost William Miller in 1970 that

> "Universities should and are moving from the isolation of their ivory towers, yet they should not and can not become a Red Cross or service station for every social problem. . . ." In the past, he believes, there was an overemphasis on discipline-oriented research; "just as there is probably the beginning of a slight overreaction now" with funds going to more problem-oriented research.[19]

Note how this article took "discipline-oriented" as the *opposite* of "problem-oriented." Similarly, in 1971 associate dean of engineering William Rambo critiqued the

> blind rush toward so-called "relevant research" . . . which ha[s] marginal value in our focus on engineering education. . . . But it is both possible and desirable to supplement our basic research in engineering with projects derived from real-world problems. . . . [I]f the problems are really of substance, the solutions will inevitably . . . interact with economics, politics, law, and so on.[20]

Again, note how for Rambo "real-world problems" are topics that go beyond "basic research in engineering" and that are inherently interdisciplinary *because of* their real-world relevance.

In private, Rambo was considerably more scathing toward interdisciplinarity than in this quote. A satirical speech that he gave at a private dinner on the eve of President Pitzer's resignation in 1970 gives some sense of his frustration that the School of Engineering faced so much pressure to solve civilian social problems and to work with Stanford's other schools. I quote it at length because it is important to understand just how much emotion interdisciplinarity and societally relevant research sparked in this era.

> As part of our program to update the undergraduate curriculum, I can report two new courses. The Freshman Seminar on "The Luddites and Creativity in Engineering" has met with only modest success, this despite an electro-historical focus with real interdisciplinary appeal. We have done much better, measured by attendance figures, with the senior-level offering entitled "Pornography in Science." Frankly, we see this course combination as a frontal attack on a two-sided problem, as a sort of intellectual bridge uniting C.P. Snow's two cultures.
>
> But this is not the end of our innovation binge! We are readying an exciting new course for Spring Quarter, as yet untitled. It follows suggestions of consultants from the other Schools; we are particularly gratified by their assurances that the proposed course format will bring our engineering students into the mainstream and thereby solve a problem which neither we nor our students knew existed—our student Identity Crisis. The course has a firm basis in pedagogy (we are told), and will make full use of advanced encounter techniques evolving naturally from Dramatic Play, Creative Playtime, Show & Tell, and Sesame Street. . . .
>
> The condition of our research still causes mild alarm. Our program meets completely this week's Research Guidelines, and draws close to a state of exquisite purity and infinitesimal size. Nonetheless, in accord with Community urging, we hope (and by next month's deadline) to have solved the assigned research problems clearing difficulties with transportation, low-cost housing, health care,

> the Post Office, pollution, ecology, Mrs. Mitchell, and population control. We are particularly mindful of the last topic, some local students having established well, if indirectly, the case for population control about 20 years ago.[21]

In public, however, Rambo had to be much more circumspect: even someone as powerful as Stanford's associate dean of engineering could do little to resist calls for interdisciplinarity and social relevance.

That's because the coalition in favor of a more wide-ranging interdisciplinarity was astonishingly diverse, not just at Stanford but almost everywhere. As Simon Schaffer puts it, "During the Cold War the virtues of interdisciplinarity were urged both by defence administrators and student radicals. . . . '[I]nterdisciplinarité' also appeared on protestors' banners in [the Paris uprisings of] May 1968."[22] At Stanford, elements of both the national security state and the antiwar left agreed that Stanford's postwar research model, as established by Frederick Terman and upheld by allies such as Rambo, was insufficiently interdisciplinary.

Recall that Terman thought of interdisciplinarity as a way to get engineers to be more like physicists. One of his big successes in that vein came when Stanford attracted one of the first eight Interdisciplinary Laboratories (IDLs) funded by ARPA (Advanced Research Projects Agency) in 1961. The IDLs were intended to create a new interdiscipline of materials science, and when they were transferred from ARPA to the NSF in 1972 (because of the Mansfield Amendment) they were renamed the Materials Research Laboratories (MRLs).[23] Yet when the NSF commissioned a review of the MRLs in preparation for the transfer, it found that:

1. Research in progress is in many cases not materials research
2. Research in progress is often not interdisciplinary
3. Materials science is stressed more than materials engineering[24]

That is, interdisciplinarity in the Terman model was achieved more in talk than in practice. And not just at Stanford—as Hyungsub Choi and I have argued with reference to Cornell's IDL, the military's dissatisfaction with academic researchers' attempts at interdisciplinarity meant ARPA managers weren't unhappy that the Mansfield Amendment gave them an excuse to shed the IDL program.[25]

Similarly, administrators at NASA—nominally civilian but closely linked to the military—informed Stanford in 1968 that they disliked "engineering doctoral programs . . . modeled after the more classical ones for scientists with research and studies directed at highly specialized work in physics rather than toward advanced engineering problems"—that is, exactly the kind of program that Terman had introduced and Rambo sought to maintain.[26] Instead, NASA preferred engineers' training to "deal with economic, social, legal, business, and management issues."[27]

Student protestors, meanwhile, understood NASA's and ARPA's position and embraced the interdisciplinary *methods* of national security R&D even though they rejected the *ends* to which those methods led. One of the most instructive—and explosive—instances in which antiwar groups took that position concerned the Stanford Research Institute (SRI), a contract research outfit founded in the late 1940s. In the late 1960s, the Stanford campus was engulfed in heated debate over whether to keep SRI and its military contracts. The SRI story has been told in more detail elsewhere, but without specific attention to the role of interdisciplinarity in the controversy.[28] Instead, earlier work has looked at antiwar groups' objections to SRI's research on chemical and biological warfare and counterinsurgency-inspired "land reform" in Vietnam. Protests, culminating in 1969, leveraged SRI's research on those topics to implicate Stanford in the Southeast Asian conflict. Opposition to SRI's defense work was widespread and was therefore central to antiwar groups' attempts to create a coalition that could force more sweeping reforms.

The SRI debate centered on two possible futures for the unit. The one favored by the Stanford administration was to "divest" SRI—that is, to make it independent of Stanford by selling it to itself. The outcome favored by protestors, meanwhile, was to keep SRI part of the university and turn it into Stanford's clearinghouse for applied, interdisciplinary civilian research. Proponents of the latter option noted that SRI's interdisciplinary methods were invaluable to the military—so why shouldn't they be used to benefit civil society instead? As the occupiers of Stanford's Applied Electronics Laboratory put it in 1969, the new SRI should adopt an

Interdisciplinary approach A program of research to benefit all peoples will necessarily transcend narrow disciplinary perspectives and consider the interrelation of many aspects of each research problem. The physical impact and psychological implications of technological progress must be continually projected. Stanford Research Institute has already developed an interdisciplinary approach to many of its projects. Our objection to the present content of SRI research does not keep us from appreciating the need for this approach.[29]

Perhaps unsurprisingly, divestment won out, as at MIT and other schools with similar institutes. That's because Stanford's administration was adamantly opposed to civilianizing SRI, ostensibly because there were few plausible funders for such an organization—meaning a civilianized SRI would either close down quickly or become a budgetary liability for the university.

That argument alone might not have been enough to sway ambivalent faculty members, however. So proponents of divestment (or of the status quo) stoked fears that civilianization would lead to social acceptability tests being applied, first to SRI's research but eventually to all Stanford departments. Opponents of the student protestors warned that groups like Students for a Democratic Society (SDS) threatened academic freedom and research quality because their plans would force scientists to follow the whims of the public, rather than the meritocratic guidance of disciplinary peers. In any case, proponents of divestment argued, all research is bound to upset someone, so the set of topics that no one would object to would be miniscule—that is, even a civilianized SRI would be a permanent locus of controversy and protest. This is what Rambo meant by "Our program meets completely this week's Research Guidelines, and draws close to a state of exquisite purity and infinitesimal size."

These appeals worked. Moderate faculty members who had earlier bucked the administration in supporting a ban on on-campus classified research (because they had seen classified research as contrary to academic openness) were now convinced to join the administration in supporting divestment because the alternative—an SRI controlled by the SDS—seemed to them equally inimical to academic freedom.[30] As the conservative Engineering Students for a Fair Hearing put it, using the heated rhetoric of the era, "A new precedent is being established . . . of political force and coercive tactics deciding what types of investigations are appropriate to a university. . . . [T]he knife is poised to strike. But the sacrifice of military sponsored research will not satisfy SDS long. Other lambs will be called for, and you may be one."[31] Or, as one engineer in the Systems Techniques Laboratory wrote to Kenneth Pitzer amid the SRI dispute,

> What will be the targets in the future? Already, businessmen (GSB) [Graduate School of Business] are looked upon with disfavor; physical education, particularly varsity competition, does not aid in solving the social problems of the world and furthermore is open to far more academic criticism than ROTC [Reserve Officers' Training Corps—another campus flashpoint]. Most of the engineering and physical science research (unclassified or not) directly benefits the Department of Defense, not always due to support by DOD, but mainly because DOD is the main user of advanced technology. Liberal arts departments can be criticized for only educating future professors of liberal arts (i.e. a closed society) rather than solving problems of society. It can be asked whether the enormous SLAC expenditures benefit the ghettos, or solve problems of pollution. It is clear that no segment of the University is immune from criticism, and thereby a target for elimination by another segment of the community.[32]

That is, plans to divest SRI succeeded because supporters of divestment whipped up fears that the alternative (turning SRI into a center for applied, civilian interdisciplinary research) would leave the *disciplines* vulnerable to attack for failing to serve all of society.

HEALING INTERDISCIPLINARITY

The SRI controversy throws the politics of interdisciplinarity in stark relief. Interdisciplinarity was not a neutral and universal good. Among both squares and nonsquares, it was seen as part of a package of reforms that would benefit some and harm others. More contentiously, some square scientists and engineers believed that those reforms would simply redistribute power from one elite to another. Yet if interdisciplinarity was, from some perspectives, a wedge dividing the Stanford community, from other vantages it was seen as a glue for keeping that community together.

Today, it seems implausible that anyone could've worried that Stanford would need such a glue—that it could fall apart, perhaps even cease to exist. But at the lowest point in 1969 to 1970, those fears were very real. When Stanford's board forced President Pitzer to announce his resignation in 1970 for failing to quell what they viewed as a student insurrection, the university seemed on the verge of an unending cycle of protest, backlash, and administrative chaos. Pitzer's successor, Richard Lyman, proved more adept at appeasing protestors while simultaneously placating the board by limiting students' ability to disrupt campus life. He was therefore able to regain control relatively quickly; but if he hadn't the university might well have become ungovernable.[33]

Yet despite the tensions among Stanford's warring factions, and despite the unequally distributed benefits of interdisciplinarity, almost everyone agreed that Stanford could be saved only if everyone on campus cultivated a more interdisciplinarity perspective. What follows is a tour of some of the voices representing those factions—all of which were saying much the same thing regarding interdisciplinarity.

Square science and engineering students: For instance, Stanton Glantz, a graduate student in Applied Mechanics, put it thus in the first issue of *Grindstone*, an "unofficial journal for airing controversial issues concerning members of the Stanford School of Engineering," in 1970:

> As an engineering student, I feel a great deal of anxiety about some of the attitudes I see in the School of Engineering as well as the possibility of violent confrontation between Stanford's technical and non-technical communities. . . . [W]hile the School of Engineering has been very upset about lack of understanding of Engineering on the part of the outside world, it has made very little effort to really understand why the outside world is so upset. . . . [If] we continue ignoring other people's issues and setting their arguments up as straw men, if we continue being self-righteous and continue to refuse to take a hard look in the mirror, I can only foresee more alienation, more broken windows, and more quarters lost.[34]

Glantz is an interesting case. He was more outspoken than most square engineers (at least as I've defined them), having helped set up

Grindstone. He also organized a 1971 student-run course that investigated the (military and nonmilitary) sources of faculty members' research support and criticized the university's overreliance on the military.[35] But, as the quote shows, he was quite square in his ambivalence toward the political options available to him; even though he bemoaned the attitudes of many engineers, he also opposed a ban on military research funding and was clearly heartsick about the tactics ("broken windows") some protestors used.

Square science and engineering faculty: Like Glantz, some professors saw interdisciplinary conversation as key to lowering the temperature among Stanford's warring factions. For instance, Michael Arbib, a professor in Electrical Engineering, in 1969:

> "SDS [Students for a Democratic Society] see clearly what they want to destroy, but . . . [s]tudents are too young to understand the complexity of what they're trying to change," he told an overflow crowd of students, faculty, and visitors, "and the faculty needs to be more sensitive to what's going on outside their disciplines. We need to educate for an adaptive society. We must learn new disciplines and learn how to stop hating, to learn how to temper ideology with concern."[36]

Contrary to Arbib, some professors held the view that SDS members *could* engage constructively, once you got to know them in a setting conducive to interdisciplinary dialogue rather than confrontation. For instance, the director of the Systems Techniques Laboratory (STL), Charles Shoens, sent this note to "all administrative officers of the university, all members of the Academic Senate and Academic Council, research staff, administrative assistants, secretaries, technicians, students of all political persuasions" in 1968:

> The really important outcome of this SDS march [on the STL], however, was that after the crowd had fragmented, after some had gone upstairs and talked with [various STL researchers] . . . much, much more was accomplished in these small discussion groups, and after talking with our staff members later, they unanimously expressed the opinions that (1) the kids were really likeable, (2) that they are sincere and (3) that after the discussions both sides felt that they were not really so far apart. All the discussions ended amicably.[37]

Of course, that was several months before students took over Shoens's lab for nine days and rifled the lab's files for evidence of its researchers' complicity in the war!

Antiwar students: Still, such episodes aside, there were indeed signs that the sides were "not really so far apart," at least with respect to interdisciplinarity. One such indicator was that the SDS, too, used similar rhetoric bemoaning the lack of dialogue among Stanford's factions. When Kenneth Pitzer began his short, chaotic tenure as president, for instance, SDS members leafletted the following statement:

> Welcome Pitzer!! On our campus, as in other communities, we are isolated from each other. We have no real community. And we, as a group, are isolated from people outside the Stanford Community.
>
> The research on [*sic*] our university is *not* in the interests of most people. *Much research is basic research with no intrinsic value of its own*! Other work supports the war or government policies which cause wars. Not enought [*sic*] scientific resources are dedicated to the channeling of Nature's resources and the disposal of Man's wastes (smog, pollution, etc.).[38]

Note how civilian, applied, and (implicitly) interdisciplinary topics were presented as the antidote to the atomization of the Stanford community. Elsewhere, the SDS explicitly argued that interdisciplinarity was the key to loosening the Establishment's grip:

> College students are trained to fit into vocational niches already defined for them by government, industry, and universities. The courses offered merely represent "choices" in a limited number of narrowly defined fields. . . . Each department has its own hardened arteries. . . . SDS people believe that students should be offered the opportunity to freely reach into other fields, and "broaden their minds."[39]

The institution itself: Yet, ironically enough, it was Stanford's establishment that had first made interdisciplinarity a central element of reform, well before the peak of campus protests. In 1967, Wallace Sterling (Pitzer's predecessor as president) commissioned a major review, the Study of Education at Stanford (SES), which issued ten volumes of reports starting in 1968. The scope of the SES review was such that for the historian, it offers an approximation of the institutional voice of "Stanford." It's notable, then, that prominent in its recommendations were new interdisciplinary programs and requirements designed to expose Stanford students to views from the other sides of campus: "science and technology for the nonspecialist; mathematics and computer science for the nonspecialist; interdisciplinary and problem-oriented studies in the humanities and the social sciences."[40]

Starting the next year, the School of Engineering incorporated these recommendations into a new and avowedly interdisciplinary curriculum:

> To avoid over-specialization and gain a broader viewpoint . . . students [now may] . . . map an "interdisciplinary" or an "innovative" study program. . . . [I]nterdisciplinary majors include Product Design (mechanical engineering and art), and Resource Strategy (civil and industrial engineering, economics, and humanities) . . . the "Technology and Society Program" and the "Applied Science" major.[41]

This kind of institutionally mandated interdisciplinarity was, of course, intended to serve the institution's needs. It would, above all, produce undergraduates who were happier (they could chart their own educational course), more committed to the integrity of the institution (they would be in conversation with all sides of Stanford), and better prepared to find jobs.

Administrators also hoped that interdisciplinarity would save the institution from the financial precarity that started to bite in the late 1960s. As the *Campus Report* noted in 1970,

> cutbacks in federal and state budgets, the general condition of the economy, showed that we were heading into a period of serious financial constraint. . . . Among the tough-minded decisions that have been made, the Vice Provost said, is to lay greater stress on interdisciplinary programs for undergraduates, such as human biology, environmental engineering, and "values, technology, and society. . . ." The University will stress a thoroughly trained undergraduate body which can be expected to go into graduate professional training in order to attack today's pressing problems. . . . Thus, "by reform of education, reduction of costs, with inspired leaders and creative faculty and students, we can bring Stanford to its full potential."[42]

Stanford administrators didn't just think that interdisciplinarity would allow "reduction of costs"—they also believed that it would help bring in new sources of funding. For example, the justification offered for a new Center for Interdisciplinary Research in 1972 was that

> The Stanford faculty has demonstrated a substantial and growing interest in the development of interdisciplinary research. Much of this is oriented toward what might broadly be called "problems of society. . . ." The increasing availability of external support insures that the faculty will be able to translate these interests into effective research activities.[43]

Administrators also saw interdisciplinarity as a way to recoup the institution's investment in its assistant professors: "Giving young faculty members 'a sense of educational power' through developing and evaluating new programs which cut across normal departmental lines can be 'the key factor in cultivating loyalty to the institution as a whole' according to Prof. [Vice Provost Herbert L.] Packer."[44]

At times, the various institutional motivations for interdisciplinarity were knitted together in a hoped-for virtuous cycle: interdisciplinary dialogue would help keep Stanford from falling apart, which was necessary to gain funding to keep the institution afloat, which was necessary for its researchers to address social problems in an interdisciplinary fashion, which would placate restive undergraduates and thus stanch the fractiousness pulling Stanford apart in the first place. Richard Lyman invoked something like this logic in recounting

> The Ford Foundation's grant of nearly $2 million [which] enabled the university to launch the most successful of interdisciplinary undergraduate majors, human biology. As I recall, what influenced the foundation . . . was the spectacle of a founding committee that included several chairs of Medical School clinical departments, not a group generally to be found putting time and energy into developing innovations in undergraduate education. Human biology increased Stanford's reputation for entrepreneurial activity in academic matters, perhaps more than any other development of that period, and showed that as an academic institution we were refusing to be stymied by the campus unrest.[45]

That is, interdisciplinarity was overdetermined—there were many justifications offered for it, cultural, economic, political, and scientific. Although not everyone agreed with all the justifications, *almost* everyone agreed that more interdisciplinarity was better. Though some, like Bill Rambo, grumbled, even they could see the advantages of a "tough-minded" interdisciplinarity that would win funding during a period of declining R&D budgets.

We can see here how reforms championed by the New Left were also taken up by their establishment foes as a way of preserving institutions—*and* how those same reforms paved the way for a neoliberal university that neither the SDS nor establishment figures like Bill Rambo had had in mind. Interdisciplinary, and especially individual, degree programs empowered students to tailor their educations so that they could tackle social problems rather than study abstract knowledge offered up by the disciplines. Over the long run, though, interdisciplinary and individual degrees weakened disciplinary departments. The emphasis on extrinsic, societal relevance devalued the intrinsic worth of departments' disciplinary knowledge. Without that intrinsic worth, departments had to offer students other inducements to major in their degree programs; the most durable of those inducements has been the supposed market value of their degrees. Similarly, interdisciplinary funding instruments encouraged faculty members to present themselves as individual experts in specific topics, rather than as members of disciplines with a collective expertise. Finally, administrators who worried that their universities might fall apart cultivated faculty members and students who were loyal to the institution as a whole and discouraged loyalty to departments or extrainstitutional disciplinary communities.

The end result was a university made up of atomized individuals, each pursuing a research portfolio or pedagogical trajectory tailored to their personal preferences, rather than pursuing a path that would integrate the individual into a collective body of expertise. As was often the case, Richard Lyman foresaw this outcome more clearly than almost anyone else. As he told the *Campus Report* in 1971, "The trick will be to assure effective linkages between so-called traditional disciplines and skills, on the one hand, and problem-oriented or experiential education on the other. Without such linkages, the contemporary American university could descend . . . into the biggest and most over-financed federation of bull sessions in history."[46]

At the time, it was the SDS, Congress, and even the national security state who wanted Stanford to emphasize "problem-oriented" research and education, with administrators such as Lyman and Rambo left to defend the "traditional disciplines." Today, such a counterweight hardly exists: university administrators are as invested in problem-oriented interdisciplinarity as anyone and view their universities as accountable to donors, funders, markets, student consumers, and various publics—but hardly accountable to bodies of disciplined knowledge. Whether that is a good thing or not is beyond the scope of this book. My point is that the conditions for that kind of university were created in the turmoil of the long 1970s, even though virtually no one said that that was the kind of university they wanted.

NEW VISIBILITY

So what actions did people take, and what *did* they say they wanted those actions to lead to? Square scientists and engineers' enactments of interdisciplinarity at Stanford in the long 1970s can be placed, more or less, in three broad categories: new visibility for extant interdisciplinary projects; new interdisciplinary projects modeled on the older, newly visible ones; and new interdisciplinary institutions. That list runs in rough chronological order: in the first instance were interdisciplinary projects that predated the student protests and budget shortfalls of 1969 to 1970 but that suddenly gained the limelight in those years; with the onset of protests, new interdisciplinary projects were rolled out hastily, often with the result that they failed after a few years; and finally new interdisciplinary institutions appeared, some of which took more time to set up but which prospered over the longer term.

One of the more striking examples of the first category was the Optacon (or OPtical-to-TActile CONverter). This was a tactile reading aid for people who would otherwise read Braille documents. With the Optacon, a user would run a photoelectric scanner over a document with one hand, and the dark-and-light patches imaged by the scanner would be turned into a pattern on a set of pins underneath the user's fingertips on the other hand. With practice, users could feel the patterns of pins and translate them into the corresponding printed letters. Rudimentary attempts at this technology dated to 1963, well before the budget crises and student protests began. John Linvill, chair of Electrical Engineering, led the Stanford side of the project, with another electrical engineer, James Bliss, joining from SRI. Linvill's inspiration, in part, was personal; his daughter, Candace, was blind and eventually became the Optacon's most prominent user. Both Bliss and Linvill, however, started with national security funding from the Office of Naval Research, NASA, and the Air Force.[47]

By 1967, cuts in national security funding were starting to be felt. Linvill, almost always ahead of the curve, could sense that he and his colleagues would need to find alternative sources of research money. Thus, in a memo (which he circulated to members of his department) responding to a query from the SES committee, Linvill outlined how interdisciplinary collaboration would allow electrical engineers to tackle problems of civil society and therefore win funding from civilian agencies:

> Stanford University can and should become more effective in studying and attacking the problems of today's society. Electrical Engineering, with its aim to bring technological tools to the solution of man's problems, is interested to join with other departments in working on these contemporary problems. . . . These problems cannot be attacked within a single discipline. Technology, by itself, is not sufficient. Multidisciplinary approaches must be used. . . . Financial support for the research must be found outside the university, however, as has been the case in our DOD and space research.[48]

Later that year, Linvill and Bliss put those words into action by submitting a proposal to the federal Office of Education for money to buy equipment for fabricating integrated circuits that would make the Optacon fast enough to compete with Braille.

That equipment was housed in the Integrated Circuits Laboratory run by another Stanford electrical engineer, James Meindl. With that equipment, and Meindl's joining as a third member of the team, Optacon development accelerated. "By April 1969 three improved reading aids . . . were being used in reading experiments with blind students."[49] Soon, James Angell, Robert White, and other electrical engineering faculty members were making use of the same ICL equipment to develop other biomedical and disability technologies: cochlear implants, blood flowmeters, blood gas analyzers, and others.

April 1969 was also the peak month for research-oriented protest at Stanford. In response, administrators moved to show that Stanford was already doing interdisciplinary, civilian, problem-oriented research that might appease the protestors. Thus, they put the Optacon on a pedestal in outlets such as the university's annual *Electronics Research Review*. Prior to 1969, the reading aid had received terse descriptions buried at the back of these annual reports. But in that year, the Optacon was given a gushing ode on page *one*, emphasizing that "Integrated circuits, principally developed to the present stage for space and military applications, are powerful tools for the solution of human problems, channeling interest of the imaginative graduate student to important social problems and providing a vehicle for interesting technical research."[50]

Likewise, when a faculty committee organized a 1971 research "exposition extending over the entire campus community. Interdisciplinary activities particularly are desired,"[51] they highlighted

> the use of holography (three-dimensional laser holography) in displaying art objects, a continuous interactive opinion poll with a computer terminal, the use of electronic devices to help the blind and to measure medical data inside the body, and a presentation of computer-composed music.[52]

Note how all these highlighted projects knit engineering and applied science into collaboration with disparate parts of the campus with which they had not previously interacted much: art, political science, medicine, and music. Indeed, the Optacon formed the basis for particularly wide-ranging collaborations between its inventors (in electrical engineering) and graduate students in Stanford's School of Education and departments of Mechanical Engineering and Psychology, as well as a neuroscientist at the Smith-Kettlewell Eye Research Institute.

For that reason, the Optacon was repeatedly and prominently invoked to advertise Stanford researchers' ability to cross disciplinary lines and (thereby) address civilian problems. The Optacon also provided evidence that Stanford was willing and able to leave the ivory tower and reach out to communities: in 1970, "Carolyn Weihl of SRI conducted a successful experiment in Optacon reader training . . . at the Monroe Elementary School in Campbell, California . . . [while] Susan Melrose, a Stanford student, taught beginning Optacon reading to a group of blind high school students . . . in San Diego."[53] Finally, the Optacon showed that Stanford's engineers could attract funding from a variety of military *and* civilian agencies: Linvill, Bliss, and Meindl sought funding from the NIH, NSF's RANN program, and the DOD's Joint Services Electronics Program, in addition to the Office of Education. In every way, the Optacon showed how Stanford's square scientists and engineers could help the school navigate through the various and conflicting demands made on them in the long 1970s.

NEW RESEARCH DIRECTIONS

It should be evident from the Optacon example that square scientists and engineers were doing interdisciplinary, societally relevant, civilian-oriented research before the long 1970s—I am not arguing that such research appeared out of nowhere. But the overwhelming sense from the contemporary sources I've consulted is that people *said* that much more interdisciplinary and societally relevant research was being called for—and being done—around 1970 than even a few years before. When people responded to those calls, they did have older examples like the Optacon to look to. Yet those examples had only recently become known to the entire Stanford community—and the main reason for that new awareness was that the university administration was anxious to uphold those examples as a model for others to follow.

In this section, I look briefly at two of Linvill's colleagues in the Stanford Electrical Engineering department, Calvin Quate and Rudi Kompfner, who tried to follow Linvill's example.[54] Quate had been one of the first PhD graduates, in 1950, of Terman's redesigned Cold War Stanford, after which he joined Bell Labs, where he worked with Kompfner for almost a decade. After returning to Stanford to join the Microwave Laboratory and the departments of Applied Physics and Electrical Engineering in 1961, Quate became interested in the coupling of electromagnetic and acoustic phenomena in crystals. That work, supported by the Navy, was oriented to signals processing for military radar and communications, though variants of this technology (surface acoustic wave, or SAW, devices) later found their way into civilian cellphones, garage door openers, televisions, and other gadgets.

In 1966, while "sitting around the swimming pool," Kompfner suggested that Quate's SAW devices could form the basis for a microscope.[55] That is, the SAW device could pulse an acoustic wave through a sapphire lens, thereby focusing the wave onto a sample—a sonic equivalent for how a lens focuses light onto a sample in an optical microscope. In the acoustic microscope, the sonic wave would then pass through (or be absorbed or reflected by) the sample, be detected by a transducer, and be converted into an image. Images generated in this way would have several advantages over optical microscope images. In particular, the acoustic microscope could image samples that were transparent to optical wavelengths of light and that therefore wouldn't show up in an optical microscope. Also, different shades in an acoustic micrograph would correspond to variations in the elasticity of different parts of the sample and would therefore tell users something about the mechanical, rather than optical, properties of the sample. A third advantage was not mentioned much initially but became important later: the acoustic microscope could image the inaccessible interior of a sample, in the same way that a medical ultrasound provides images of the interior of a human body.

Slowly, Quate followed Kompfner's suggestion, diverting money from his Navy grant for initial investigations. However, even in this early phase he and Kompfner framed the microscope in terms of interdisciplinary collaborations and civilian applications. Kompfner later remarked that "the original intention behind the Stanford work was to make an instrument . . . [for] studying biological objects."[56] That limitation to biological samples is curious, because earlier proposals for an acoustic microscope (which Quate and Kompfner knew about) were oriented to characterization of crystalline materials used in military technologies.[57] That is, despite counterexamples (and their own previous military funding), Quate and Kompfner at this point preferred to articulate the acoustic microscope in terms of its interdisciplinary, civilian relevance.

Soon, Quate began pushing for Electrical Engineering to make a joint appointment with the medical school, because that was where he hoped that the acoustic microscope would be used. However, Linvill, despite his and Meindl's own collaborations with biomedical researchers, warned Quate

> that the Biophysics Lab has not succeeded in their efforts to form a joint program with the School of Medicine; and, furthermore, the Administration is acutely aware of this failure at the present moment. Is the School of Medicine the only alternative? I would think a connection with the Biology Department could be just as rewarding and allow for the same sort of interacting research programs.[58]

Nevertheless, Quate began talking with medical school faculty members, one of whom suggested the nonprofit John A. Hartford Foundation as a funder to wean acoustic microscopy from Navy support.

In 1969, Quate obtained a two-year, $170,000 grant from the Hartford Foundation, followed by a three-year $300,000 grant conditional on Quate's working more closely with biomedical researchers. Over the next few years, Quate did make progress along those lines: his group used the acoustic microscope to image samples prepared in the Stanford departments of surgery, pathology, psychology, radiation oncology, hematology, anatomy, medical microbiology, and biology, as well as from biomedical researchers at the Mayo Clinic; University of California, Irvine; Albert Einstein School of Medicine; University of North Dakota; National Cancer

Institute; US Department of Agriculture; and the Faculty of Medicine in Montpellier, France.

Yet when Quate tried to leverage those samples into a grant from the National Institutes of Health, he was told that he had not achieved the right degree or kind of interdisciplinary collaboration. Even Hartford's program director chided Quate in 1976:

> Although you have had some collaboration from medical scientists in various fields, it has been scattered, transient and too inconclusive to be of value in obtaining NIH support. It is also apparent from the criticism of the NIH review group that you have still not put together a specific, detailed biomedical research protocol that within a given period would develop convincing answers as to the microscope's potentials. For the same reason, a similar proposal would not pass our scientific review and evaluation procedures. I surmise that at least some of the trouble with NIH lies in your understandable lack of experience in devising a sound and workable research proposal in the alien field of biomedical research. This is an art in itself and must surely require the active collaboration, dedication and commitment of a leading medical research scientist—the more prominent the better.[59]

Thus, starting around 1974, Quate began to look beyond biomedicine for new collaborators, and hence for new kinds of samples to image with the acoustic microscope.

Kompfner, meanwhile, had retired from Bell Labs in 1973 and took up a half-time position at Stanford to work more closely with Quate. Together, they began to question their starting assumption that the acoustic microscope was most relevant to the study of biological materials. I am hesitant to say that changes in funding and public sentiment *caused* that shift; Quate's explanation was that "as this instrument evolved we found that it was much more versatile" than originally imagined, because in reflection mode it could image crystalline materials, whereas in the original transmission mode it was more suitable for imaging biological materials.[60] Still, it is fair to say that Quate put more effort into using the microscope to image organic materials when the funding and political environment tilted in favor of biomedical topics in the late 1960s, and then put more energy into using it to image *inorganic* materials when the funding and political environment tilted in a different direction from 1974 onward.

That the turning point came in 1974 was a result of both push and pull. By then, Quate felt pushed away from biomedical applications because of the disinterest of potential collaborators and funders in those fields. But he was also pulled toward applying the acoustic microscope to inorganic samples by the OAPEC oil embargo in the fall of the previous year and by the deterioration in the economy that followed the embargo. The embargo led the Nixon administration in the summer of 1974 to undertake a massive review, announced as Project Independence, of ways to increase domestic production of energy while cutting consumption, with the stated aim of making the US energy independent by 1980.[61] Included in the package were a number of changes in federal R&D, such as the creation of the Energy Research and Development Administration (ERDA) and the Solar Energy Research Institute (SERI), as well as the reorientation of much of the NSF's RANN program toward energy topics.[62]

Quate was clearly paying attention to the federal shift toward energy. In 1974, he and another Stanford applied physicist, Ted Geballe, responded with a proposal to RANN to develop materials for superconducting power transmission lines.[63] Later, Quate turned acoustic microscopy in this direction as well by applying it to determination of carbon content in coal formations.[64] He also worked with SERI to apply the microscope for characterizing silicon photovoltaic cells.[65]

Over the longer term, though, the energy crisis was less important for Quate's research than the nation's economic anxieties. Starting in the mid-1970s, the newspapers began to fill with worries about economic competition from Japan across a range of manufacturing industries: consumer electronics, automobiles, shipbuilding, semiconductors. Semiconductors, in particular, caused concerns within the US R&D establishment after Japan's government announced a major program in 1975 to bring Japanese semiconductor firms up to the global state of the art. That announcement prompted significant reorganization of parts of the US research system and reallocation of research funding.[66]

Luckily for Quate, microelectronics was a topic where he could form collaborations around the acoustic microscope that remained within the boundaries of his home disciplines of electrical engineering and applied physics—that is, collaborations that avoided the frustrations he had experienced in dealing with biomedical researchers. Thus, he soon developed a new application for acoustic microscopy: quality control inspection of integrated circuits. Quate easily acquired sample integrated circuits, and sometimes funding for research into that application, from firms with nearby sites in Silicon Valley: Hewlett-Packard, Avantek, IBM, and Fairchild Semiconductor.[67] He also noted to federal funders that microelectronics researchers—unlike life scientists—were eager to build their own acoustic microscopes or buy them when they became commercially available.

If the microelectronics applications of the acoustic microscope allowed Quate to return to familiar disciplinary territory, they also allowed him to return to a familiar funder: the military. From 1971 onward (and especially after the end of conscription in late 1972), campus protests waned sharply, as did stigmatization of military funding. Slowly, Quate moved the acoustic microscope back under the military's umbrella by claiming that it could detect microcracks in battlefield equipment (e.g., jet turbine blades or reactor shielding in nuclear submarines).[68]

Still, money remained tight, and Quate prioritized applications that could combine funding from *both* military and civilian sources. That included microelectronics, which the military viewed as a mission-critical technology. The military wanted specialized, exotic integrated circuits with extreme capabilities (e.g., radiation hardening or very high speed for cryptography), and it wanted to purchase them from American firms that could presumably be better trusted to keep such technology secret. But the Pentagon also wanted to purchase conventional integrated circuits from plants located in North America, so that supplies could not be disrupted by Soviet submarines in case of war. Thus, military and civilian agencies experimented with cooperative support for the domestic microelectronics industry. In 1976, Quate obtained a grant to develop acoustic microscopy for microelectronics characterization from one such program, the Innovative Measurement Technology for the Semiconductor Industry initiative, jointly funded by the National Bureau of Standards and DARPA.[69]

With that support, as well as a later cooperative funding scheme between the NSF and IBM, Quate was able to develop the acoustic microscope into a reliable tool for nondestructive testing and quality control in semiconductor manufacturing. Microscope manufacturers started to market their own versions, which were widely adopted by semiconductor firms in the 1980s. Contrast semiconductor manufacturers' evident interest in the acoustic microscope from 1976 onward with what happened in 1974, when Quate had asked Varian Associates to market an acoustic microscope for biomedical users:

> Their response to our [Varian's] inquiries has been uniformly negative. They point out that the acoustic microscope—while it has certain attributes that are attractive—does not yet have the resolution that is achievable with the optical instrument. It, therefore, cannot be compared on a one-to-one basis. The training and background of such people as pathologists and histologists cannot be easily transferred in a way that will allow them to interpret the acoustic micrographs with the same facility that is now done with the optical photos. They further argue that the work to date has been done in an electronics laboratory with people possessing a background in physics and electronics. No one in biology or in medical research has carried out a significant piece of research with this new instrument. Until that is accomplished the size of the market is in question.[70]

Thus, the acoustic microscope shows that the near-universal consensus that applied research oriented to societal problems *had* to be interdisciplinary was wrong. Quate initially adhered to the consensus view and tried to take an interdisciplinary path toward societal relevance. But that route led only to frustration and failure. When he tried again, this time following a path that stayed within his own discipline, he succeeded. True, the kind of societal relevance that he achieved—aiding the domestic semiconductor industry—was not what most reformers were pushing for circa 1970. But it was the kind of societal relevance that civilian federal agencies were looking for from the middle of the 1970s onward.

His experiences led Quate to sour on the reforms to the science system that had encouraged interdisciplinary collaboration with biomedical researchers in the first place. In 1979 he wrote the director of the NSF, Richard Atkinson, to compliment him on the new joint funding program with IBM and to complain about earlier reforms:

> there has been great improvement [at the NSF] over the past few years in the areas that are of interest to me. The IBM program is much better than its predecessor—RANN. . . . You can see it in the people, Dick. Their spirit is good. They now talk about the evolution of programs and the extended planning. That is in contrast to the revolution and abrupt changes in course that took place in those days before you took over as Director.[71]

This is one of the clearest windows into what does and doesn't encourage square scientists and engineers to conduct "responsible" research. Quate was willing to try working with biomedical researchers and to vie for RANN funding in the early 1970s—clearly he was not immune to calls for a more socially relevant research paradigm. But that paradigm was only partially, hastily, and reversibly put into place. Incentives were established for physical and engineering scientists to seek collaborations with life scientists but not vice versa, and funding priorities were unstable and unpredictable. In that atmosphere, Quate and many of his peers gave up and went back to funding sources and collaborators that were more stable and familiar.

Quate should probably not have been surprised by the frustrations of interdisciplinarity. As we've seen, Linvill cautioned him that cooperation with biomedical researchers would be harder than he imagined. Yet Quate forged ahead anyway, feeling confident that he would succeed in getting biomedical researchers and even clinicians to adopt his microscope, and feeling unhappy when they didn't. That is, indeed, one of the hallmarks of square scientists and engineers' passage through the long 1970s: a high-modernist optimism that they could solve any problem and that their solutions would self-evidently be attractive to any audience—followed by pessimism and disengagement when that turned out not to be the case.

Nevertheless, the mark left on Quate by the long 1970s was not entirely fleeting. He gained a lasting taste for patenting and licensing his work. He learned to secure funding from philanthropies, corporations, and civilian agencies in addition to the military services. And he acquired a cautious yen for interdisciplinary collaboration, particularly with the life sciences. Thus, when he switched his lab to work on a new tool, scanning tunneling microscopy, in the 1980s, he wrote enthusiastically in his notebook that STM could ensure that "someday we should return to cells."[72] The same could be said of James Meindl in this chapter and Virgil Elings in the previous chapter: all three moved from almost no interest in biomedical applications before roughly 1967 to significant interest in the 1970s, back to low levels of collaboration with the life sciences in the 1980s, and then a return to biomedicine after the end of the Cold War. As I argued in chapter 1, the long 1970s *can* be seen as a brief interruption followed by a return to normalcy; but they can *also* be seen as establishing a new normal that was interrupted in the Reagan era and resumed in the 1990s.

NEW INSTITUTIONS

The third enactment of the new interdisciplinarity at Stanford was the creation of institutions to unite disciplines and connect them with resources and audiences. As figure 3.2 (from Andrew Nelson) shows, the onset of the long 1970s initiated a boom in the founding of interdisciplinary centers at Stanford: the number of such centers doubled in the year of peak student protest, and then continued expanding at roughly seven times the pace after 1970 compared to before. Many of these centers were anchored in the social sciences because as Jamie Cohen-Cole has shown, an ideology of interdisciplinarity already gripped American social science in the 1950s.[73] Specifically with respect to Stanford, Beatrice Cherrier and Aurélien Saïdi have shown that the school had a long tradition of ties between engineers and economists.[74] Thus, it should be unsurprising that when the university founded a Center for Interdisciplinary Research in 1972, it used an existing interdisciplinary social science center, the

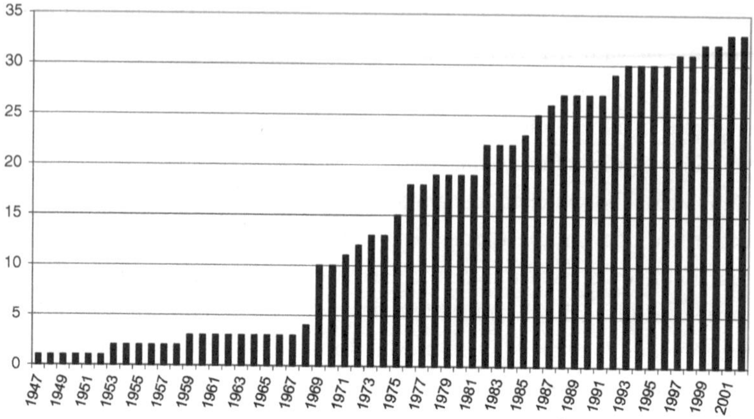

Figure 3.2
Number of degree-granting interdisciplinary centers at Stanford. *Source*: Data compiled by, and graph courtesy of, Andrew J. Nelson.

Institute for Public Policy Analysis, to nucleate a broadening of "the scope of interdisciplinary research activities to include engineering, the physical sciences, the professional schools and humanities, as well as economics and the social sciences."[75]

Yet some of Stanford's most successful new interdisciplinary centers focused on *humanistic* application of and reflection on *engineering* expertise. As mentioned earlier, in 1971 Stephen Kline, in mechanical engineering, assembled a diverse coalition for the new Values, Technology, and Society program. Although Kline's relationship with campus activists was often tense, reform-minded faculty members still "counterpoise[d] the brilliantly-conceived program in Values, Technology and Society to which Kline and [Walter] Vincenti are devoting so much creative energy" "against the unregenerateness and glacial metamorphosis" other colleagues showed in the face of calls for change.[76]

The institutionalization of interdisciplinarity didn't happen only through the founding of new centers: there were a few existing labs that began to broaden their disciplinary scope as they turned toward more civilian-oriented problem areas in the late 1960s. As with individual researchers, the new emphasis on interdisciplinarity in these existing centers was stimulated in part by a decline in funding from the military. A particularly salient example was the Stanford Artificial Intelligence Laboratory (SAIL), which was founded with ARPA funding in 1963 as the Stanford Artificial Intelligence Project.[77] By 1968, SAIL had leveraged the growing interdisciplinarity of its more than one hundred members to expand its funding base to include NASA, the NSF, and the National Institute of Mental Health, though the Pentagon still contributed half to two-thirds of its annual funding.[78]

That ahead-of-its-time interagency/interdisciplinary attitude, plus SAIL's ownership of significant computing power, led to one of Stanford's most profitable and influential new interdisciplinary institutions. In 1964, an undergraduate SAIL programmer, David Poole, was playing the tuba in the Stanford orchestra near John Chowning, a percussionist and graduate student in the music department. Poole heard that Chowning was interested in computer-generated music and suggested that Chowning use SAIL's computers. Les Earnest, SAIL's deputy director, was supportive, and Chowning became a fixture in the lab. Over the next decade, Chowning acquired enough international stature, funding, and local supporters to spin off from SAIL and found the Center for Computer Research in Music and Acoustics (CCRMA—pronounced "karma," naturally).

From the beginning, CCRMA drew favorable publicity for Stanford in venues such as *Newsweek* and *Rolling Stone*. Notably, every popular article about CCRMA in the 1970s—and there were dozens—mentioned the word *interdisciplinary*. Most articles dwelt on how wide-ranging the center's collaborations were. That, in turn, allowed CCRMA researchers to bridge the two main forms of interdisciplinarity in contention at Stanford: on the one hand, Terman's paradigm in which physics would teach other disciplines its approach to basic research; and the newer paradigm in which disciplines would cooperate to demonstrate the relevance of academic research to civil society. As Chowning and colleagues put it in a 1974 joint proposal to the NSF and the National Endowment for the Arts,

CCRMA's "rich interdisciplinary environment" would allow musicians to "produce fundamental knowledge" for the "science of music," while at the same time "creative minds from diverse disciplines" would together "produce works of art which reflect the scientific-technological riches of the present."[79]

Unusually for a music center, CCRMA also approached industry. Indeed, contacts with firms were one thing that engineering faculty members could bring to CCRMA in return for the opportunity to work on something new and expressive. For instance, in 1974 James Angell (one of the Stanford electrical engineers forging collaborations with the medical school) received a Research Development Fund (RDF) grant for work on "digital systems for the generation of musical sound." The RDF was a pot of money provided by the NSF (matched by Stanford's own funds), which the university used "to stimulate innovative research endeavors by both junior faculty and faculty who were redirecting their research interests."[80] Note the criterion "redirecting"—this money was intended for those who were changing their research portfolio; that is, in most cases those who were moving away from dependence on military funding and toward applied, civilian-oriented interdisciplinary projects. Angell used his RDF to support Dick Moore, a graduate student in electrical engineering and research assistant at CCRMA who later joined the music faculty at University of California, San Diego, and directed UCSD's equivalent of CCRMA. Together, Angell and Moore also submitted a proposal "to Intel Corporation, who have told us informally that they would be happy to provide us with any amount of . . . integrated circuits."[81]

Chowning, meanwhile, had started patenting ideas for electronically generating sounds in the mid-1960s, eventually striking gold with his invention of frequency modulation synthesis in 1967. That patent was licensed by Yamaha in 1975 for their line of electronic instruments and became one of Stanford's most profitable patents. Over time, as Andrew Nelson has shown, CCRMA created an interdisciplinary environment in which the unique demands presented by musicians elicited innovations from engineers that turned out to have wide applicability and commercial value.[82]

Yet despite the fame and later fortune that he brought to the school, Chowning, who had taken an assistant professorship at Stanford after receiving his PhD, was initially denied tenure in 1973. Accounts of his tenure case point to a classic problem of interdisciplinarity: musicians couldn't judge his electronics, and electrical engineers couldn't judge his music.[83] Probably some of Stanford's squarer engineers also found Chowning a bit too groovy to take seriously—including Quate, who was associate dean of the School of Humanities and Sciences from 1972 to 1974 and played a role in denying Chowning tenure.[84] Curiously, though, Quate himself filed a patent disclosure in 1974 for an "electronic device for converting written music to audible sound"—perhaps another indication that friction between squarer and groovier researchers was more about style and presentation of self, and *not* primarily about research topics, because on those there was substantial convergence.[85]

In any case, when Chowning challenged the tenure decision, Stanford agreed to review it, but only reversed its decision *after* Yamaha sought his help in implementing the patents that they had licensed from Stanford, and after other schools had offered Chowning tenured posts. The significant external funding for CCRMA that Chowning brought in probably also helped, though surprisingly it does not seem to have been determinative. Yet the fact that Chowning was promoted to full professor in 1978 shows how far his brand of interdisciplinarity traveled from precarity to unassailability in less than a decade.

CROSSING DISCIPLINES, ENTERING MARKETS

Here we begin to see how the new forms of interdisciplinarity that came into being around 1970 facilitated Stanford's transition from a Cold War university to an entrepreneurial university. Links to industry weren't, of course, completely new in the long 1970s. The university long had links to local entrepreneurs such as the Varian brothers and William Hewlett and David Packard. The postwar rise of Stanford's School of Engineering was fueled in large part by recruiting industrial researchers. The department that I've discussed most in this chapter, electrical engineering, was utterly dominated by people who had previously worked in industry. Bell Labs, in particular, was the former employer of Linvill, Quate, and Kompfner, as well as the notorious Nobel Laureate semiconductor physicist William Shockley, who had coinvented the transistor there before moving back to his hometown of Palo Alto. Shockley's very public eugenicist views were a frequent target of campus protests.[86] Shockley aside, ongoing contact with Bell Labs was incredibly valuable to Stanford researchers. Chowning, for instance, learned computer music from Bell Labs' John Pierce and Max Mathews, whereas in 1970 Quate called on Pierce for help getting the Hartford Foundation to fund his acoustic microscope research (to which Pierce replied "is [the microscope] fire resistant and bomb proof? Or, are things now quiet at Stanford?").[87]

Still, even if Stanford was friendly with industry well before the long 1970s, there is some consensus that those years did mark some kind of turning point in Stanford's orientation to commerce. Before then, the university recruited personnel (and their knowledge) *from* industry, while sending its graduates *to* industry.[88] The disciplinary scope of those exchanges was generally rather limited. Also, in the pre-1970 paradigm, Stanford was careful not to conduct research that could be interpreted as competing with industry, and it was generally hesitant to aid faculty members who wanted to patent their research or form start-up companies.[89] Even experienced industrial researchers such as Linvill and Quate patented virtually nothing while at Stanford before 1969—but did start patenting in the long 1970s. It's notable, therefore, that 1969 was the year that Stanford formed its Office of Technology Licensing (OTL), which has since become famous for its role in the emergence of the biotechnology industry, and in particular for its help in patenting the recombinant DNA techniques that ensured the success of Genentech and other early academic biotech start-ups.

There is, therefore, a considerable literature on the OTL and its founder, Niels Reimers. Much of that literature does not acknowledge that 1969 was a peculiar year for Stanford.[90] A few studies, particularly by

Doogab Yi and Jeannette Colyvas, do note the backdrop of calls for societally relevant research but without drawing much of a connection to the student protests and fears of Stanford's imminent demise.[91] And I, too, am hesitant to draw too direct a connection—maybe the OTL would've been founded even in the absence of protest. I am, however, comfortable arguing that in its early years the OTL was, like CCRMA, one of the new Stanford institutions that drew on the idealistic language of societal relevance in justifying its existence. For instance, Colyvas quotes a Stanford vice president saying, in 1970, "I am pleased by the high societal value of inventions we have licensed [such as] a potential cure for viral infections and a potential ecologically safe insect control."[92]

The early OTL also, and relatedly, invoked interdisciplinarity as a prerequisite for societal relevance. For instance, when the OTL negotiated a licensing agreement for Quate's patents on the acoustic microscope, it emphasized that technology's interdisciplinary biomedical applications and framed its microelectronics applications secondarily and purely in terms of their relevance to medical and disability technologies:

> A license agreement for commercial development of a microscope that uses ultrahigh-frequency sound waves to see into living cells and other materials has been signed with Stanford by the American Optical Co. of Buffalo, N.Y. The new device, called a scanning acoustic microscope, was developed by scientists at Stanford. It may prove useful in early detection of tumors and other medical diagnosis, as well as for searching out flaws in such manufactured items as semiconductors and integrated circuits. . . . The resolution is fine enough to study human cells, and the acoustic image shows features that the optical image does not. By comparing the two, diagnosticians should be able to tell more about cell functions and disorders. Since ultrasonic waves can penetrate materials, they also can be used to show defects in microscopic integrated circuits. The new integrated circuit technology already has provided the world with life-saving heart-pacers, hearing aids, minicomputers, and many other benefits, though it is still in its infancy.[93]

That announcement was from 1975—by which time it was already known that the biomedical market for the acoustic microscope was disappointingly small, whereas semiconductor manufacturers were showing real enthusiasm for its application to nondestructive testing of commercial integrated circuits. The fact that the press release downplayed the microscope's real monodisciplinary achievements and emphasized its disproven interdisciplinary appeal shows that the early OTL was focused on presenting itself, at least for public consumption, as an instrument for aiding society rather than for generating revenue—and that it viewed the interdisciplinarity of the inventions it patented as key to their societal relevance.

Like most experiments, the OTL wasn't a complete success at the start. Quate, for one, grew increasingly frustrated with the licensing agreement that OTL had negotiated with American Optical. Linvill, Meindl, and Bliss seem to have been unimpressed by Reimers and managed the patenting of the Optacon and other biomedical technologies themselves. The OTL's struggles in its early years, when it was dealing with technologies such as pesticides and acoustic microscopes, contrast strikingly with its later success in biotechnology. That's probably why little of the literature on the OTL delves into its more varied and interdisciplinary efforts in those first few years. And by the time biotechnology took off, the protests and counterculture were a distant memory, so the ways in which they shaped the OTL are hard to locate.

Thus, the OTL has come to be remembered as a harbinger of post-1980 entrepreneurial Stanford, rather than a creature of the long 1970s. We should, however, see the OTL's founding in the context of the crises besetting Stanford in the late 1960s, with its early operations defined by the need to connect with the discourses of interdisciplinarity and societal relevance that arose in those crisis years. As with Virgil Elings and David Phillips's start-ups in chapter 2, the OTL was initially framed as a means of getting socially useful technologies out of the academic lab and into the world, where they could actually be of use to society. That justification was, in fact, crucial in persuading a reluctant Stanley Cohen to allow the OTL to shepherd his and Herb Boyer's recombinant DNA research through the patent process.[94] But once that patent became worth millions of dollars and seeded an industry worth billions, profit became the only justification for the OTL worth invoking; the original justifications relating to societal relevance could be discarded from memory.

Moreover, by the time recombinant DNA was patented, opposition to military research had been pushed out of the mainstream, and the turbulent days of 1969 to 1970 were becoming ancient history. Even as early as 1980, Donald Kennedy—Lyman's successor as Stanford's president—could give a major speech on the quandaries of universities' getting involved with the biotech industry in which he dated civil society's doubts about basic research to 1976—leaving out *any* mention of the SDS or the heated debates about basic research that had roiled his campus just ten years earlier.[95] It's unclear whether Kennedy really forgot that era or purposefully ignored it; but it is clear that by 1980 he could leave it out with little fear that his audience would object.

That said, the debates of the early 1970s never disappeared entirely. The organization of research rarely changes wholesale; new approaches are simply layered on old ones. As a consequence, *contestations* of the organization of research rarely go away; new critiques simply appear alongside old ones. Let me close by pointing to a particularly clear example of this layering from 1983. That year, Stanford's Committee on Research (C-RES)—then chaired by James Meindl—confronted this docket of issues:

- Review of "Weapons-Related" Research at Stanford. . . .
- Copyright Policy. . . .
- Internal Audit of Faculty Consulting Activities. . . .
- Indirect Costs. . . .
- Employee Patent Agreement Forms. . . .
- Pending Matters . . . potential conflict of interest situations in university-industrial relations.[96]

Here we see a bundle of controversies, each rooted in a different era of Stanford's past: the early Cold War; the antiwar protests of the late 1960s; and the entrepreneurial Stanford that emerged over the course of the 1970s.

The "indirect costs" item, for instance, points to an enduring vestige of the Cold War research paradigm. Indirect costs are those incurred by a university's research infrastructure and are what a grant's overhead rate ostensibly pays for. Securing federal grants and their overhead to finance rapid expansion was the core of Stanford's early Cold War strategy. Yet although indirect costs are often described by the shorthand of "paying to keep the lights on," in fact they are not always so utilitarian. The 1983 C-RES discussion was related to Stanford's move to raise its overhead rate precipitously to make up for the shortfalls in federal research funding that colored most of the other events in this chapter.[97] That led to a windfall that the university spent with abandon. As a result, in the early 1990s the Navy accused Stanford of using indirect costs to pay for items not obviously necessary for research, most famously upkeep on a yacht and flowers and antiques for Donald Kennedy's home—a scandal that would result in Kennedy's resignation.

The "weapons-related" issue on the C-RES agenda, meanwhile, stemmed from objections made by researchers at the Stanford Linear Accelerator Center to being forced to participate in the lab's contract research on, among other things, topics related to radiation from nuclear weapons. This, clearly, represented a vestige of the old objections of the SDS and of reform-minded members of the Stanford community such as Stanton Glantz. Yet in 1983, this case elicited no sit-ins, no arson, no protests—just some debate within the C-RES committee and a report stating that the objectionable research should be evaluated on its scientific merit rather than its possible relevance to weapons.

Of the other subjects on the C-RES docket, the copyright policy, consulting activities, patent forms, and university-industrial relations items all centered on the growing pains of the new entrepreneurial university. The new modes of market participation that had entered the university over the course of the long 1970s raised real concerns that academic research would be kept secret from the public, that faculty members would use their students and their public funding for private gain, and that commercial incentives might harm the quality of Stanford's research. Activists interested in reforming academia were not raising such concerns very loudly in the late 1960s. By the 1980s they were, though. I have tried to show that the criticisms of monodisciplinarity and military funding that flourished in the late 1960s pushed academic researchers toward the marketplace, where they encountered a new set of criticisms in the 1980s. That is, some of the responses to the crises of the beginning of the long 1970s *became* the crises of later eras.

Other responses to the crises of 1969 to 1970 are now taken for granted, with their roots in those years forgotten. That's certainly the case with interdisciplinarity. Maybe it doesn't matter that today's deans and provosts don't know that some of their predecessors regarded interdisciplinarity as a crude and dangerous weapon of the New Left. What does matter is that today's deans and provosts don't know what happened over the course of the 1970s as American universities rushed to embrace a wide-ranging form of interdisciplinarity. Stanford's experience in that decade shows that the consensus view of interdisciplinarity—then and now—is too simplistic. In particular, the enactments of interdisciplinarity that unfolded in this chapter demonstrate that some real-world problems are more amenable to a monodisciplinary approach, and that some attempts to force researchers into interdisciplinarity collaborations can be counterproductive or have problematic unintended consequences.

My point isn't to condemn interdisciplinarity, but rather to encourage reflection on its variety, psychology, politics, and evolution over time. There is no one interdisciplinarity, and no form of interdisciplinarity is a timeless and universal good. Interdisciplinarity can be a force for good, but any time that it is promoted we should ask who stands to benefit and who stands to lose from the erosion of the traditional disciplines. The answers to those questions won't be static: interdisciplinarity's opponents today will be its supporters tomorrow and vice versa. But how we think about interdisciplinarity today and in the future has been conditioned by the upheavals of the long 1970s—even if, in most cases, we cannot (or choose not to) remember that.

4

NOTHING FAILS LIKE SUCCESS: FROM MOON TO EARTH AT NASA

Most readers of this book will know that on the evening of July 20, 1969, Neil Armstrong hopped from the *Eagle* Lunar Excursion Module onto the moon's surface, to the cryptic misutterance, "That's one small step for [a] man, one giant leap for mankind."[1] Indeed, a significant portion of "mankind"—and a much larger proportion of the American public—was watching him. Yet many outside the United States were unimpressed with the moon landing, and even many Americans (particularly people of color) thought that the time and money spent on it exemplified the nation's misplaced priorities.[2] Americans as a whole were generally positive toward the moon landing on that day, but public opinion was (as Matthew Tribbe and Roger Launius have shown) much more ambivalent about the Apollo Program both before and after July 20, 1969.[3] The singular moments when Armstrong set foot on lunar soil and when the Apollo 8 astronauts circled the moon the previous December have left a mark on both popular culture and the discourse of public intellectuals.[4] But those singular moments stand somewhat outside the longer flow of public and elite opinion about the value of America's space program.

Nevertheless, Armstrong's hop was in preparation for a long time. The National Aeronautics and Space Administration was created in 1958, and its astonishing growth over the next decade made the moon landing possible. NASA wasn't created specifically for that purpose, of course; but it *was* created with the intent of preventing the Soviet Union from gaining a monopoly on space achievements, including human exploration of the moon. NASA's founding occurred in an atmosphere of panic over the launch of Sputnik and a series of Soviet space triumphs and humiliating American failures.[5] Even after NASA's creation, the Soviets continued to reach notable space milestones before the United States at an alarming rate. In terms of unpiloted craft, those milestones included firsts in the following: artificial satellite, animal in space, uncontrolled landing on the moon, pictures of the far side of the moon, flyby of another planet (Venus), controlled landing on the moon, and artificial satellite of the moon. In terms of piloted craft, the early 1960s saw the Soviets launch the first man and woman into space and the first two- and three-person spacecraft, as well as conduct the first spacewalk and the first simultaneous two-craft piloted space operations. As a result, the language of a "race" and of being "behind" suffused discussions of America's space objectives from the late 1950s until the return of Apollo 11 in the summer of 1969.

BUILDUP AND DRAWDOWN

In September, 1962, a month after the Soviet tandem missions (Vostok 3 and 4) mentioned in the previous paragraph President Kennedy journeyed to Houston to visit the new Manned Spacecraft Center (renamed the Johnson Space Center in 1973—I will refer to it as JSC throughout this chapter). While in Houston, he gave a since much-quoted speech at Rice University reifying the "race" with the Soviets and America's current place in it: "we are behind, and will be behind for some time in manned flight. But we do not intend to stay behind, and in this decade we shall make up and move ahead" by "putting a man on the moon" before the Union of Soviet Socialist Republics (USSR).[6] Although Kennedy had earlier made similar commitments, that speech act in Houston created an existential benchmark for JSC and for much of the rest of NASA—a benchmark that was met the instant Armstrong's foot touched the lunar surface.

Meeting that benchmark required expenditures that are hard to fathom today. In 1965, 4.3 percent of the total federal budget went to NASA—an agency that didn't exist before 1958![7] In 1966, NASA's funding peaked at 4.4 percent of the federal budget, or roughly 0.7 percent of US GDP. Much of that money was spent on the rapid construction of semipermanent infrastructure such as the gigantic Vehicle Assembly Building at Cape Canaveral. By 1967 most of that infrastructure was in place, and therefore expenditures dropped rapidly. By 1975, NASA's budget had stabilized at around 1 percent of the federal budget (or roughly 0.2 percent of GDP). It would stay close to that level until the late 1990s, when it would again decline to approximately 0.5 percent of the federal budget (0.1 percent of GDP) today.

Much of NASA's spending in the 1960s took the form of contracts with aerospace companies such as North American Aviation and Grumman. Thus, the rapid decline in NASA's budget had an immediate effect on the military–industrial complex. As George Low, NASA's acting administrator, reported to Congress in 1971:

> Since the peak NASA employment level established in early 1966, NASA contractor employment has shown a sharp decline. Estimated total employment on NASA programs in early 1966 was 420,000. The comparable figure in June 1970 is 161,000.[8]

Historians have documented the employment crisis among engineers in the early 1970s, usually with reference to cutbacks in defense spending caused by a drawdown in US involvement in the Vietnam War.[9] But NASA's cuts also contributed significantly to engineers' employment woes. Importantly, NASA's cuts were almost inevitable—once infrastructural elements such as rocket test stands and vehicle assembly buildings were built, NASA didn't need to build them again. Maintaining that infrastructure did take money and personnel, but not at nearly the scale required to build them in the first place. So most of the engineers who designed and constructed that infrastructure had to be either put onto new projects or let go.

The US space program's retrenchment had a similar effect in the sciences. Historians have shown that NASA's money and unusual requirements stimulated rapid gains across a wide variety of disciplines, from the obvious (astronomy, planetary geology, climate science) to the less so (cybernetics, ecology, medicine, botany, surface science, systems engineering, etc.).[10] Those achievements—especially in the less-obvious fields—were most associated with the era of NASA's rapid growth and slowed with NASA's budgetary retreat. Across much of American science, particularly the physical and engineering sciences, researchers made unprecedented strides with NASA's help up to 1966, then had to adjust when that patronage began to fade.

Some did so by finding other patrons to allow them to continue the work that NASA had earlier supported. In chapter 3, for instance, we saw how John Linvill and James Bliss's initial Optacon research was funded by NASA, but then in the late 1960s they had to turn to other agencies to keep the Optacon going. Other researchers reacted to NASA's cuts by dropping the topics that had won them funding from the space program; yet the easiest way forward for this group was to find new ways to apply the space age *skills* that that funding had helped them develop. This group therefore went in search of new but related topics, often ones in keeping with the cultural experimentation of the times. Again from chapter 3, we saw how John Chowning first learned the fundamentals of computer music during his time in the NASA-supported SAIL before founding his own computer music center. Other electronic music innovators such as Don Buchla and Alan R. Pearlman worked on nonmusical NASA projects during the agency's budgetary inflation, then turned to developing their synthesizers when the space program contracted.[11] Across many fields one can find similar figures who gained fundamental insights while working with NASA—insights that they then had to adapt outside the context of space once NASA became a less generous patron.

EXISTENTIAL SUCCESS

This dynamic, in which an organization or initiative makes an outsized contribution to a number of fields during its heyday, then makes a rather different contribution during their decline (by releasing researchers to do other things), is a recurrent feature of American science and engineering.[12] Sometimes, this dynamic occurs because the organization or initiative breaks apart after failing to accomplish what it set out to do. For instance, Robin Scheffler argues that the biotechnology industry benefited from the cadre of young researchers who were recruited into Nixon's War on Cancer and then set loose when the "war" effort bogged down.[13] In other cases, though, organizations break up because they *succeed* in their existential goals. One can see this quite clearly with many of the organizations set up during World War II, which released a generation of scientists back into (more or less) civilian life with a new set of skills, contacts, and questions earned from their wartime work. And in yet other cases, the distinction between success and failure is less clear, but the dynamic is the same. The collapse of AT&T's Bell Laboratories in the 1990s, for instance, gave a one-time boost to many fields that recruited ex–Bell Labs staff. Yet Bell Labs didn't disintegrate because it failed—its existence was simply no longer sustainable in a changing economic, political, and regulatory environment.

NASA falls quite easily into the category of organizations that succeeded to the point of failure—a dynamic that I have already introduced under the label of *existential success*. The thumbnail history of the space race that I've sketched shows why I apply that label to NASA's difficulties in the long 1970s. The space program, after all, was founded with the existential aim of denying the Soviets' monopoly on space achievements. By 1962, one achievement in particular—putting "a man on the moon"—had come to define the piloted side of the space program. In 1969 that existence-justifying benchmark was reached. Although the Soviets continued to make piloted space firsts after that—particularly in long-term human occupation of low earth orbit—American observers no longer cared in the way that they had a few years earlier. The US piloted space program had succeeded in its existential mission—but with the Soviets definitively outrun, the US piloted program had no *new* mission in place to define and justify its continued existence.

On the unpiloted side, the Soviets kept pace for a bit longer (to about 1975) in firsts that American observers deemed important. After that, NASA ran through a series of unpiloted achievements (photos from the Martian surface and flybys of Jupiter, Saturn, Uranus, and Neptune) that were widely heralded in the United States. The Soviet unpiloted program continued making gains—for example, a series of landers on Venus—but after 1975 most American observers took the Soviets' unpiloted achievements to be narrower and less impressive than the US program's. Thus, although the space race had some reality as a guide to action in the 1960s, by the 1970s that was no longer the case, at least in terms of US views of the Soviets. Even as early as August, 1969, President Nixon's Space Task Group could say

> The attitude of the American people has gradually been changing and public frustration over Soviet achievements in space, an important force in support of the Nation's acceptance of the lunar landing goal in 1961, is not now present. Today, new Soviet achievements are not likely to have the same effect of those in the past.[14]

If anything, in the 1970s NASA started to view Europe as a more relevant international space power than the Soviets, both as a collaborator and competitor.[15]

Apollo 11 and its aftermath mean that NASA offers an especially stark instance of existential success. For the purposes of this book, though, I take NASA as representative of—or a synecdoche for—the wider, less clear-cut existential success of postwar (and especially post-Sputnik) US science and engineering more generally. That is, the arc of the US research system from the end of World War II until the late 1960s overcame the same more-or-less existential challenge as NASA; thus the US research system (in general) faced many of the same problems that NASA did (in particular) during the long 1970s.

Indeed, the research system as a whole was structured by a perceived race against the Soviets that closely resembled the space race—not least in that both races had more or less been won by 1969. In the early Cold War there were a number of science and engineering fields in which the Soviets appeared to be "ahead" of their US counterparts, or where the US "lead" appeared to be small and growing smaller.[16] In particular, Soviet gains from the late 1940s to the mid-1960s in fields most closely related to nuclear strategy—building different kinds of nuclear weapons, different means of delivering them to a target, and different means of preventing the enemy's bombs from reaching their target—led to panic, purges, and budgetary largesse on the part of US policymakers.

Fears that the Soviets were or would soon be ahead in weapons-related fields were deftly exploited by American opportunists such as Edward Teller to get funding for far-out projects such as a nuclear airplane, a nuclear spaceship, and nuclear environmental modification and earth-moving.[17] But the two superpowers also raced each other in fields that were less obviously related to weapons production—especially ones where there was some easy metric of who was ahead. For instance, the US and Soviets leapfrogged each other in building ever-larger particle accelerators of various types through the 1950s and 1960s—though, again, after 1970 the United States regarded Western Europe, not the USSR, as its main competitor in accelerator design.[18]

By the 1970s, the number of fields in which American scientists, engineers, or policymakers cared about a possible Soviet lead had dwindled. Usually that meant that US actors now viewed the Soviets as hopelessly lagging: as Kremlinologist Thane Gustafson put it in 1980, "by any measure . . . U.S. scientists lead their Soviet colleagues in most disciplines, and in many there is simply no competition."[19] Admittedly, there were a few fields in which a Soviet lead was still acknowledged in the 1970s, but these were now largely written off as unimportant.

Probably no domain better captures this attitude than that of high-speed circuitry. By the early 1970s virtually no one believed that the Soviets would ever catch up to American firms in the speed of solid-state integrated circuits. National priorities in this area therefore shifted to ensuring that Japanese and Western European firms did not pull ahead and that Soviet and Eastern European intelligence services could not access and reverse-engineer American circuits.[20] There were, however, varieties of high-speed circuitry for which the Soviets were still competitive with US firms and researchers, yet without causing American observers much concern. Some of these were cutting edge but rather limited areas, such as molecular electronics and superconducting electronics.[21] But in one ostensibly low(ish)-tech area, Soviet capabilities completely eclipsed the United States: high-quality vacuum tubes. Even today, the (now extinct) Soviet Union is still the source of the vacuum tubes that audio enthusiasts prize and comb the world for![22] Yet in the United States the vacuum tube was mostly regarded as a relic of a bygone age, something that belonged in a museum rather than in a useful piece of equipment. Soviet dominance in vacuum tubes therefore caused no sleepless nights in the United States after 1970.[23]

Does this state of affairs qualify as existential success? The label is a slight exaggeration, less easily applied to American science and engineering as a whole than to the particular case of NASA. Yet I believe that we can still usefully (if cautiously) use the concept. The US research paradigm that policymakers such as Vannevar Bush and Lloyd Berkner set up in the late war/early postwar period was not existentially predicated on competition with the Soviets at first, but it was designed to keep American science and engineering capacity in a high state of readiness in case of a future global conflict.[24] The Soviets must always have been prominent in the list of potential foes against whom US scientists and engineers could be mobilized. With the Soviet testing of a fission weapon in 1949 (and perhaps even before), preventing or blunting Soviet technoscientific advantage became an overriding if not existential priority for much of the US research system.

As with the space program, meeting that priority required enormous expenditures. And, as with NASA, in the early phase infrastructure building *temporarily* boosted those expenditures. As Hyungsub Choi and Brittany Shields have shown, until the mid-1960s the federal government was quite willing to earmark research funds for very expensive brick-and-mortar building projects.[25] Once a laboratory is built, though, it doesn't need to be built again. Over the long term, maintaining buildings isn't cheap, but in terms of annual budgets, maintaining an old building is a far smaller expense than constructing a new one. Thus, money for laboratory construction was one of the first things to disappear when federal R&D funding was retrenched starting in the late 1960s. From then on, if universities wanted a new building, they had to rely on alumni, philanthropies, companies, and internal accounts to cover construction costs. Universities eventually became quite good at soliciting donations for new buildings, but that hard-won expertise was still rudimentary in the 1970s. Thus, laboratory construction—which had exploded in the early 1960s—nearly halted during the long 1970s. Even finding money for maintenance of labs, lab equipment, and lab buildings became a struggle, such that by the early 1980s experts worried that the deterioration in research infrastructure was hindering American science.[26]

The US research system's retrenchment in the long 1970s was to some extent inevitable. Presumably, federal R&D budgets could not have continued to rise at post-Sputnik rates forever. But few predicted that the political will to continue those increases would evaporate as early as it did—many American scientists and engineers seem to have been truly caught off guard. You could ascribe that tipping point to economic malaise or the budgetary pull of the Great Society and the Vietnam War, yet those factors could only have been decisive once the Soviet threat was no longer seen as overwhelming other considerations. By the late 1960s, national priorities *other* than the struggle against communism had come to the fore, thanks to the convergence of several developments: demographic, economic, and cultural change; the civil, environmental, and other social movements; decolonization; and the first steps toward détente. The Cold

War wasn't over, of course, but the urgency of staying ahead of the Soviets had diminished. Those parts of American society that had been mobilized against the Soviets—including American scientists and engineers—were seen as having fulfilled their existential task. Thus, they could now be given other tasks, or find new tasks for themselves. This book is filled with examples of scientists and engineers doing just that.

TOO SUCCESSFUL TO EXPAND, BUT TOO BIG TO FAIL

Some scientists and engineers perceived that sea change early, others much later. To continue the nautical metaphor, some got out ahead of the tide, some moved with it, and some fought it. Among the latter were those NASA personnel who made the moon landing possible but did not see—or wish to see—that Apollo 11 was an end rather than a beginning. In particular, one of the main architects of Apollo—Wernher von Braun—laid out plans for the moon landing to lead to continued expansion of the space program and continued elevation of its aims: a permanently crewed space station, a permanent base on the moon, and a crewed trip to the Martian surface and back.[27] After July, 1969, this "von Braun paradigm" continued to exist, but only within what the physicist Freeman Dyson dubbed the "paper NASA" of long-range planning.[28] It also continued to resonate with enthusiasts outside NASA, especially among the generation of young men who had grown up with von Braun's 1950s magazine articles and television appearances and NASA's 1960s missions, and who became captivated in the 1970s by von Braun's visionary successors, most notably Gerard K. O'Neill.[29] As we'll see, NASA and its vendors made opportunistic use of O'Neill's proposals for space "colonies" and the enthusiasm that they generated, but most people in the space program recognized that O'Neill's designs required laughably high expenditures (and leaps in technical ability) and so would never come to fruition.

Administrators and engineers within what Dyson called the "real NASA" therefore understood that existential success left the von Braun paradigm with no traction. The Soviets seemed to be in no position to put a human on Mars or a permanent settlement on the moon, and Congress could never be convinced to continue expanding NASA just to stay in a race where the other runner had dropped out. Indeed, given the public mood and the state of public finances, Congress and the Nixon administration were eager to *shrink* the space program, not expand it. Thus, all that NASA's administrators could hope for was to contain the scale of cuts in budgets and personnel.[30] For that, one of their most powerful arguments was that NASA's sunk costs were too large to be thrown away. Communities all across the nation (read: voters across many, many Congressional districts) benefited from employment at nearby NASA facilities or NASA vendors. Their other main argument was that the moon landing demonstrated the competence of NASA personnel in a world hungry for technical expertise, and that some societally beneficial use could therefore be found for such talent.

So at the moment when Armstrong's foot hit the moon, almost no will existed for NASA to continue the von Braun paradigm. Indeed, there was little will even to continue the moon landings. President Nixon pressed to cancel the Apollo 16 and 17 missions and had to be personally persuaded that doing so would provide no budgetary surplus. At the same time, few policymakers wanted NASA to immediately close up shop. The medium-term debate therefore concerned what NASA should do next. After considerable confusion and horse-trading, that discussion eventually settled on the space "shuttle" as NASA's next marquee piloted initiative. As John Logsdon and others have shown, though, NASA was unable to resist other parts of the government's imposing their interests on the shuttle program. The final shuttle design was therefore a mish-mash that satisfied no one.

With the shuttle under discussion and Apollo winding down, NASA administrators of the early 1970s had to figure out how to shrink the space program, in a semicontrolled fashion, down to a sustainable size. The ensuing contraction set up one outcome of NASA's existential success: an outflow of personnel who went on to do other things. Most visible in that diaspora were former astronauts who moved into jobs for which the relevance of their previous experience was opaque: politician, artist, chief executive officer, diplomat, even a vice president of a National Football League team.[31] But the astronauts were by no means alone. Engineers left NASA and its vendors in droves, creating a political headache for the Nixon administration. As one strategist noted

> About 100,000 aerospace workers are now unemployed in Southern California. . . . [T]he Administration is not seen as doing anything. . . . [D]emocratic Senators and Congressmen from Southern California are "cutting the Administration apart" by showing concern.[32]

Many of those unemployed engineers moved into new industries, into government, and even into new disciplines, some with help from Nixon administration "projects to retrain displaced aerospace engineers and scientists in bio-medical, environmental engineering, and State and local administrative fields."[33] As we'll see, these administration programs that were targeted at individuals were replicated at the organizational level with projects that redirected NASA itself into biomedicine, environmental engineering, and state and local administration.

Budgetary contraction meant many people were pushed out of NASA and the aerospace industry; but existential success meant other organizations also *pulled* NASA and aerospace industry personnel toward them. Everyone could see that NASA personnel were effective—they had come from behind to accomplish their organization's existential mission against daunting odds. What organization would not want people of that caliber? And because that existential mission had been completed with no new one to take its place, there was little to keep personnel at NASA. Budget cuts meant some personnel were simply dismissed, but many *chose* to leave NASA to seek new challenges or because other organizations offered them attractive inducements. The NSF's RANN program, for instance, was run during its heyday by Alfred Eggers, a former NASA Deputy Associate Administrator for Advanced Research Technology.[34] In this chapter we will encounter some other members of the NASA diaspora in the Department of Housing and Urban Development (HUD) and elsewhere.

This aspect of NASA's existential success demands further research—we need to know more about where NASA people went and what organizations, industries, and fields were changed by their presence. It's more difficult to apply this aspect of existential success to the rest of the US research system in the long 1970s, but nevertheless we should also pay more attention to the mobility of non-NASA scientists and engineers during those lean years. We know a bit about what people chose to research when the pressure to conform to national security aims lifted. But we know less about the organizations that recruited such people precisely because their professions had proven their competence in the struggle against communism. My guess is that if we followed scientists and engineers who switched fields or industries during the long 1970s we would find interesting fingerprints in areas such as computing, video game development, and finance.

Those who stayed at NASA and its aerospace vendors, meanwhile, experienced a mirror-image version of existential success. Their organizations had succeeded; few people wanted those organizations to disappear entirely; yet almost no one wanted those organizations to continue as they had before. Other organizations saw that NASA and its vendors could accomplish preset goals and therefore tried to make their own organizational priorities part of the space program's next set of goals. The Nixon administration, for its part, believed that NASA should accede to such demands. Per Lawrence Goldmuntz (a scientist in the White House Office of Science and Technology):

> NASA should assume responsibility for certain technical developments that would be within their areas of competence and that would enrich the civilian life and economy . . . in certain new areas of transportation . . . communications and information processing for improved mail handling, in urban crime and alarm and location . . . in weather modification and water resources. These areas are the normal responsibilities of some of the civilian mission oriented agencies, but they are not being accomplished with the attention they deserve because of preoccupation by the mission oriented agencies with the more immediate responsibilities. It is suggested that such projects be funded through the mission oriented agencies—HUD [Department of Housing and Urban Development], DOT [Department of

Transportation], EPA [Environmental Protection Agency], [Department of the] Interior—but accomplished by NASA. This would tend to assure overall project relevance while making use of NASA facilities and personnel.[35]

Some NASA personnel welcomed this turn toward relevance. NASA administrators could view collaboration on projects not directly related to spaceflight as a "'gap-filler' or 'insurance' program . . . to reduce the manned [space]flight gap" and keep NASA's infrastructure and staff (and its vendors' engineers) occupied while waiting for shuttle operations to begin.[36] But we shouldn't view NASA's attitude in entirely cynical terms. We cannot discount the likelihood that some scientists and engineers and perhaps even upper administrators viewed these projects as an enjoyable opportunity to work on something different, with interesting technical challenges and visible benefit to society.

JOHNSON SPACE CENTER AND JET PROPULSION LABORATORY

Both external pressure and internal priorities, then, led NASA and its vendors to pursue a variety of projects that brought the concerns of civilian, terrestrial society into contact with space-age expertise in the long 1970s. In this chapter I will concentrate on projects associated with the Johnson Space Center outside Houston, Texas, and the Jet Propulsion Laboratory (JPL) outside Los Angeles, California. These facilities were particularly ensnared by existential success: JSC by the success of piloted missions and JPL unpiloted ones. Readers should not take JPL and JSC as entirely representative sites, though geographically they nicely stand for the regions (California and the South/Southwest) where NASA and the aerospace industry were most concentrated. Still, there was considerable variation in how NASA's facilities experienced the predicaments of the long 1970s, such that no sample of two would capture the full range. Some NASA sites, such as Cape Canaveral, faced dilemmas similar to JSC and JPL's but were more constrained in dealing with them. One—but only one!—facility, the Electronics Research Center in Cambridge, Massachusetts, was actually closed.[37] And a few facilities experienced the long

1970s quite differently: the Ames Research Center in the San Francisco Bay area, for instance, thrived by cultivating stronger ties with industry and universities while at the same time lending its reputation to O'Neill and the space colonization community.[38]

That said, many of NASA's facilities did see major programs shut down and responded in ways similar to what I will describe for JSC and JPL. Mark Bowles, for instance, has written about NASA's nuclear reactor facility at Plum Brook, Ohio, which was built in 1961 to generate neutron flux to test designs for a nuclear airplane and then a nuclear spaceship before becoming a more mundane but very useful tool in materials research.[39] The reactor program was then summarily ended in mid-1973, leaving NASA to desperately seek new partners such as the Cleveland Clinic and the EPA. Those attempts failed and the reactor was mothballed, leaving the Plum Brook site to struggle through the 1970s by retooling to support, among other things, NASA's programs in terrestrial alternative energy. Plum Brook was part of the Lewis Research Center, which similarly partnered with local agencies to monitor pollution levels in the Cleveland area and obtained funding from the EPA to develop a more efficient, less polluting car engine.[40]

That path through the long 1970s was mirrored at JSC and JPL, though with slightly different organizational logics. And it is at the level of organizational logic that this chapter will focus: the sources that I offer were largely written in a bureaucratic tone that effaced personal motivations. In places I will speculate about the people behind those sources, but for the most part the sources force me to ask what was going on with the *organization*—and not the individual—at the time that the source was written. My other chapters bring out individual voices more clearly, so we can use those chapters to infer what individuals at NASA thought. Conversely, we can use this chapter to get a sense of the organizational logics that constrained and enabled individual expression within the organizations profiled in my other case studies.

At JSC, the overriding organizational context was the gap between the Apollo and shuttle programs. Staff at JSC understood the 1970s as a (painful) transition period, but one with some kind of end in sight. In the early 1970s that transition was made bearable by convincing the Nixon administration to allow the piloted program to adapt Apollo hardware for other purposes: first for three missions to the Skylab space station (itself a kitted-out third stage of the Saturn V rocket) and then the joint US–Soviet Apollo–Soyuz Test Project.[41] Meanwhile, JSC's activities gradually increased pace over the course of the decade under the umbrella of the piloted program's next existential effort: the space shuttle.

Readers should therefore picture JSC's experience of the 1970s as a reasonably fast ramp down *from* full capacity in the Apollo era, superimposed on a slow ramp up *to* full (though lower) capacity in the shuttle era. That left a significant period, though, when JSC's engineers were not fully occupied and were therefore in danger of being laid off. Although the number of full-time–equivalent NASA employees (or civil service "man-years" in contemporary lingo) at JSC declined only by about 13 percent between fiscal years 1969 and 1974, *contract* employment (i.e., people working at JSC but employed by aerospace and consulting companies) shrank by more than a third before starting to recover.[42] During that time, administrators looked for short-term projects that would tide JSC engineers over from Apollo to the space shuttle so that they wouldn't permanently lose highly trained personnel (and thus slow down or endanger the shuttle program). As the following sections will show, projects similar to what we saw at Stanford and UCSB in previous chapters fit that bill and therefore briefly became more common at JSC before being squeezed out again as the shuttle program intensified.

In contrast, the situation was more open ended at JPL. That facility was technically run by the California Institute of Technology (Caltech) on NASA's behalf, so academic administrators (and to some extent even students) exerted some influence over JPL's direction. Even at conservative Caltech, some of the national debate about military funding of academic research filtered through. As a result, the institute released reports on the JPL–Caltech relationship in 1970 and 1976, with particular focus (as at Stanford) on the propriety of allowing classified research at a (nominally) academic institution.[43] Thus, JPL had to deal with the short-term ups and downs of academic politics, but it also had to surf the much longer waves of unpiloted space exploration. Unpiloted missions can last much longer than piloted ones. In fact, Voyager 1—launched in 1977—is still an active mission. Thus, missions that entered the pipeline when there was still political will to deny the Soviet monopoly on space firsts could potentially outlive that political will by many decades. Indeed, JPL's most dramatic missions of the 1970s were set in motion before that decade.

During the 1970s, however, the lab's mission queue became much shorter.[44] There was a real possibility that the will to undertake unpiloted space exploration would disappear completely and that JPL—if it were to continue to exist—would need to find some other mission. As we will see, JPL went a long way in exploring other organizational objectives, and for a time it seemed as though other missions—particularly terrestrial energy conservation and alternative energy—would take up as much of its attention as space exploration. In the early 1980s, though, those new organizational objectives were quashed by the Reagan administration, and JPL returned to space exploration as well as to its military–industrial roots.

Despite their differences, therefore, JSC and JPL followed similar arcs: from existential success in beating the Soviets, to a period of uncertainty characterized by collaborations with other organizations in aid of societal relevance, followed by the establishment of a new existential mission and the dissolution of projects aimed at solving civilian, terrestrial problems. To the extent that problems of civil society continued to occupy these organizations after 1980, it was in the mode familiar from other chapters: the societal problems considered most worth solving after roughly 1980 were those set by the market, to be approached in collaboration with industry and *not* in partnership with nongovernmental organizations (NGOs) or state and federal agencies associated with the social safety net. As in other chapters, we will see that the long 1970s introduced new ways for scientists and engineers to work with industry, but that initially their aims were not overtly commercial. It was only at the end of the long 1970s that those new practices were reconfigured in ways that can today be seen as an accommodation to neoliberalism.

FROM THE MOON TO THE EARTH

Of course, in some sense NASA was always supposed to contribute to solving problems of civil society. The agency's organic charter contained a vague reference to applications of "aeronautical and space science and technology . . . within and outside the atmosphere," which headquarters used to justify opening a Technology Utilization Office in 1962.[45] Through the 1960s this office was best known for annual reports enumerating dozens of supposed "spin-offs" of space programs. Yet, as Bruce Seely argues, early on NASA seems to have measured "utilization" by how widely its documentation circulated, rather than by whether its discoveries ended up in civilian technologies.[46]

Around 1970, though, headquarters-sponsored "applications teams" were "diversif[ying] into other, nonmedical, public-sector areas" such as "air and water pollution, fire safety, housing and urban development, transportation, law enforcement, criminalistics, the postal services, and mine safety."[47] There then seems to have been a fairly rapid transition in the early 1970s from headquarters pushing the field sites to create application teams to "the Field Centers themselves submit[ing] proposals based on their own perceptions of the relevance of their in-house skills and capabilities to public-sector problems."[48] At the same time, external actors seem to have become more interested in working with NASA on terrestrial applications. For instance, industrial requests to NASA's Technology Utilization program went from ten to twenty per year in the late 1960s to seventy to ninety per year in the early 1970s.[49] Similarly, reports from this era offered long lists of other government agencies with whom the Technology Utilization program collaborated: the "Department of Health, Education, and Welfare; the Department of Housing and Urban Development; the Bureau of Mines; the Department of Transportation; Environmental Protection Agency; the Law Enforcement Assistance

Association; the National Bureau of Standards; the Veterans Administration and others."[50]

In keeping with the Nixon administration's ideological commitments, many of these projects aimed to alleviate market failure by ushering a prototype to the point where a private entity could profitably manufacture and sell a commercial version. As Lawrence Goldmuntz's 1971 memo outlined,

> These projects cannot be undertaken by the private sector because of their long range character requiring extensive development and/or because national standards must be developed before a market can aggregate. Nevertheless once the research and development is completed and once a national demonstration is accomplished, substantial domestic and foreign markets would materialize for development by the private sector.[51]

This mode of working was aided by a new patent policy in 1972 that allowed NASA to grant exclusive licenses on its patents nine months after filing rather than two years after issuance.[52]

An illustrative example of how this worked in practice was a test of whether the lunar rover could be adapted for use by people with lower-body paralysis conducted at the JSC in 1974 to 1975. As Goldmuntz would have encouraged, the project's leaders saw it as an attempt to resolve a market failure:

> It is unlikely that specialty equipment manufacturers can afford to develop a high-quality, well-engineered product such as the proposed Lunar Rover controller for automobiles. This is an opportunity for NASA, as a public agency responsible for the technology, and the VA [Veterans Administration], as a public agency responsible for the care and service of a large segment of the physically impaired population, to sponsor the development of such a system.[53]

As the quote implies, the lunar rover conversion project partnered NASA with the VA—but also with the US Department of Transportation, as well as a number of Texas state agencies and NGOs that worked with people with physical disabilities. These kinds of complex collaborations with organizations that represented possible users or sources of expertise relevant to terrestrial applications of space technology were quite common.

The focus on disability technologies was also quite typical. As we saw in chapters 2 and 3, projects to aid people with visual, hearing, or mobility impairments were common responses to calls for civilianization of the R&D system in the years around 1970.

One of the more colorful examples of this market failure approach was a program in the mid-1970s to adapt astronaut food for delivery to elderly shut-ins by mail, volunteer, or social worker. The program started in 1974

> when Anne Kohler, Director of the Texas Governor's Committee on Aging Research Utilization Program (now a part of the Texas Department of Public Welfare) made a request of NASA: could the space agency help to improve nutritional services for the elderly, particularly those in areas not reached by meals-on-wheels programs? As it turned out, NASA's experience in developing a shelf-stable, nutritious, easily transportable meals system provided the foundation for design and development of a meals system intended to meet the special needs of the elderly.[54]

Eventually, the project expanded to include the JSC, the Texas Research Institute of Mental Sciences, United Action for the Elderly (an affiliate of Meals on Wheels), and researchers from the University of Texas (both the School of Public Affairs in Austin and the UT Medical Branch in Galveston). The Army's Natick Research and Development Command also provided some support for the project, though this was not always publicly acknowledged.[55] The cost was around a quarter million dollars ($2 to $3 million in today's dollars), about half paid by NASA; the Texas Department of Public Welfare and the Ford Foundation also contributed money. Some of this funding went to an aerospace contractor, Martin Marietta, to develop the "meal system for the elderly" and run field tests.

Programs such as Meal System for the Elderly illustrate how issues and groups that were spotlighted by the era's political debates and social movements were incorporated into NASA's post-Apollo search for relevance. In the 1960s, Lyndon Johnson's Great Society programs had (briefly) made the alleviation of poverty in rural and urban areas and among people of color and the elderly a federal priority. Even though many of Johnson's programs were gone or in decline by 1974, those groups were still the ones that Meal System for the Elderly was intended to aid. Admittedly, none of NASA's publications about the meal system mentioned social justice or civil rights. Yet almost half the participants in the project's field trials were beneficiaries of the USDA Food Stamp program; two thirds were women, 44 percent were African American, and project reports identified another 29 percent as "Mexican-American."[56] Project reports suggested that the same meal system could benefit the nonelderly poor, people with physical or mental disabilities, and prisoners. All these were groups whose interests were promoted by the civil rights movement and the new social movements of the late 1960s and 1970s.

Perhaps unsurprisingly, NASA viewed the concerns of these movements through a paternalist prism. For instance, an official report suggested that the system could be useful to "some urban elderly [who] are within walking distance of stores, but fear to venture out of their homes."[57] A few sentences later the same report observed that

> Some also criticize the food stamp program, because it does not guarantee that recipients will buy food items that contribute to a balanced diet. Such a criticism might be quelled if a balanced, nutritious meal such as that developed by NASA were available for use by food stamp participants.[58]

Nor should it be surprising that attention to the needs of prisoners, the poor, people of color, and those with disabilities waned once NASA handed the project over to commercial manufacturers. At least two companies commercialized technologies developed within the Meal System for the Elderly program: Sky-Lab Foods and Oregon Freeze Dry Foods. In the end, though, these companies saw greater potential for profit from selling meals to campers or homeowners preparing hurricane kits than to people of limited means. These companies did market to some elderly users, as the original NASA program hoped. For instance, Oregon Freeze Dry advertised its Easy Meal as "Space Age Food, developed for Adults of Retirement Age."[59] Yet as figures 4.1 and 4.2 indicate, Easy Meal's imagined elderly customer was whiter and more middle-class than the Meal System for the Elderly's field trial participants.

That Meal System for the Elderly gave rise to a consumer product at all is probably a testament to the short leap from astronaut food to a

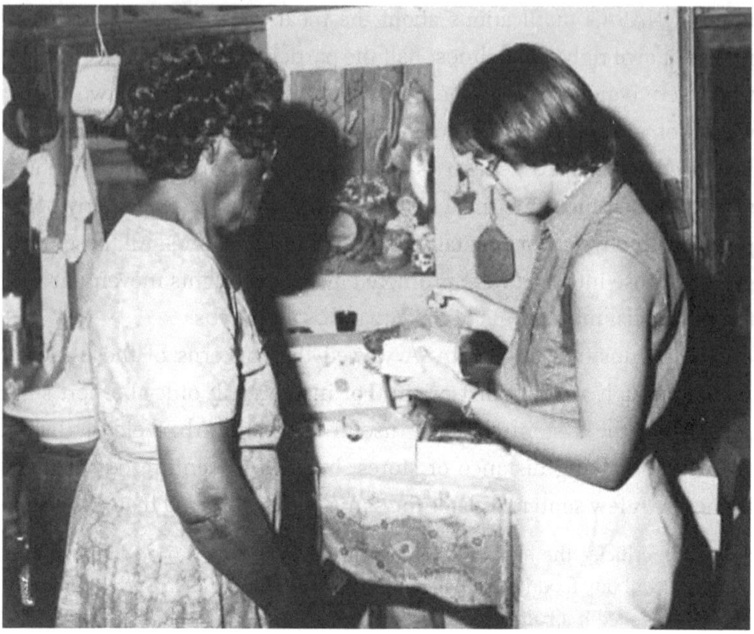

Figure 4.1
Fieldworker (right) demonstrating the Meal System for the Elderly in the home of a trial participant (left). *Source*: Technology Utilization Program, *Meal System for the Elderly*, JSC-11191 (Houston: Johnson Space Center, 1976), 9.

terrestrial analogue. Indeed, many of NASA's forays into public sector problems in the early 1970s involved fairly straightforward translations of aerospace technologies. For instance, NASA partnered with the Federal Aviation Administration to develop better air traffic control technologies, and with ERDA to create new designs for wind turbines (which are in some respects just giant propellers) and fuel-efficient airplane engines.[60] In other cases, though, NASA's relevance to solving social problems was thought to stem from its abstract expertise in systems engineering and "space age management" rather than a specific technology.[61] And in yet other cases NASA's engineers argued for the societal relevance of space technology by drawing debatable analogies between spaceflight and what

Figure 4.2
Publicity shot used in advertising for the Easy Meal line of products sold by Oregon Freeze Dry Foods. The company collaborated on the Meal System for the Elderly, as noted in its ads. The company's participation in the project was also noted by NASA, for instance in the publication this shot is taken from. *Source*: Technology Utilization Division, *Spinoff 1978: An Annual Report* (Washington, DC: NASA, 1978), 91.

they believed to be the typical situations of the people they claimed to be trying to help.

For instance, NASA staff with expertise in biomedicine and communications were deeply involved in a number of efforts to improve the delivery of health care to poor rural communities, especially those with a large proportion of ethnic minorities. These projects were justified on the basis of an explicit analogy between the remoteness and low population density of outer space and that of rural areas. As one proposal put it,

> The medical information obtained from an abnormal electrocardiogram on a crewman in space may be identical to that obtained from an electrocardiogram on a patient in a remote location on earth. A high resolution, slow scan TV presentation of an x-ray picture . . . from a space base crewman might be the same as that obtained from a farmer's x-ray taken and developed in a remote area health services unit on the ground and transmitted to a medical center. . . . After considering these parallels . . . NASA is considering the development of a demonstration program for a remote area health services system on earth as a necessary step to the agency's development and verification of a remote health services capability for space.[62]

This proposal later evolved into the STARPAHC (Space Technology Applied to Rural Papago Advanced Health Care) project, sited on the lands of the Tohono O'odham Nation in southern Arizona and begun in 1973. ("Papago" was at the time the common non-Indigenous term used for the Tohono O'odham.) Because the leader of STARPAHC, JSC's Sam Pool, became prominent in the telemedicine community, this project is frequently mentioned in histories of that technology, both by medical practitioners and by anthropologists and historians.[63] So far, though, STARPAHC has not been placed in the context of the space program's existential success and contemporary calls for the civilianization of NASA and of American R&D more generally.

Yet there are at least two reasons to place STARPAHC in exactly that context. First, Pool and his collaborators explicitly did so themselves. The proposal quoted earlier framed the project in terms of national needs articulated by the highest civil authorities:

> We also believe this earth based remote health services system demonstration project to have significant potential for general public benefit as well as for future space systems development. President Nixon pointed out in his message to Congress in February 1971 that, "For some Americans—especially those who live in remote rural areas or in the inner city—(health) care is simply not available." The Department of Health, Education and Welfare and the American Medical Association have endorsed the idea of "the provision of access to quality health care for every American. . . ." as a major national goal."[64]

The other reason to view STARPAHC in the context of NASA's existential success is that it was not unique. NASA was involved in nontelemedicine projects that treated communities on Tohono O'odham land as analogous to space capsules; in particular, in 1978 NASA's Lewis Research Center built an experimental solar farm to power the village of Schuchuli, Arizona.[65] With respect to telemedicine, NASA was involved in (or proposed as a partner for) STARPAHC-like efforts in rural areas of New Mexico, Nevada, Alaska, Missouri, northern Alabama, and the Permian Basin region of West Texas.[66] As with STARPAHC, these were economically disadvantaged areas with either majority-minority or substantial minority populations whose interests were being vocally represented in the 1970s by the civil rights movement, the American Indian Movement, the Chicano Movement, and other social justice movements of the day.

And, again, all these efforts did (or planned to) partner NASA and its aerospace vendors with nonspace organizations to make space technology societally relevant. In some of these projects, NASA led the collaboration. In STARPAHC, for instance, JSC personnel headed a team that included Lockheed; the Department of Health, Education, and Welfare; and the Tohono O'odham nation and its Executive Health Council. Similarly, for the Schuchuli solar farm, NASA Lewis Research Center personnel brought together support from the Department of Energy, the US Public Health Service, and the Tohono O'odham. In other projects, though, NASA seems to have been drafted in. For instance, a Remote Area Health Services Research Demonstration and Evaluation Project in New Mexico was run by the Physical Science Laboratory at New Mexico State University on behalf of the state's Health and Social Services Department, with

NASA and the Air Force's School of Aerospace Medicine brought in once the program was running. Either way, the 1970s were an era in which NASA and its vendors could not go it alone but instead needed (and had difficulty resisting) partnerships with organizations representing civil society.

SPACE AND THE CITY

Both Meal System for the Elderly and STARPAHC (and other telemedicine projects) operated primarily in rural areas. Yet urban problems were of at least equal concern within NASA for much of the 1970s. In fact, many of the publications issued by NASA's public sector projects pictured urban and rural areas as similarly amenable to interventions with aerospace technology. Sometimes this supposed similarity led to arguments that solutions to rural problems would also find application in cities and vice versa. As the researchers leading STARPAHC put it, "Consider groups who live in remote areas or, for that matter, in the inner cities who should have access to local health services and to the health care establishment."[67] At other times, the concentration of poverty in rural and urban areas was taken to mean that city and countryside would compete for scarce resources. As a report on the remote health care project in New Mexico put it, "More is heard of ghetto problems, but rural health care has suffered particularly because most young physicians . . . and their wives prefer to settle in urban areas. . . . Even ghetto practice is often near splendid medical centers."[68]

The sentiment is grating, but the author was correct that public discourse concerning urban problems was mounting rapidly in this era. Recall the Ngram from figure 1.5, where we saw a rise in mentions of "urban problems" and an "urban crisis" starting around 1964 and peaking in 1968 or 1969. Ngrams are a crude measure, but this periodization maps reasonably well onto the prevalence of talk about urban problems at NASA (with a little bureaucratic lag). It also maps onto aerospace firms' explorations of, for instance, new markets in public transportation and other urban infrastructure. Rohr Aircraft, for instance, became an important maker of city buses and subway systems in the early 1970s; later in the decade Rohr, and its mass transit products, were bought by a much larger aerospace firm, Grumman.[69] Other military–industrial companies tried to follow Rohr's lead, sometimes with poor results, as with Westinghouse's failed attempt to build controls for the Bay Area Rapid Transit system (today a classic case in engineering ethics).[70] Most companies dabbled for a few years, then pulled out. A good instance is Boeing's participation (with JPL oversight) in the Personal Rapid Transit system in Morgantown, West Virginia—which, though successfully built, never led to hoped-for replications in other US cities nor to any long-term business for Boeing.[71]

Even more than NASA, aerospace firms had to deal with their association with the war in Vietnam *and* budgetary cutbacks. Demonstrating the societal relevance of their expertise was an obvious answer to both problems; no form of societal relevance was more valorized in the early 1970s than relevance to the urban crisis. A good indicator is a 1970 pamphlet issued by the Aerospace Industries Association entitled *Aerospace Technology: Creating Social Progress,* in which "urban affairs" were the first site of "social progress" presented (figure 4.3).

In Nixon's America, though, there was considerable disagreement as to what urban problems were and how they should be addressed.[72] From the point of view of many in the New Left and the civil rights movement, the urban crisis was caused by institutions—including the institutions of American science and engineering such as NASA—that had a vested interest in bringing about, or at least doing nothing to reverse, the second-class citizenship of most city dwellers, especially those of color.[73] On the other hand, many Nixon voters no doubt believed that the urban crisis stemmed from what they saw as growing disregard for authority among African-Americans, young people, and the antiwar Left.[74]

From the point of view of NASA and aerospace companies, though, these differing explanations for the urban crisis simply meant that there was an even wider array of research topics by which they could make

Figure 4.3

Cover and page 2 of *Aerospace Technology: Creating Social Progress* (Washington, DC: Aerospace Industries Association, undated but probably 1970). *Source*: Permission granted by the Aerospace Industries Association.

themselves relevant to the pressing social issues of the day. Thus, on the one hand, NASA (particularly JPL) and aerospace firms were deeply implicated in Nixon's "law and order" rhetoric aimed at harvesting the votes of white Democrats who disliked both the civil rights and antiwar movements. JPL had projects to help local law enforcement agencies communicate better with each other (using NASA's communications networking expertise) and to develop fingerprint recognition software for the FBI (using image processing software originally developed for the planetary probe program).[75] In 1974, JPL researchers even invented an Automated Drug Identification device to detect "drugs in blood or urine samples taken from suspected narcotics users" at the request of the Los Angeles Police Department.[76] JPL was hardly alone, though; for instance, its drug identification device was similar to a "portable device for detecting extraordinarily small amounts of heroin in a urine sample" developed by a short-lived NASA–New York City Applications Project in 1973.[77]

On the other hand, NASA also positioned itself relative to the "urban crisis" in a way that would put the Nixon administration in a good light with the nominally more liberal Nelson Rockefeller and George Romney wing of the Republican party. JPL, for instance, ran a multiyear "four cities program" in which the city managers of four municipalities (Anaheim, Pasadena, San Jose, and Fresno) were each teamed with a JPL engineer and an engineer from an aerospace firm (TRW, Lockheed, Aerojet, and Northrop) to see how aerospace and systems engineering might be applied to their cities' issues.[78] The AIA similarly coopted talk of an urban crisis in several different ways. On the one hand, its 1970 pamphlet bragged of AIA members' efforts in "school community relations . . . job training . . . city programming and budgeting . . . [and] justice, welfare studies."[79] The *first* illustration in that pamphlet, under "Urban Affairs," showed a group of young men—at least one of them seemingly African American—receiving vocational instruction from an (apparently white) aerospace engineer. The *second* illustration, however, showed two police officers surrounded by banks of computers under the caption "Information Systems."

At JSC, urban applications actually made their way onto the organizational chart in the form of an Urban Systems Project Office (USPO). This group was formed in 1972 with a staff of thirty-five, including both civil servants and contract personnel from Lockheed and Boeing.[80] The unit was NASA's stake in a project headed by HUD, which also included the Environmental Protection Agency and Oak Ridge National Laboratory (then part of the AEC). The NASA–HUD collaboration seems to have come about in part because HUD Secretary Romney had earlier poached his Assistant Secretary for Research and Technology, Harold Finger, from NASA. In the late 1960s Romney apparently hoped that Finger (and his NASA contacts) could bring space-age expertise to HUD while securing NASA's help in future collaborations. However, the HUD–NASA collaboration slipped from Romney's control when it was coopted by the New Technology Opportunities program, a high-profile Nixon administration initiative put forward in 1971 by William Magruder (a special assistant to Nixon) with help from Edward David (Nixon's science adviser) and others.[81]

The main focus of the NASA–HUD collaboration was the Modular Integrated Utility System (MIUS) project, which aimed to apply NASA expertise to making residential utility systems cheaper, more efficient, and friendlier to the environment. Again, as with STARPAHC, the promoters of MIUS invoked somewhat debatable analogies to tie their work to the national priorities set out by the administration:

> In his February, 1971, address on environmental quality, President Nixon said: "A great deal of our space research has been toward creating self-sustaining environments, in which people can live for long periods of time by reprocessing, recycling, and reusing the same materials. We need to apply this kind of thinking more consciously and more broadly to our pattern of use and disposal of materials here on earth." Programs have been conducted to develop semiclosed ecological systems to support crewmen for long-term space missions. The spacecraft for which these systems have been developed is, in many respects, a housing unit which must contend with many of the logistic supply and ecological problems we face on earth today.[82]

As the quote indicates, urban and environmental problems were, at this point in the 1970s, still closely aligned—a conjunction that environmental historians are starting to contend with.[83]

Indeed, this quote will sound familiar to environmental historians, who have documented a number of ties connecting the environmental movement with architecture and space capsule design.[84] Yet such studies have so far mostly focused on rather groovy groups such as the New Alchemists and the Biospherians and their plans for habitats for the few. The MIUS project, by contrast, was *very* square and aimed at an architecture for the many. Admittedly, there are indications that at the beginning of the project, some NASA engineers imagined their space technology embedded in futurist luxury homes à la John Lautner.[85] But relatively quickly they reoriented the project toward a type of large garden apartment complex commonly found in working-class neighborhoods in the Houston metropolitan area near the JSC.[86]

In applying their expertise to such mundane settings, however, MIUS engineers ran into an equally mundane but stubborn problem: lack of trust from other organizations, including both their erstwhile partners at HUD and potential competitors in the utility industry. Throughout the program, NASA engineers complained that "HUD doesn't really grasp our approach to the problem. Whatever the 'hang up,' we need to understand and correct it. If we don't it will be most difficult to justify" further funding.[87] JSC staff also thought that "HUD has been less than candid on their program funding availability."[88]

For its part, HUD viewed NASA as oblivious to the sensitive politics of housing, particularly the need to represent MIUS "as an alternate means of supplying services, to be employed by existing utility companies, and not an alternate to those entities."[89] It's likely that HUD faced pressure on this point both from the utilities themselves and from the Nixon administration. As one NASA engineer noted, the utilities "have heard of MIUS and the consensus seems to be 'The government is in competition with us,' especially the electric people."[90] Nixon's powerful

Office of Management and Budget, meanwhile, questioned the need for MIUS at all, arguing that "development and demonstration of subsystems is already underway in the private sector and in EPA and HUD" and that the HUD–NASA–AEC project should therefore be given low priority.[91]

To get around objections from utilities and to placate Nixon officials, NASA proposed to experiment with MIUS prototypes in situations where it would not compete with the private sector. For instance, in 1972 JSC's USPO conducted a "Mini-MIUS Feasibility Study" in partnership with the National Bureau of Standards and the EPA in an isolated village in Alaska that was not served by utility companies and that had access to only limited amounts of electricity supplied by diesel generator.[92] Similarly, John Ehrlichman (Nixon's Chief Domestic Adviser) argued that although MIUS was

> something that the private sector probably can and should do . . . there are two ways that we might finesse the problem and still get the job done. One was to use moral suasion with some of the major utility companies or equipment producing companies, such as GE, and encourage them to conduct the experiment. The other was to conduct the experiment the next time the Federal Government builds some large installation. Perhaps such an experiment might fit into plans for a major rehabilitation of a defense base.[93]

We'll see again in chapter 5 that efforts to convert military–industrial R&D to address civilian problems often ran afoul of established civilian actors—and that a common proposal for overcoming such obstacles was to fall back on the military as a captive market.

But an even more fundamental issue led to the cancellation of MIUS and closure of the USPO: money. During NASA's peak years of the mid-1960s, its engineers could request almost any amount of money to help them build components for a one-of-a-kind space vehicle that would be used once and thrown away. As Diane Vaughan showed in her study of the 1986 Challenger disaster, that mindset persisted at NASA long after the agency's budgets declined.[94] HUD, by contrast, had never received sufficient funding, much less the superabundance that NASA experienced in the 1960s. HUD therefore needed MIUS to yield something cheap and scalable to thousands of housing units.

It was only after the project started that the two agencies realized that their different budgetary histories led to divergent expectations. As a 1981 report on MIUS put it,

After about 6 months of intensive design work [i.e., late 1972 or early 1973], the USPO team made its first presentation to HUD, and was shocked to find out that HUD did not intend to use advanced technology but, instead, planned to build MIUS systems using only existing equipment. HUD set a new rule for the project—only bondable catalog-available hardware could be used in the MIUS development."[95]

From that point on, JSC staff complained that they were "constrained by the guidelines which require that we consider only off-the-shelf hardware. This precludes the inclusion of solar space heating and fuel cells and other attractive technologies in our design studies."[96]

In other words, for JSC engineers, the scheme was too *low-tech*, and they continued adding in higher-tech, designed-from-scratch elements. But for HUD, even the initial design was too *high-tech*; they would never be able to get developers to include it without a government subsidy, which HUD was never in a position to supply. From HUD's vantage, NASA's continued involvement only made things worse! These disagreements, plus the increasing pace of the shuttle program, meant that JSC director Chris Kraft began looking for ways to extricate NASA. With the New Technology Opportunities program canceled and one of its proponents, Edward David, forced out of the Nixon administration in 1973, Kraft saw his chance. The next year he abruptly cancelled future MIUS planning and started winding the project down to a close in 1976.

As USPO Deputy Manager Harold Benson reported it, Kraft reminded project personnel that

they were working for the Manned Spacecraft Center and that what they were working on was not in any way related to manned space. Dr. Kraft went on to say that the JSC has the best engineers in the world, and the world is full of problems; if he would allow his engineers to go to work on all the world's problems, they would fragment the agency and the NASA's mission.[97]

By that logic, though, the USPO should never have been formed. Yet it was formed, and it received resources and personnel so long as urban problems were a major focus of the Nixon administration and so long as the shuttle was only in the planning stages at JSC. In 1970 to 1971, as the Apollo program was winding down, much of the American public and government believed that "NASA has a competence which should be used to complement HUD's activities in developing new housing."[98] It was only later that a project aimed at improving multifamily housing no longer seemed desirable within America's space agency.

CRISIS CREEP ONE: URBAN TO ENVIRONMENT AND ENERGY AT JSC

JSC's USPO, JPL's Four Cities Program, and similar projects of the early 1970s briefly focused NASA's attention on the urban crisis. By 1973, though, that attention was wandering, especially to environmental and then to energy issues. Before then, energy and environment were seen to a significant extent as subordinate *parts* of the urban crisis, rather than as separate issues. As the MIUS prospectus put it in 1971,

> The expanding population, the decay of our cities, and the continued demand to increase the American standard of living have created a shortage of acceptable housing which can be afforded by all U.S. citizens. However, providing these new home units should be considered an opportunity as well as a problem. This new generation of homes not only should be as economical, livable, and durable as possible, but also should be designed to improve fire safety, to minimize environmental pollution, and to conserve water, energy fuels, and electrical power.[99]

After 1973, though, use of the keyword "urban" dropped away and environmental and energy projects came to the fore in their own right. There was still an urban dimension to some of these projects—for example, studies of the aerodynamics of subway trains in tunnels—but their "urban-ness" was deemphasized in favor of their relevance to energy conservation, pollution reduction, and other topics.[100]

At JSC, energy and environmental research flourished in the mid-1970s—but in a relatively chaotic, uncoordinated way. There were a number of small efforts, some in partnership with local aerospace companies or social groups. For instance, the Rotary Club of Space Center spawned an Earth Awareness Foundation that partnered with JSC to assist a group of local high school students in a water quality monitoring project.[101] JSC engineers themselves built a "trailer-mounted automated water monitoring system" in partnership with Boeing and the Gulf Coast Waste Disposal Authority.[102] JSC was also brought into a solar energy project that involved HUD, the Bureau of Standards, and the NSF, as mandated by the Solar Heating and Cooling Demonstration Act of 1974.[103]

Two energy and environmental efforts did gain some longer-term traction at JSC, one involving the "real NASA" and one the "paper NASA." The former was the Earth Resources Survey Program Office, which served as JSC's stake in the Landsat remote-sensing project. The Landsat program has already received quite a bit of attention from historians for two reasons: first, it shows how NASA responded to the growing environmental movement; and second, Landsat's later privatization shows how commercial objectives increasingly permeated the agency from the late 1970s onward.[104] I would simply add that there were ways in which Landsat linked to the concerns of the long 1970s beyond environmentalism. For instance, JSC proudly drew attention to Landsat's use on behalf of indigenous communities.[105] Similarly, JSC partnered with the University of Texas School of Public Health on a remote sensing project to monitor how changes in land use affected public health—a topic that incorporates but goes beyond environmental monitoring, and which is also rather difficult to imagine getting much attention during Landsat's later, commercial phase.[106]

The "paper" JSC, meanwhile, was deeply involved in one of the wildest, most hubristic proposals of the era: a solar power satellite (SPS). This would have been a permanent crewed solar farm in space, collecting photovoltaic energy and converting it into electromagnetic radiation to beam to earth, where it would be converted into electricity. The idea has a long history, but its modern form is usually credited to Peter Glaser, an engineer at the Arthur D. Little Corporation, who gave a much-cited talk on the subject in 1968. In 1971 Glaser wrote a proposal for an SPS that somehow reached William Magruder (among others).[107] The idea then seems to have languished until it was taken up by Gerard O'Neill as a way to justify the enormous expense of building giant space "colonies." Patrick McCray has documented how O'Neill and other "visioneers"—supported by technophile impresarios such as Stewart Brand—publicized the SPS concept.[108] The interest that they stoked led to pressure on NASA to investigate the idea, and in 1976 JSC issued a call for proposals for an SPS "system definition study." That study was awarded to Boeing, which then undertook a wide-ranging examination of the engineering requirements of a hypothetical SPS.[109]

As Jeff Womack argues, JSC leadership saw the SPS largely as an opportunity to keep people occupied and to think through engineering issues of likely relevance to projects that NASA was more seriously contemplating (such as what later became the International Space Station).[110] There is little evidence that NASA was enthusiastic about the prospect of beaming vast amounts of energy into the earth's atmosphere, most likely aimed at points in proximity to urban areas (such that a very small error in beam alignment could result in costly damage or even fatalities). Nevertheless, JSC's initial study soon morphed into a more ambitious assessment led by the Department of Energy.[111] This project, in which NASA continued to be involved, persisted even after the shuttle began flying—a testament, as Womack argues, to the bureaucracy's ability to hijack crisis talk to ensure the bureaucracy's own continuing work.

CRISIS CREEP TWO: JPL'S "DUAL MISSION"

At JPL, R&D oriented to public-sector projects, and then alternative energy and energy conservation, was centrally coordinated much more than at JSC. In 1969, the lab formed a Civil Systems Projects Office, which housed a growing portfolio of work in biomedicine, urban affairs, criminology, environmental monitoring and remediation, alternative energy, and assorted other topics such as an automated system for taking attendance in schools![112] A few, broadly similar projects were also paid for from the Director's Discretionary Fund, but for the most part the Civil

Systems office oversaw JPL's overtures to societal relevance. That meant that JPL's public sector work could be steered more effectively than the scattered efforts at JSC.

Thus, when JPL signed a memorandum of understanding in 1975 with ERDA (later renewed when ERDA became the Department of Energy [DOE]), the Civil Systems office turned sharply in the direction of energy research. Other projects were mostly set aside, and by 1978 the office had been renamed Energy and Technology Applications.[113] Under the memorandum of understanding, ERDA–DOE referred a varied portfolio of projects to JPL. Some were short-term, rather mundane contracts, such as testing the fuel efficiency of modified engines placed in AMC Pacers and Volkswagen Rabbits.[114] Rather deliciously, the site where these "test vehicles" were driven was a runway at Edwards Air Force Base, a location better known for the X-plane flights chronicled in *The Right Stuff.*

The largest and most enduring projects operating under the agreement with ERDA–DOE, though, were in solar energy, particularly the Low-Cost Solar Array Project and the Thermal Power Systems Project.[115] We will return to JPL's solar activities in chapter 5 when we look at a residential solar energy system developed by Texas Instruments; that's because "the Department of Energy has assigned the Laboratory as the lead center for the national photovoltaic program" in 1978.[116] Thus, anyone wanting or receiving federal funding for solar R&D in the late 1970s had to contend with JPL's experts. As we'll see in chapter 5, not everyone was happy about JPL's role as the American solar community's obligatory passage point.

Indeed, JPL sometimes gave people plenty to be unhappy about. Moving from space into civil systems required the lab to acquire expertise with which its personnel were unfamiliar and sometimes uncomfortable. For instance, ERDA commissioned a series of reports on coal mining that were, in part, technological, but which also involved sociological studies of coal mining communities and even historical and philosophical studies of the acceptance of new technologies.[117] These forays into new territory were sometimes scientifically unsound and sometimes politically inept—and sometimes both. Perhaps the most glaring example is a 1975 report on "alternative automotive powerplants" entitled "Should We Have a New Engine?" (JPL SP 43–17).[118] Although this research was funded by Ford Motor Company (with some assistance from Chrysler), it provoked such heated opposition from across the entire car industry that JPL later had to issue a "Compendium of Critiques of JPL Report SP 43–17."[119]

The potential for JPL's civil systems researchers to venture out onto interdisciplinary thin ice elicited considerable concern from Caltech faculty, who were afraid that their reputations would be damaged or that they would have to clean up JPL's mess. Caltech—with its small and until 1970 all-male student body of scientists and engineers—was never home to an organized anti-JPL movement similar to the anti-SRI movement at Stanford.[120] But divestment initiatives on other campuses did spur a 1970 report on Caltech–JPL relations that recommended some mild reforms, even if it generally "found the JPL connection sound and worthwhile."[121] The 1970 report didn't settle matters, however. "As the Civil Systems Operations at JPL have expanded, conflicts have arisen between JPL and the [Caltech] Faculty over the question of the competence of JPL to undertake work in certain areas," leading to a new study of JPL's status in 1976.[122]

From Caltech's side, the concern was that JPL staff were unqualified to pursue civil systems work that was so far removed from missiles, space tracking, and planetary probe technology. JPL staff didn't disagree, per se—they simply viewed civil systems R&D as necessary given the lab's budgetary situation, and appropriate given the nature of engineering knowledge. As a former JPL engineer wrote to the Caltech Ad Hoc Study Group on JPL/Campus Interaction in 1976:

> Currently, much effort at JPL is being directed toward programs pertaining to the so-called civil systems sector. These activities involve issues and problems of a socio-economic and health care nature. These are certainly a large step away from past experiences with space and military matters. Again, a Caltech faculty member views the Laboratory's movement in this direction with much unease and concern. Why is there concern and uneasiness? . . . [T]he civil systems program involves personal interaction with the public on issues which are complex and difficult. . . . Caltech today is primarily a body of scientific talent. It had at

one time some considerable engineering skill, but little today. . . . Unfortunately, some of the Caltech faculty is alarmed by these steps and arrogantly demands, "What are those clowns at JPL up to now? We must monitor what they are doing closely or they will wreck our reputations." I question who should be monitoring whom in a situation such as this, because by instruction, training, and experience, engineers are expected to work with the public, certainly to a greater degree than is the scientist.[123]

Not that this engineer's views should be taken at face value: Caltech still had plenty of engineers, such as John Pierce (Rudi Kompfner and Calvin Quate's former Bell Labs colleague and one of John Chowning's computer music mentors). Although Pierce seems not to have worried that JPL would harm Caltech's reputation, he did believe "If JPL cannot in any large and useful way be successful in energy without a change in form, then either JPL will have to drop energy or it will have to change form."[124]

For a time, it seemed as though JPL *would* "change form" to accommodate energy R&D. In 1978, when Civil Systems changed its named to Energy and Technology Applications, energy R&D constituted "about one-seventh of JPL's total activity" or possibly closer to one-fifth of JPL's "manpower distribution."[125] The lab's director, Bruce Murray, was at that point broadcasting that "JPL is well along toward the goal of a dual mission, in energy and space."[126] Yet JPL's commitment to its new mission in energy was light, and based more on institutional survival than on a clear vision for renewable energy. As one Civil Systems manager put it, "JPL's decision to favor energy over biomedical systems seems based primarily on which gets the most federal budget, which makes sense only if one expects JPL's competence in an area to be marginal"—a view supported by Caltech faculty members who worried that "quality is quite uneven" in the Civil Systems program.[127]

When Ronald Reagan took office in 1981, therefore, JPL's shift toward energy R&D was halted. Reagan was opposed to federal agencies competing with private interests, and he particularly disliked federal support for alternative energy and energy conservation. He had also run on a platform of reintensifying the armed standoff with the Soviet Union. As we will see again in chapter 6, the new administration slashed federal alternative energy funding, including at JPL. By 1983 "reduced federal priorities in energy [resulted in] a decidedly smaller effort in Civil Programs," with DOE work down to 8 percent of JPL activities, down from a high of more than 14 percent.[128] To avoid mass layoffs Murray made up for the losses by seeking military contracts. As he told his management committee in 1982, "the only *really* urgent matter . . . is: what is our best strategy for achieving rapid growth of DOD business, and what actions do we need to take to support it?"[129]

EXISTENTIAL SUCCESS, THE LONG 1970S, AND SPACE HISTORY

And with that, the long 1970s came to an end at JPL. There, and at NASA's other facilities, the public sector projects I've surveyed left few traces. A few endured in some form—most notably Sam Pool's experiments with telemedicine—but even major efforts like MIUS or JPL's Low-Cost Solar Array Project had disappeared almost completely by the mid-1980s. The cultural and political changes of the long 1970s did leave an imprint on NASA itself—the space program became more diverse, as symbolized in particular by an astronaut corps that began to include women, non-whites, non-Americans, and more civilians. These changes have indeed drawn much attention from historians.[130] But their connection—or lack thereof—to NASA's brief interest in civilian, terrestrial social relevance has not been much explored, with the notable exception of Neil Maher's *Apollo in the Age of Aquarius*.[131]

My analysis does not significantly complicate Maher's. Like him, I want to put the "who" of NASA in the 1970s in the same frame as the "what." That is, like Maher I argue that the space program's shifts in personnel that were inspired by the civil rights movement and second-wave feminism are not separable from its ventures into alternative energy that were inspired by the environmental movement. A nice emblem of this confluence is a 1975 newsletter from NASA's Goddard Space Flight Center in Greenbelt, Maryland, which juxtaposed an article about a collaboration with the Greenbelt municipality to develop energy conservation and solar heating technologies; a photo essay on Goddard's observation of Black History Week; and a profile of John Bogert, a blind computer programmer and user of a commercial version of the Stanford Optacon.[132] These different strands of the 1970s were entangled at the time, even if some of them left few traces in the organization—or in the organization's historical memory—today.

What this chapter adds to Maher's argument is the concept of existential success as a way of understanding the near-inevitability, and the contours, of NASA's predicament in the long 1970s. It is hard to imagine a scenario in which NASA could have avoided severe budget cuts in the early 1970s. Some outflow of personnel to other organizations—and the export of space-age expertise to those organizations—was therefore close to inevitable. It is perhaps easier to imagine NASA avoiding entanglements with the era's socially relevant research agenda, and certainly much of NASA did remain aloof from civil society. But at the time many actors were trying to guide NASA in that direction, and for many people within NASA engagement with public sector projects was the best (or only) option available.

I have also argued that NASA's predicament in the long 1970s casts light on the situation of American science and engineering more generally. I have therefore drawn out parallels between, for instance, the Meal System for the Elderly project and some of the Master of Scientific Instrumentation theses surveyed in chapter 2 or the acoustic microscope in chapter 3. Those parallels are probably clearest when talking about the origins of socially relevant projects; but the concept of existential success also tells us something about why these projects mostly petered out. Socially relevant R&D flourished in a time when one existential mission had ended and a new one was not yet in place. By the end of the 1970s, though, NASA had been given a new set of priorities—priorities that did not include nearly as much room for urban, environmental, or energy research.

Those new priorities were varied and sometimes contradictory: to support remobilization for the Cold War; to improve diplomatic relations with Europe, the Pacific Rim, the Soviet bloc, and elsewhere; to assist private industry in the commercialization of space; and to give taxpayers an accountable return on the use of "their" money.[133] Together, those new priorities allowed NASA to justify a fairly stable budget; but none of these priorities uniquely justified NASA's *existence* in the way that the space race had. NASA's history since July 20, 1969, has therefore seen a fairly steady if gradual decline in funding and public faith.

The same applies to American science and engineering more generally. From the end of World War II until the late 1960s, there was a clear mission justifying the government's investment in (and the public's deference to) American science and engineering: to contribute prestigious discoveries, militarily valuable technology, and mobilizable personnel to the struggle against communism. That mission lost priority in the late 1960s—that is, it no longer functioned as an existential mission. Through the early 1970s American scientists and engineers bumped from one new existential mission to another: urban problems, environmental problems, the energy crisis, global economic competition. By the late 1970s some of the same priorities that took hold at NASA had begun to restructure American science and engineering more generally. But, again, those new priorities were sometimes contradictory, and none could uniquely justify the enterprise as a whole to its various beneficiaries. After 1970, no one thought that they were getting everything that they wanted in return for their support of American scientists and engineers. The history of the American R&D enterprise since July 20, 1969, has thus been one of long-term decline in public funding and public faith as well.

5

MISTAKING THE SUNSET FOR THE DAWN: JACK KILBY'S SOLAR BOOM AND BUST

This chapter looks at Texas Instruments (TI), one of the most influential microelectronics companies in the world from the late 1950s through the 1990s. I approach TI through the later career of its star employee, Jack Kilby, who among other things shared the Nobel Prize in Physics in 2000 for his coinvention of the integrated circuit. The theme of the chapter is change, and specifically the awkward politics of change for buccaneering companies reliant on square engineers like Kilby (and vice versa). Change was a constant at TI; the company famously started out in oilfield services, then branched into military technologies, semiconductor manufacturing, and finally consumer electronics. Nor was TI alone in making such major changes to its core business: some of the most well-known names in semiconductor manufacturing—Samsung, Philips, Motorola, Ericsson, Nokia, IBM, Philco, and Toshiba, for example—did not start out in that industry.

As they entered the tumultuous 1970s, companies like these had to decide whether to change again—and if so, in what direction to change. As with NASA in chapter 4, I will draw out ways in which a particular organization's strategies for dealing with change help us to better understand the decade's changes in American science and engineering more generally. Everyone involved in America's R&D system at the time wanted to know how far it would and could change from the early Cold War paradigm, and in what direction. Looking back, we need to know why some systemic changes were pursued and then dropped but others persist to the present. Like the other organizations in this book, TI offers enlightening examples of major strategic changes that were abandoned as well as ones that endured.

Specifically, we will look at a residential solar energy system that TI and Kilby developed in the 1970s and then mothballed in the early 1980s. As an alternative energy source, solar power has long attracted interest from those pursuing "alternative" politics and lifestyles.[1] That was as true in the 1970s as before and after. But for a few years after the 1973 oil embargo, interest in solar power was remarkably widespread across the political and cultural spectrum. Through the rest of the decade, pillars of the national security state like TI, and conservative individuals such as Kilby, attached themselves to solar power. They justified their own—and their patrons' and potential customers'—interest in that technology in terms of cost and national interest, rather than in the environmental or antiestablishment terms that attracted countercultural solarists. If we can understand why companies like TI and individuals like Kilby turned toward solar power in the first place, why they formed some coalitions around the technology but not others, and why they swerved away from solar power later, then we gain some insight into what conditions make square scientists and engineers (and their employers) more socially responsible or less so.

SQUAREST OF SQUARES

One of those conditions, I'll argue, is squareness itself. So before getting to the case study of TI's solar energy system, I first want to show how this chapter's central figure—Jack Kilby—relates to the category of square scientists and engineers. As noted in this book's introduction, the category has fuzzy boundaries and significant internal variability. In earlier chapters I have counted both the acerbic Bill Rambo and the idealistic David Phillips as squares even though they would've disagreed on most of the debates I've discussed. The common characteristics that lead me to label all three—Rambo, Phillips, and Kilby—as squares is their "small-c" conservative presentation of self and their publicly ambivalent or reticent stance toward the politics of the day.

Of course, demographic affinities—their status as white, middle-class, middle-aged men—helped Phillips, Rambo, and Kilby to cultivate a square, broadly quietist presentation of self. None of them were apolitical, but all three expressed their politics quietly, if at all, and their public actions were often only ambivalently related to their private political views. That kind of quietism was a privilege afforded by their demographic status. But demographics aren't destiny. Other white male scientists and engineers used similar privileges to cultivate more provocative public personae instead. Gerard K. O'Neill, who appeared in the previous chapter, is a good example. If anything, O'Neill's background in instrument building meant that he began the long 1970s on the square end of his field (high-energy physics) relative to flamboyant theorists such as Richard Feynman and Murray Gell-Mann. But once he began promoting his utopian visions for space habitation, O'Neill's appearances on the *Tonight Show* and fervent youth following made him rather outlandish relative to his square colleagues. The line that O'Neill crossed, from avoiding the limelight to seeking it, was fluid and contestable yet keenly felt by many squares. One aim of this chapter is to map out where at least some squares drew that line and how they related to those on the other side of it. We'll see that their attitudes toward nonsquare colleagues—or at least colleagues with an outsized public profile or links to the counterculture or the outspoken left—shaped the outcomes of the squares' ventures in the long 1970s.

Jack Kilby is thus the central figure of this chapter because he so thoroughly exemplified squareness. It is hard to imagine someone to whom the label could more aptly be applied. Kilby's public presentation of self was notoriously buttoned-down—to the point that his only biography thus far (a breezy hagiography written by his friend Ed Millis) is entitled *Jack St. Clair Kilby: A Man of Few Words.* There, Kilby is depicted as almost parodically middle American right from the start:

> Life in [Kilby's parents'] household [in Great Bend, Kansas] was probably typical of small-town midwestern U.S.A., but some of the family customs seem quaint by today's standards, or lack of standards. The evening meal, dinner, was a sit-down affair with a white tablecloth and napkins and food brought in from the kitchen. The plates were served by Father Kilby to each family member. This was not just Sunday dinner, or Thanksgiving, but every evening. Conversation during this time was encouraged, but was not to include controversy.[2]

"Conversation encouraged, but not to include controversy" is a better motto for square scientists and engineers than any I could coin.

This is perhaps as good a moment as any to acknowledge that I grew up in a similar milieu to Kilby (the same Midwestern state, in fact), though a half century later. Kilby was of roughly the same generation as my parents, and Millis's picture of Kilby's childhood dinners doesn't deviate much from my mother's recollections of meals among the small-town Midwestern bourgeoisie in the 1940s.[3] So my references to Kilby's squareness are in no way meant as a slight but are rather grounded in a kind of autoethnographic recognition of how his squareness informed the lines that Kilby drew between himself and others.

Once Kilby left Kansas, his adulthood continued in much the same square vein as his childhood. He received a bachelor's degree in electrical engineering from the University of Illinois, interrupted by a stint in India assigned to the Office of Strategic Services (forerunner of the CIA) during World War II. He then took a job with a Milwaukee firm, Centralab, developing techniques for making miniaturized, printed circuits. While there Kilby married another Illinois alum, Barbara Annegers, and they began a family. After eleven years in Wisconsin, the Kilby family moved to suburban Dallas so that Jack could join TI's rapidly growing Semiconductor Research and Development Lab. There they settled into what Millis depicts as unexceptional white Texan suburbia, with Jack immersed in photography and woodworking in his spare time and Barbara becoming an accomplished ceramic artist. That life, however, was compartmentalized from Jack Kilby's work: "People that he worked with for years knew very little about Jack and his home life."[4]

At work, meanwhile, Kilby's name and identity became entwined with TI's thanks to his sudden burst of inventive activity in the first two months after he arrived in 1958. Although his coworkers were all on holiday that summer, Kilby had yet to accumulate vacation time.[5] He thus had the semiconductor lab to himself and could concentrate on solving the microelectronics industry's most pressing problem: the "tyranny of numbers."[6] This crisis arose from the growing number of components (transistors, capacitors, inductors, resistors) in high-performance circuits, especially those made for the military. With discrete components soldered together, failures at the solders became all too frequent and the circuits were uselessly short lived. In the late 1950s the military services and their vendors were exploring ways to "integrate" circuit components to eliminate this problem, and a number of people proposed variants of the eventual solution—monolithic integrated circuits with components and interconnects all made from a single chunk of material—at about the same time.[7] Kilby's tinkering in 1958 led to the first realization of at least some aspects of what became the standard form of integrated circuit—though a long patent battle has resulted in the convention that Kilby "coinvented" the integated circuit with Robert Noyce, then of Fairchild Semiconductor in California.

In part because of that patent dispute, TI has put a great deal of money and energy into enshrining Kilby's 1958 breakthrough in historical memory, down to securing a Texas State Historical Marker commemorating the "demonstration of the first working integrated circuit," unveiled by the state's governor in 1988. (I've heard a TI veteran caustically note that Fairchild's equivalent California state historical marker is for the less punchy "first commercially practicable integrated circuit."[8]) As a result, there is a curious asymmetry between Kilby and the aggressively extroverted Noyce. What they both did in 1958 has been retold many times; but of their lives *after* 1958, only Noyce has been memorialized as *The Man behind the Microchip*, the title of Leslie Berlin's biography.[9] No scholarly—or even full-length—biography of Kilby exists, even though he (unlike Noyce) later won the Nobel Prize. In my experience, historians of microelectronics usually have some vague knowledge of Kilby's achievements at TI in the 1960s—for example, ushering the first handheld electronic calculator to market—but not to nearly the same degree that they are familiar with Noyce's post-1958 contributions.[10] This isn't to say that Kilby *deserves* more attention; my point is simply that TI's campaign to secure priority for its intellectual property, plus Kilby's squarely taciturn profile, means that his later life is a cipher compared to his actions of 1958.

Thus, it is not well known that Kilby was a salaried employee of TI for only slightly longer than he was an employee of Centralab—roughly twelve years. Technically, his departure from TI in 1970 was a leave of absence, but there is little indication that he ever intended to go back, and in fact he never did. Millis ascribes Kilby's exit from TI to mounting managerial duties and a consequent disengagement from the technical puzzle solving that he enjoyed.[11] To recoup that creative outlet, Kilby set himself up as a consultant and independent inventor, free to define his own priorities with no pressure from management. The break with TI was not complete, however—Kilby still filed monthly reports to Fred Bucy, a TI vice president (later CEO), and he negotiated a generous deal in which TI paid him a retainer in return for right of refusal on his inventions. He also represented TI on numerous committees, particularly national commissions assessing military policy toward microelectronics research, development, and manufacturing. TI reaped real gains from Kilby's influence over those committees and the insights that he acquired from them. Those factors, plus the reputational advantages of associating Kilby's name with TI, were probably why the company supported Kilby's new venture. Certainly, Bucy did not pursue commercialization of Kilby's inventions for the first few years, so it is unlikely that is the reason that he was kept on retainer.

It's difficult to place Kilby's post-TI venture in historical context. Was he simply an engineering consultant who tinkered on the side? Was he a relic of an archetypal character in American technology, the Edisonian *inventor*? Or did he represent the dawn of a new figure, the disruptive *innovator*? The conventional story has long been that independent inventors became scarce after World War II, but Eric Hintz has shown that companies like TI maintained fruitful relationships with independents throughout the postwar years—so Kilby may have seen his continuing relationship with TI in that way.[12] Yet the 1970s were also a point at which the US corporate R&D system was shifting. Research managers at companies like TI were increasingly aware that they struggled to commercialize in-house inventions, and they therefore needed to externalize some of their R&D efforts. By the end of the decade the microelectronics and biotech industries had established an enduring model for such externalization: fund university researchers, industrial consortia, and small start-up companies and then absorb any intellectual property that they developed.[13] But as Matthew Wisnioski has argued, that model was accompanied by a rhetoric of making "every American an innovator"—that is, encouraging radically decentralized innovation that (with the right policies) could be made to flow into industry.[14] It's possible, then, that TI management saw Kilby's invention shop as an experimental model for externalizing and then capturing innovation.

In any case, Kilby's departure from TI left him in limbo. He continued to be that squarest of squares: a white, straight, suburban, Midwestern-turned-Texan, fairly conservative (but publicly politically reticent), decidedly patriotic, military veteran and family man. But for almost a quarter century he had also been that postwar icon—the *company man*—only now he was without a company. Like NASA after Apollo 11, Kilby had already done the thing that he would always be most famous for—and now he had to find something else to do.

TINKERING AND TEACHING

For the first few years, Kilby's new venture did not yield much. True, he generated a reasonably impressive list of patent disclosures, but not for anything world-changing: a reliable flashlight, a robbery immune cash register, a battery charger, a thermal printer, a display for electronic clocks and watches, and a portable paging system with visible display (of the sort widely adopted in the 1980s).[15] A few of Kilby's projects, though, speak to a more visionary outlook. For instance, an "electronic checkwriter" that he patented in 1975 (filed in 1973) foresaw digital handling and storage of financial transactions on a personal, portable device, thereby paving the way to the "checkless society" that has come into being only in recent years.[16]

More intriguingly, Kilby invested significant energy, time, and money in developing a personalized, interactive electronic teaching machine. Here is how he described the device in 1974:

> Basically the unit consists of a conventional audio cassette playback device, a group of twelve switches and integrated circuit logic to compare the student's answer to an answer recorded on the tape. In operation the student will listen to a question which has been recorded on the tape. When the question is over, the machine will stop. The student then sets his answer on a bank of switches.[17]

This teaching machine seems to have been important to Kilby: there is much more correspondence about it in his papers than about his other inventions of this period. Indeed, Millis speculates that Kilby's attachment to the teaching machine helps explain why he left TI: he invented an early version in 1964—on his own time—that TI nonetheless refused to let him patent in his own name.[18] As Millis notes, TI did eventually market a tremendously successful teaching machine, the Speak & Spell, which Kilby's device closely resembled in appearance (though the Speak & Spell's digital signal processing capability was radically different from, and more pathbreaking than, Kilby's invention).

Electronic teaching machines are the kind of technology that exemplifies the common interests that still could have united American scientists and engineers of the long 1970s even when they were divided by cultural and political affiliations. On the one hand, teaching machines were a vibrant site of countercultural activity, particularly in the San Francisco Bay Area. In fact, the Portola Institute, publisher of *The Whole*

Earth Catalog, was founded primarily to stimulate educational computing.[19] The trivial distance (a quarter mile) between Portola and Douglas Engelbart's group at SRI (Stanford's former contract research arm) aided Engelbart's team in developing technical aspects of personal computing such as the mouse and the graphical user interface. Yet proximity to Portola also encouraged Engelbart's group to indulge in countercultural visions of human potential liberated—and human cognition augmented—by forming a close pedagogical relationship with a computer from childhood onward.[20]

Engelbart's group and the Portola Institute both played an important (if sometimes exaggerated) role in the emergence of personal computing as we know it today. They were by no means alone, though, in the enormous hope that they invested in educational computing. Similar technologies were also a focus of attention among the conventional liberals in the Johnson administration's Office of Education.[21] And they were a hobbyhorse of a few conservative inventors like Kilby, possibly because of growing interest from their patrons in the US military. For instance, Eugene Kleiner (one of the cofounders, along with Robert Noyce, of Fairchild Semiconductor) founded an educational computing company, Edex, in 1962 before selling it to a defense contractor, Raytheon, in 1965. Other military–industrial firms like Litton Industries and Hughes Aircraft entered the field around the same time.[22] By the time Kilby left TI, however, the Mansfield Amendment had forced some square educational computing researchers—for example, the Artificial Intelligence Group at MIT's Media Lab—to shift from military to civilian funders (and consequently toward a more child centered line of work).[23]

Perhaps because of such widespread interest in the technology, Kilby thought that there was a ready market for his teaching machine, and he tried to persuade a number of companies to develop and market it: Westinghouse, Xerox, Singer, Telex, IBM, Bell and Howell, McGraw-Hill, and Sintra. He also hoped that William Magruder—whom we encountered in chapter 4 running Nixon's New Technology Opportunities program—would connect him to people in the federal Office of Education who might be interested.[24] The Office of Education and several of the companies that Kilby contacted (e.g., Sintra and Bell and Howell) already had or would later have educational computing projects—yet none wanted to collaborate with Kilby or license his invention.

HOW TO TRAIN YOUR ELECTRICAL ENGINEERS

So Kilby's efforts in educational computing did not lead directly to anything (though one can speculate that they gave some impetus to his friends who later developed the Speak & Spell). Yet the teaching machine did mark the start of Kilby's growing interest in bringing his technical expertise to bear on educational innovation. By the early 1970s he had become particularly focused on incorporating hands-on experience in fabricating integrated circuits into undergraduate electrical engineering curricula. Kilby's efforts in this area were stimulated in part by his correspondence with a former TI colleague, Jay Lathrop, who had moved to Clemson University's electrical engineering department in 1968 to be closer to his elderly parents.[25] One of the recurring topics in their letters was Lathrop's attempts to introduce integrated circuit fabrication into his department's curriculum. Around the same time, Kilby developed ties with the electrical engineering department at Texas A&M University, where he promoted a similar incorporation of training in integrated circuit fabrication into the undergraduate degree program. Eventually (1978–1984) Kilby would serve as a Distinguished Professor of Electrical Engineering at A&M, and today that department still hosts the TI Jack Kilby endowed chair.

Kilby's dive into reforming undergraduate electrical engineering pedagogy later facilitated his move into solar energy. His efforts in this area also tell us something about the attitudes toward his peers that he brought with him when tackling a new technological arena—in particular, his greater regard for peers who were allied with military–industrial coalitions rather than civilian ones, and his low regard for peers who sought out the limelight. So it's worth delving into this aspect of

Kilby's post-TI career even though that means skipping slightly ahead in time.

The background for Kilby's new interest in academic electrical engineering was that by 1970 integrated circuit technology had stabilized as *the* platform for mass-produced microelectronics. That is, although Kilby and Noyce's 1958 invention was later taken as a (literal) landmark, through the 1960s other microelectronics technologies were still more practical than integrated circuits for many applications. From the early 1970s onward, however, it was obvious that integrated circuits would predominate in most commercial microelectronics products, whereas alternatives to the integrated circuit would endure only in niche applications.[26]

That realization had two consequences relevant to Kilby's perspective on education. First, companies like TI increasingly wanted to recruit college graduates who were already familiar with integrated circuits and how to make them. At the same time, academic electrical engineers (many, as we saw in chapter 3, former employees of companies like TI) wanted to keep up with industry and make their research relevant to their corporate peers. After 1970 it was obvious that they could do so only if they had ready access to integrated circuit fabrication equipment. Until then most academic electrical engineers made do by scrounging time on industrial equipment, but in the 1970s more and more universities founded microelectronics centers to house their own tools, with which they could train students and facilitate faculty research.[27] The ICL at Stanford, founded in part to advance Optacon development, was an influential early model for such centers. Kilby wanted A&M to set up something along those lines; to that end he corresponded with people who were trying to establish similar facilities, such the University of Minnesota's Microelectronics and Information Sciences Center.[28]

Second, everyone could now see that innovation in microelectronics primarily meant innovation in silicon integrated circuit manufacturing—and that "innovation" in that context largely meant continued miniaturization of components and integration of greater numbers of components onto a single chip. Whichever firms managed to push miniaturization and integration furthest would dominate the market. Various governments therefore initiated projects in the 1970s to boost the capabilities of their nations' semiconductor firms in miniaturization and integration. Japan's Very Large Scale Integration (VLSI) project, announced in 1975, is the most famous and perhaps the most successful of these, but a number of other countries moved in the same direction, especially after Japan's effort was launched.[29]

In the United States, the Pentagon started playing a similar role much earlier, in the 1950s. The US military took that early lead because integrated circuits were suitable for the unusual and extreme requirements of battlefield technologies long before they were practical for most civilian customers. At TI, contracts with the US Air Force and coordination by an Air Force grant officer, Dick Alberts, made it possible to move from Kilby's single, kludgy prototype of 1958 to mass production of integrated circuits in the 1960s.[30] That transition to mass production allowed economies of scale, which meant that over the course of the 1960s, integrated circuits did become cheap enough for increasing numbers of civilian customers. By the beginning of the long 1970s, companies like TI were even incorporating integrated circuits into *consumer* products such as watches. That is, the integrated circuits market civilianized, and it did so at exactly the same time that military support of semiconductor R&D became more controversial, both on university campuses *and* among some of TI's corporate competitors. Indeed, the two trends were to some extent interrelated. Yet Kilby and TI remained committed to the Pentagon's patronage and mission—a commitment that informed Kilby's approach to both educational innovation and solar power.

Opposition to military support of semiconductor R&D took two main forms, both of which Kilby disdained. First, there were semiconductor manufacturers who thought that the military market came with too many burdensome strings attached. Under Robert Noyce, Fairchild had been one of the first semiconductor manufacturers to step away from military contracts for this reason.[31] Other Silicon Valley firms then followed suit—to Kilby's dismay. As he put it in a 1978 report to Fred Bucy,

DoD continues to pursue the VHSI [Very High Speed Integrated Circuit] project. . . . [A] problem is Bob Noyce, perhaps supported by a few of the smaller West Coast houses. I gather that he feels that people capable of working on the program are in short supply, and that the DoD effort will divert them from more productive tasks. Noyce has always felt that DoD support of R&D has been unproductive, and that industry—or Noyce—could do a better job of directing it. His view is a short term one, and ignores the fact that the supply of trained people can be increased and that the needs of the military are not being well met by the present semiconductor industry.[32]

Thus, for Kilby, it was the duty of the semiconductor industry to ensure that the needs of the military *were* met, even if that meant incurring some extra costs.

Second, *classified* academic research was the subject of intense and broad-based opposition on many American campuses in the late 1960s. Even many academics who had enjoyed military patronage joined that opposition, with the result that many universities banned on-campus classified work. Yet, if anything, the national security state became *more* stringent in classifying microelectronics R&D over the course of the 1970s. Many of Kilby's monthly memos to Fred Bucy raised concerns about the Pentagon's export controls on semiconductor products and manufacturing know-how, because these had the potential to undermine TI's profits. Kilby was unmoved, however, when universities voiced similar worries about restrictions on the open flow of information. As he told Bucy in 1981,

I am chairing a NRC [National Research Council] panel on the "Impact of VHSIC," a study requested by the DoD. Most of the areas are quite straightforward, but we do have a few problems on the impact of information restrictions on the universities. I am not very sympathetic to the university position. The universities who are complaining got out of classified work ten years ago. If they really want to work on military systems, even if the systems are built on silicon they should be prepared to accept some restrictions. A major part of the problem is caused by delusions of grandeur on the part of the faculty. Many of them feel that with their new microelectronic capabilities that they can build anything. They will not be asked to build VHSIC chips. They should be able to contribute in the rather basic research area, but this is not covered by the VHSIC regulations. It is unfortunate, but the problem has become a matter of principle.[33]

Again, Kilby clearly believed that the military should get to pay the piper and should therefore get to call the tune—and that universities that had allowed antiwar students to run riot should have to bear the consequences.

The person whose "delusions of grandeur" that Kilby had in mind was almost certainly Caltech's Carver Mead, one of the leaders both in VLSI research and in the emergence of academic microelectronics centers. Mead was closely associated with Noyce and was a regular consultant for Intel, the second and ultimately more famous firm that Noyce cofounded. So it's possible that the patent dispute with Noyce colored Kilby's views of Mead. But Mead was also a charismatic promotor of his own ideas—that is, a type of not-very-square academic that Kilby disliked. As Kilby reported to Bucy in 1980,

> I have been asked to serve as Chairman of an NAS panel to review the impact of the VHSIC program for DoD. We have had several meetings, and at the last heard from Carver Mead, who seems to represent the only real opposition to the program. From a technical standpoint just what Mead is proposing is not too clear, but he is very explicit on one point. He feels that vertical integration in the semiconductor industry is bad, and very much wants DoD to build a state-of-the-art fab facility, or silicon foundry. The fact that DoD (ARPA) has already agreed to fund one such facility at Stanford does not seem to deter him. Mead seems to have a messiah complex, and feels that he alone has seen the light and fought the good fight.[34]

Two months later, Kilby expressed relief that "the Carver Meade [*sic*] situation has quieted down . . . [probably because] Meade felt that he had achieved his purpose of securing recognition by all as the father of VLSI, declared a triumph, and withdrew."[35]

Obviously, Kilby's skepticism may have been colored by the rivalry between Intel and TI and the differences in their business strategies (with Intel, like Mead, rejecting vertical integration and TI embracing it). Yet, the fact that Kilby nominated Gordon Moore, Intel's very square cofounder, for membership in the National Academy of Engineering suggests that his dislike of Mead and Noyce was more about personality than corporate or personal rivalry.[36] In any case, whatever his feelings, Kilby couldn't simply dismiss Mead's contributions to both research and teaching related to integrated circuits, but he also couldn't pass up the chance to one-up Mead. As he told Bucy in 1981,

> Carver Meade [*sic*] had the annual meeting of his disciples a couple of weeks ago.... To his credit, the courses that he started at a half dozen schools have attracted a great deal of interest, and something over a thousand students have been exposed to the course.... Texas A&M is still very enthused about the TI program, and is making a major commitment of capital equipment to enhance it. Their spending will be concentrated on CAD equipment and on testing, which will let them do a better job of designing circuits and evaluating them.[37]

"Better" here almost certainly means better than Mead, because Kilby had earlier explained,

> We have chosen to structure the course somewhat differently than the Carver Meade [*sic*] course, with more emphasis on the basic principles of IC design and less on VLSI. It is intended to be a two semester course, which should give enough time to do a good design and have time to evaluate it and find any errors, which I think is a significant lack in the Meade [*sic*] course.[38]

I'm offering these extended quotes not to slight Kilby *or* Mead, but rather to convey the emotionally loaded language that Kilby used in talking about Mead. One of the arguments of this book is that the turmoil of the 1970s generated strong emotions, even among the most ambivalent square scientists and engineers. Affect tells us something. As we'll see, Kilby's visceral reactions to outspoken and avowed agents of change like Mead and Noyce similarly defined his vision for solar power—he wanted to prove himself a better engineer than such people. But his attitude toward the likes of Mead and Noyce also *limited* his vision for solar power, in that he couldn't bring himself to cooperate with such people.

BORN OF NECESSITY

Let us, then, examine that vision in detail. As far as his papers indicate, Kilby had given no thought at all to solar energy before 1973. He would, of course, have known the basic physics and engineering underlying most solar energy systems, but I can find no indication that he viewed solar energy as a locus for his own inventing activity prior to the OAPEC oil embargo in the fall of that year. Nor should that be surprising—the 1973 embargo was a wake-up call for many. Although there is a debate whether a real embargo was ever put in place and whether the "energy crisis" was orchestrated by the oil industry, there can be no denying that most Americans experienced precipitous increases in gasoline prices combined with severe gasoline shortages in late 1973 and into 1974.[39] In the short term, the Nixon administration scrambled to respond with a mix of price controls and rationing. Eventually Nixon announced Project Independence, a long-range plan for conserving energy, finding new sources of energy, and making existing sources of energy (e.g., oil from the North Slope of Alaska) more accessible. Recall that we saw in chapters 3 and 4 how the embargo, Project Independence, and the ensuing reorientation of the RANN program toward energy caused researchers at Stanford and NASA to suddenly prioritize alternative energy and energy conservation. Kilby and TI were no different.

With his attention primed by the embargo, Kilby hit on an idea at the beginning of 1974 for how his expertise in semiconductor physics could help solve the energy crisis.[40] Over the previous few months he had been in regular correspondence with Lathrop about the latter's pedagogical reforms at Clemson, so when inspiration struck Kilby naturally reached out to his friend and colleague. Lathrop seems to have been positive toward Kilby's idea from the start, and so in early May Kilby offered Lathrop a substantial consulting fee to come to Dallas to collaborate on the project over the summer.[41] Then, sometime before the fall of 1974, Kilby and Lathrop brought a third collaborator into the project, W. Arthur "Skip" Porter, an electrical engineer at Texas A&M. Porter's 1970 PhD dissertation had been on dislocations in single crystal silicon, and it was his expertise in silicon's material properties, as well as his labs at A&M's Institute for Solid State Electronics, that attracted Kilby and Lathrop.[42]

Porter's expertise was necessary because Kilby's system obtained its purported competitive advantage as much from silicon's material properties as its electrical characteristics. Most silicon-based solar energy systems were at the time constructed from large sheets of silicon. In such a system, photons from the sun excite electrons in the silicon, which then travel through a circuit to create an electric current. Making the sheets was costly, difficult, and quite energy intensive—disadvantages that Kilby proposed to overcome by processing silicon with a seemingly simple eighteenth-century technique for making shotgun pellets: a "shot tower." In such a tower—here provided by Texas A&M—a metal or semiconductor is melted and poured through a sieve, creating droplets of molten liquid of a specific size. As the droplets fall, surface tension causes them to form highly regular spheres. By the time they reach the bottom, the droplets have frozen into tiny solid spheres.

Obviously, you can't get much solar energy out of a ball of silicon approximately 250 microns in size. (TI presentations depicted a dozen or so of the frozen droplets lying on the head of a pin.[43]) The *efficiency* of such a small pellet, however, is significantly higher than that of a thick solar panel because almost all of the pellet is close to its surface. In a conventional solar panel of the 1970s, much silicon went to waste because solar photons could not penetrate deep enough to encounter most of the bulk material. With Kilby's pellets there was no bulk material, and hence little wasted silicon. So Kilby imagined combining millions of the pellets in a single modular panel to give an exceptionally high efficiency.

Instead of wiring each individual pellet into a circuit and generating electricity directly, Kilby thought to bathe them in a liquid, the molecules of which could be dissociated by photovoltaic energy. A few such liquids were discussed over the course of the project, but hydrogen bromide was the most commonly used. Photons striking the silicon pellets would create an electric potential, which in turn would break down the fluid medium: for example, hydrogen bromide would decompose into hydrogen and bromine.[44] The hydrogen and bromine would then be siphoned off and stored, and when electricity was needed they could be recombined in a fuel cell (an energy conversion device with a chemistry similar to controllably burning rocket or combustion engine fuel—hence the name). Electric current would flow out of the system, and the recombined

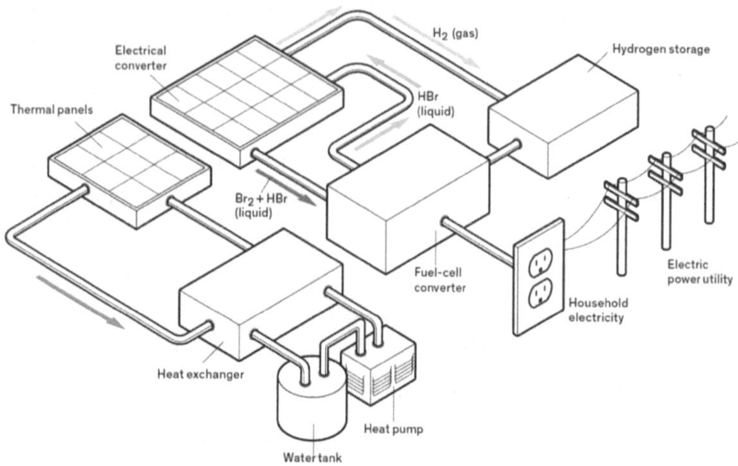

Figure 5.1

Schematic of the Texas Instruments Solar Energy System. *Source*: Image courtesy of Josh McKible.

hydrogen bromide would flow back into it to be reused again. As figure 5.1 shows, the system also included lower-tech thermal panels for generating and storing heat.

The most obvious benefit of this system is that the times when electricity is generated can be separated from the times when the sun is shining. Hydrogen bromide could be broken down by photovoltaic energy during the day, then recombined in the fuel cell to create electricity whenever doing so would be cheaper than drawing from the electrical grid. Kilby and his colleagues also claimed a few other advantages, though these were more disputed. In particular, they believed that the tiny shot pellets would "provide a couple of orders of magnitude more power per pound of silicon than anything I have seen proposed so far."[45] Manufacturing silicon shot also looked, initially, to be cheaper and less energy intensive than traditional photovoltaic panel manufacturing, during which the silicon must be brought to its (very high) melting point several times. Kilby's system also embedded the silicon in a series of narrow tubes instead of large flat panels. Because the tubes operated independently from each other, Kilby claimed that a defect in one of the tubes would not diminish the system's overall efficiency by very much, whereas a defect in one part of a monolithic silicon solar panel would make the whole panel nearly useless and dramatically lower the efficiency of the entire system.

In April 1975, Kilby formally presented his idea to TI's Fred Bucy: "I was glad to have had a chance to go through my scheme for solar energy conversion yesterday. As I am sure that you could tell, I am very enthused about it."[46] From that point on, Kilby kept Bucy regularly updated on the team's progress—a sign that TI was indeed interested in the technology. Then in July of 1975, TI agreed to fund Kilby's team through the rest of the year and to "assume full responsibility for development and exploitation of this program" at the beginning of 1976, with the proviso that Kilby could

> continue parallel development efforts after December 31, 1975, for a period not to exceed two years and for a cost not to exceed $60K per year which cost will be advanced by TI and will be considered as support advances under the existing TI/Kilby agreement. Any inventions and patents obtained on these inventions resulting from work done during the parallel development efforts will be exclusively licensed to TI at no additional cost to TI.[47]

In return, Kilby received a substantial increase in his monthly retainer plus an extra consulting fee, plus a guarantee of payments for the use of his patents (patents that TI also promised to shepherd through the patent office), plus royalties if the system went into production.

I'm outlining TI's agreement with Kilby in some detail to highlight its unusually generous nature. As Kilby himself noted, this agreement was quite different from his earlier ones with TI, not just because TI was actually interested in the invention but also because it envisioned that he would continue to have a role with the technology in tandem with TI's in-house efforts. The solar energy system agreement differed from earlier ones because TI and Kilby

> both felt that this idea was of real importance, and that it needed special attention. I then proposed to continue the work at the universities [Clemson and Texas A&M]

> until summer [of 1976], at which time the project could be moved into TI and a small facility established for it. This plan was accepted by TI and is now well under way. The additional effort has moved the idea along, and no major flaws have been uncovered. I really believe that by the end of the summer [of 1976] we will have a rather convincing demonstration of the concept.[48]

This shows, I think, that solar energy struck Kilby's imagination—*and* TI's business planners—in a way that none of his earlier inventions had. TI saw an opportunity for profit—an entirely new, booming, and untapped market to move into. Kilby likely saw solar energy partly in the same terms—he was in no way averse to making money off his idea. But the enthusiasm of his correspondence also shows an affective commitment that probably went further than just dreams of hefty royalty payments.

MANY WAYS TO BE SOLAR

Most likely one factor motivating both TI and Kilby—though possibly for different reasons—was the fact that they were not alone. By 1975 ideas for commercial solar energy systems were becoming abundant, to the point that it had become possible to speak of a solar energy "industry." TI, with its expertise in the material—silicon—at the heart of most of those solar energy systems, did not want to miss the boat. Kilby, I believe, also wanted to show that he had come up with a better technology than other inventors of solar energy systems.

The industry that Kilby and TI were attempting to join—or, really, to help create—was fractured and complex, however, in ways that Kilby and TI never entirely mastered.[49] The different parts of that industry could've provided grounds for coalition forming that might have sustained the technology through rough times. Kilby and his colleagues did seek coalitions with certain kinds of solarists—but they rejected making common cause with promotors of solar technology who did not resemble the square engineers that Kilby and TI were used to dealing with. In particular, Kilby and TI never showed any interest in allying with the growing cohort of (mostly) young people for whom solar energy fit neatly into a vision of socially relevant science and technology. These were the kind of people who, for instance, entered the UCSB MSI program for combined "tangible" (i.e., pecuniary) and "esoteric" (i.e., idealistic) reasons. Indeed, several of the MSI program's student projects concerned solar energy.

For many young, reformist solar enthusiasts, one of solar energy's main selling points was its potential for decentralization: one could use solar power to "get off the grid," to run your commune or farm as you saw fit without being at the mercy of giant, faceless utilities. Another—overlapping—group of reformist solarists were attracted to the technology because it seemed cleaner and more environmentally (and politically) friendly than coal, oil, or uranium. For some solarists these motivations were mutually reinforcing, but for others they were quite distinct though not mutually exclusive. At least for a time, different kinds of reformist solarists could cooperate even if their interpretations differed. Supporters of the appropriate technology movement associated with E. F. Schumacher, for instance, no doubt appreciated that (some) solar energy systems could be constructed with a small environmental footprint, but they were also happy to work with farmers who cast the technology more as a means to save money and preserve their autonomy.[50]

Some of these idealistic scientists and engineers formed their own start-up companies or science-oriented communes. Indeed, the line between company and commune was sometimes blurry, as exemplified by the colorfully named New Alchemy Institute, founded by Nancy Jack Todd, John Todd, and Bill McLarney in 1969. The New Alchemists ran experimental farms—which they called "arks"—on Cape Cod and Prince Edward Island, where they developed technologies for alternative energy and sustainable agriculture and architecture.[51] The New Alchemists were well-known in environmentalist circles: they hung out with the likes of Schumacher and Stewart Brand, and Prime Minister Pierre Trudeau visited the Prince Edward Island ark. Their example shows that moderately countercultural alternative energy promoters were capable of knitting themselves into quite influential coalitions. Yet it's very difficult to imagine Kilby taking such coalitions seriously or countenancing anything done by someone calling her- or himself a New Alchemist.

Other idealistic solarists joined government agencies that were developing their own solar technologies or that were tasked with evaluating solar energy systems developed by others. We'll see that Kilby and TI had little choice but to interact with some of these agencies, particularly JPL and SERI (the latter part of the Department of Energy and the former under contract with that department). Indeed, the director of SERI, Denis Hayes, exemplified this kind of idealistic solarist, having been one of the lead organizers of the first Earth Day in 1970.[52] It's no surprise, then, that Kilby was dismissive of SERI and everyone associated with it.

Not all idealistic solarists were young. Schumacher, for instance, was born in 1911. More closely intersecting with Kilby's orbit was the autodidact labor organizer turned tech entrepreneur Stanford Ovshinsky, who was born in 1922. Kilby would've already known about Ovshinsky's discovery of switching behavior in amorphous (i.e., noncrystalline) semiconductors in the late 1950s, something that many mainstream microelectronics researchers refused to accept. But Kilby also would've known Ovshinsky for his relentless, flamboyant promotion of that discovery and of the company, Energy Conversion Devices (ECD), that he built around it. No doubt Ovshinsky, with his left-wing politics and working-class roots, had to promote his ideas in ways that a more establishment figure like Kilby did not.[53] But of course for Kilby such promotionalism was one more reason to shun Ovshinsky.

Ovshinsky's announcement in 1977 that ECD was going to market easily manufacturable solar panels powered by thin (and hence light) sheets of amorphous semiconductors—perhaps even printing them roll-to-roll like a newspaper—put him in direct competition with Kilby and TI's similarly easily manufacturable solar energy system. Unsurprisingly, Kilby's assessment was that Ovshinsky was a fabulist:

> Ovshinsky and ECD continue to make the news. He has been talking to the *Wall Street Journal* and the financial community about his work, and has said that they are now building square foot amorphous cells with 7% efficiency. . . . We have

checked with some of his people, and it would seem that these stories are incorrect. . . . I think that the stories have been triggered by his need for more money. ARCO [Atlantic Richfield, an oil firm] has told him that they will cut off his funds at the $8 million level, which he will reach by the end of the year. He will probably find another sponsor. Anyone who has been able to raise $75 million on pure promises should not be underestimated, but it looks like a good chance to sell short.[54]

Again, as with Kilby's dismissal of Hayes and environmental solarists, Kilby's intense dislike of Ovshinsky reflected a worldview in which those who came to solar energy from the left or who promoted it too flamboyantly (by Kilby's standards) were probably technically incompetent and certainly not worth engaging with.

The TI program did engage, however, with some *square* solarists, especially those with backgrounds in the military–industrial complex whose fortunes had become more precarious thanks to the decline in defense R&D funding. Some of these people had developed expertise during the early Cold War that now became plausibly relevant to terrestrial, civilian alternative energy: in areas such as expertise relating to solar panels for spacecraft; silicon for microelectronics (but now adapted for photovoltaic energy); advanced airfoils used in fighter jets (but now adapted for wind turbines); construction of large movable structures that (like the dishes used in satellite tracking stations) could follow the wind or sun; and superconducting materials for transmitting power over long distances. Especially after the oil embargo and the announcement of Project Independence, many of these people could see clear incentives to repurpose their military–industrial expertise for alternative energy R&D.

A *few* of these people came to share some of the more countercultural interpretations of solar energy and to advocate that civilian, rather than military–industrial, networks take the lead in developing solar technology (much as Noyce advocated for civilian primacy in development of integrated circuits). We've already seen some examples in this book of such idealistic awakening among former participants in the military–industrial complex. Estimating their numbers would be impossible, but nothing in my exploration of this topic makes me think that such converts were a very large proportion of the square solarists. And nothing in Kilby's papers gives any indication that he thought highly of such people. Quite the contrary—former military–industrial colleagues who collaborated with countercultural solarists or advocated civilian control of solar innovation were, in Kilby's view, not very good engineers. In particular, one such former TI colleague, Paul Maycock, became something of an obsession for Kilby, Lathrop, and the TI team. As director of the Department of Energy's photovoltaic program, Maycock worked closely with Hayes and SERI; hence, Kilby and colleagues frequently expressed concern that Maycock was somehow spying on the TI program on behalf of his new allies.[55]

By contrast, Kilby had few qualms about solarists who remained (in his view) square enough: people for whom solar energy offered a means to make money and perform a patriotic duty, who either saw solar energy's environmental and decentralizing potential as secondary or rejected those interpretations of solar energy entirely, and who avoided association with countercultural solarists. As we'll see, he had deep doubts about the competence of engineers who advocated for civilian primacy in solar innovation; but he had far fewer qualms about the competence of solarists who remained rooted in the military–industrial complex. Some of these people were inside government, and others worked for aerospace firms like McDonnell Douglas (which at one point claimed to have built the "world's largest functioning solar thermal plant").[56] Indeed, one of the distinctive features of the military–industrial complex was the revolving door between government and industry, a fact that allowed firms like TI or McDonnell Douglas to quickly—but often only superficially—attach themselves to federal support for solar energy.

Kilby was also willing to entertain collaborations with square engineers in nondefense companies, so long as those companies didn't approach solar energy in a green or countercultural way. In the latter half of the 1970s there was an abundance of such companies. The trade magazine *Solar Engineering*, for instance, published monthly "stock listings of solar-related companies," featuring firms from an astonishing variety of industries: utilities, chemicals, microelectronics, aerospace, glass, mining, construction, scientific instruments, and others. The oil industry, in particular, was well represented—most of the major oil companies had in-house solar power efforts and/or partnerships with solar power start-ups, such as Atlantic Richfield's funding of ECD.[57] In that context it is worth noting once again that TI was originally an oilfield services company and was still active in that business in the 1970s. Even if Kilby was associated with the part of the company that had little to do with oil, he would've had connections in that industry, if only from being an influential high-tech figure in one of the world's great oil capitals. Later, we'll see that oil firms were prominent among the potential partners—as well as the respected competitors—that he had in mind as the TI project unfolded.

THE AMORPHOUS CRYSTAL BALL

Kilby's orientation toward the square parts of the nascent solar power industry, and away from its more countercultural corners, was not simply a matter of whom he preferred to work with or whose judgment he trusted. It was those things; but those preferences were also baked into the design of the technology itself. The TI effort—initially dubbed Project Illinois, after Kilby's alma mater, and later the Texas Instruments Solar Energy System (TISES)—was not designed for a farm, a commune, or an apartment building but rather for use in a single-family home. Moreover, the system's cost would've been high; in 1983 it was projected that the average system would cost $8,390 when TISES first reached the market in 1988—that's roughly $21,500 in 2019 dollars.[58] Thus, as the consultants Booz Allen Hamilton put it, "'High-end' new home market is the likely entry for TISES"—that is, mansions of approximately 3,000 to 5,000 square feet.[59]

In one sense, this reflected a tried-and-true strategy: develop a new technology at high cost for initial marketing to the wealthy, then use their purchases to achieve economies of scale that drive down costs and make the technology affordable for middle-class consumers. Kilby and TI's own experience with integrated circuits was a successful instance of this strategy, with the military playing the role of the wealthy early adopter. Yet in a number of respects this strategy was a poor fit for the TISES, because Kilby and TI consistently promoted the system in terms of cost savings. People who buy mansions often aren't very worried about saving on their electricity costs and show "generally lower concern with energy efficiency."[60] Unlike many countercultural solarists, mansion owners aren't very attuned to breaking the influence of the utility or fossil fuel industries. But they do want a home that they and others will regard as beautiful—whereas every indication is that they would've regarded the TISES as exceptionally ugly. Finally, though mansion owners aren't usually seen as an environmentalist constituency, there are plenty of wealthy environmentalists. TI could've chosen to market TISES as a means for mansion owners to compete with each other to be seen as helping the planet, yet there is no hint whatsoever that TI even considered that possibility.

So why did they market TISES to mansion owners but only on the limited selling point of cost savings? The answer lies in Kilby and TI's "sociotechnical imaginary," of which the TISES formed a part.[61] In the outline of that concept offered by Sheila Jasanoff and Sang-Hyun Kim, a sociotechnical imaginary is made up of the combined features of a technology and the social order in which it is embedded; specifically, those features that the parties to that technology and social order deem most salient. The collective imaginary guides technological choices in the present—a social group steers the technology one way or another on the basis of the group's understanding of the technology's features and those features' affordances for the social order in which the group is embedded. But the imaginary is also about the coproduction of the technology with a *future* social order: groups try to steer a technology to make certain preferred social arrangements more likely to come into being, just as they try to steer the social order to make certain preferred technological achievements more likely to take root.[62]

For Kilby and TI, the salient features of solar power were its lower cost relative to embargo-era oil and its potential to make the United States energy independent, and *not* its potential to undermine powerful utilities and fossil fuel conglomerates or ameliorate pollution or allow communes to be self sustaining. Kilby's imaginary for solar power guided design choices and preferences regarding what kind of people to market to or to collaborate with (and vice versa). Those design choices had a basis in certain features of the technology; but, crucially, Kilby's imaginary also encouraged or even required him to *neglect* other aspects of the technology. The logic behind accentuating some technological considerations and ignoring others was underwritten by Kilby's projection of a particular future social order. In that imagined future, TI's system would be at the center of an energy-independent society made possible by square scientists and engineers like Kilby. In Kilby's future America, no credit for the country's energy transformation would go to leftie environmentalists like Hayes or self-promoters like Ovshinsky; and the influence of such people would be so limited that any resistance to TISES from that direction could be safely ignored.

We can be fairly certain that Kilby and TI's vision of the future was defined by cost and national security factors rather than environmental or countercultural considerations because they commissioned quite explicit roadmaps for how to make TISES part of that kind of future. In late 1976 or early 1977, TI asked consultants (probably from Booz Allen Hamilton) to construct scenarios narrating three paths for bringing TISES to market: "Texas Instruments [develops TISES] alone," "Texas Instruments takes a partner," and "development with governmental support."[63] Such scenarios were at the time a relatively recent business innovation, having been developed in the 1950s at RAND and other think tanks as a means of projecting the social (dis)order that would be coproduced with various nuclear weapons designs and strategies. As Bretton Fosbrook has shown, by the early 1970s the technique had leapt from the community of defense intellectuals into the corporate world, particularly via Royal Dutch Shell.[64] TI, with its connections both in oil and in the military–industrial complex, was the kind of company in which scenario planning was adopted early and earnestly.

For scenarios to be persuasive, though, they must be plausible; and the grounds of plausibility are rooted in the audience's sociotechnical imaginary. *None* of the TI scenarios referred to solar energy's environmental or other countercultural or anticorporate affinities. We can infer, then, that Kilby and the TI team simply did not see the plausibility of the scenarios as hinging in any way on those factors. The *only* advantage of solar energy presented in the scenarios was its price relative to embargo-era oil—so the only plausible future that Kilby and the TI team could imagine was one in which people chose TISES overwhelmingly on the basis of price. Presumably they understood that there were constituencies for solar power that were more motivated by environmental or countercultural concerns. But in Kilby and TI's sociotechnical imaginary, solar power systems designed with those considerations in mind would secure such a small market share that the scenario writers could ignore them.

One thing that the scenarios could not ignore was TI's need for approximately $30 million to $50 million ($134 million to $224 million in 2019 dollars) in funding to bring the technology to market. Given that scale of investment, the "TI alone" scenario seems to have been dismissed fairly quickly. One reason that that scenario was deemed problematic was that it imagined the TISES technology leaking out to competitors. In that scenario a character, transparently based on Paul Maycock, moves from TI to ERDA and uses his inside knowledge to convince ERDA to give contracts for TISES-like solar power systems to other companies. Whatever its basis in reality, that plot device was plausible to TI managers. Thus, they decided that to develop the technology fast enough to keep ahead of competitors who were privy to TI's secrets, the project would need more money up front than the company itself could risk.

Based on the scenarios, therefore, TI sought a partner with deep pockets, preferably one that would impose few conditions on TI and that could not use its partnership to put itself in competition with TI. The latter requirement eliminated manufacturing firms such as General Electric, but not nonmanufacturing firms such as oil companies. In any case, in the strained economic conditions of the 1970s, almost the only entities with deep enough pockets were either governments or oil companies with large cash reserves from the post-1973 spike in prices. Thus the scenarios quickly narrowed TI's choice down to three partners: Royal Dutch Shell (Exxon and other oil majors were thought to have already advanced far enough in solar power technology to compete with TI); the US Department of Energy; or the Saudi government/royal family. With respect to the latter, the scenario recommended pursuing Prince Mohammed Al-Faisal because:

> He had been technically trained in the US . . . and had just raised $100 million from a group of friends to tow icebergs from the Antarctic to the coast of Saudi Arabia to serve as a source of fresh water. His project had been held up for the development of a technique for coating the icebergs, and the money was available for other ventures.[65]

The author of the scenarios believed that TI would have the most access to funds with the fewest strings attached if it partnered with the prince. The author also noted the risks of Saudi funding (e.g., the possibility that a war in the Middle East would end the partnership with TI) and thus recommended that TI first weigh those risks before deciding (by February 15, 1978) whether to find a private partner or "initiate solicitation of DOE funding."

THE MILITARY–INDUSTRIAL–SOLAR ENERGY COMPLEX

That plan was apparently put into practice, because in January 1978 TI indeed initiated solicitation of DOE funding. Partnering with an oil firm or Saudi Arabia was not taken off the table completely, but those options were temporarily put aside. The problem with seeking help from the DOE, though, was that neither Kilby nor his TI colleagues had much respect for that organization. As the scenario author put it,

> The solar photovoltaic program at DOE has recently been redirected to an early attainment of its goals through a technically unsound approach. This approach has been strongly identified with Dr. [Henry] Marvin, as being personally developed and sold by him. . . . Any alternate programs seeking substantial funds will

be personally viewed as a competitive threat by Dr. Marvin. . . . DOE personnel generally have a poor understanding of the requirements for commercialization of technology.[66]

The "development with governmental support" scenario also claimed that DOE personnel were likely to view TI as "a difficult company to work with" because "none of the DOE personnel had enough industrial experience to appreciate the importance of trade secrets."[67] Note that this latter point neglects that two of Kilby and TI's DOE bogeymen, Paul Maycock and Hank Marvin, did in fact have industrial experience (Maycock with TI and Marvin with General Electric).

It's reasonable, therefore, to assume that the scenario's objections to DOE were rooted in something other than department personnel's inexperience or "unsound" technical expertise. Most likely the problem was that Kilby and TI just didn't view civilian government agencies as trustworthy partners. That stance wouldn't be unexpected for rather conservative engineers working in the military–industrial complex and living in a very conservative part of the country. But ideology isn't exclusive of other explanations. In particular, Kilby and TI simply had more and better contacts—friendships, even—with Department of Defense personnel, and as a consequence were used to much more generous contracts and intellectual property arrangements than they believed they would get from a civilian agency like the DOE.

The scenario author therefore recommended that TI short-circuit the regular civilian staff of the DOE and instead try to negotiate directly with the department's more Pentagon-friendly leadership. At the time the scenarios were written, the Secretary of Energy was James Schlesinger, previously Secretary of Defense from 1973 to 1975. Schlesinger's deputies—the people the scenario recommended that TI negotiate with—were Dale D. Myers, a former NASA and Rockwell engineer, and Robert D. Thorne, an AEC employee since 1955. These were all square men with national security credentials that TI could trust. In the end, yet another DOE administrator, John Deutch, took the role originally envisioned for Myers or Thorne. Because Deutch was a former civilian systems analyst for the

Pentagon, a member of the Defense Science Board and the Army Science Advisory Panel, and "an influential member of the DOE management" with the ear of the Secretaries of Energy and Defense, his credentials would've been just as reassuring to TI.[68]

The final link in the chain was George Heilmeier, the director of DARPA from 1975 to 1977. Heilmeier had also worked in other Pentagon roles since 1970, and before that he was an influential research manager at RCA.[69] In December 1977 Heilmeier joined TI as Chief Technical Officer, and the *next month* he journeyed back to Washington to begin lobbying the DOE to support the TISES program. Upon his return he reported,

> during my visit to Washington on January 30th, I had planned to see Greg Canavan [an Air Force officer, probably involved with fusion research at Los Alamos National Laboratory], White House Fellow in the Department of Energy, to get an informal view of how the current DoE solar energy program was perceived by the highest levels of the Department. When John Deutsch [*sic*], DoE R&D chief, learned of my impending visit from Canavan, he directed that the meeting be a one-on-one (Deutsch and Heilmeier). During our discussion, Dr. Deutsch made the following points:
>
> - DoE has more money than good ideas.
> - He is extremely unhappy with the content and management of the DoE solar energy program. The technical sophistication of the managers is low, and he doesn't feel that they have a strategy (few truly innovative ideas).
> - He feels that the current DoE projections of $.50/watt by the mid-80's [*sic*] is unrealistic given the strategy of simply mass producing cells based on current concepts.
> - He stated that the current DoE patent policy is as flexible as that of DoD in *theory* but not in practice. He feels that overly restrictive patent practice by DoE may be inhibiting industry, but he would like to have some examples to make the case. Deutsch has been requested by Secretary Schlesinger to take a strong role in DoE patent policies including decisions on waivers.
> - Deutsch has $25M in the budget under his *direct* control for "ARPA-like" program. None of these funds have been committed.[70]

These notes showed that Deutch shared TI's low regard for DOE's civilian solar power experts and that he was willing to use his authority to cast the department's solar energy activities in a Pentagon mold, both in terms of funding ($25 million for "ARPA-like" program) and in terms of intellectual property protections for military–industrial firms like TI.

The Heilmeier-Deutch summit soon achieved the result envisioned in the scenario: a four-year, $18 million contract with the DOE, under which "The government received a royalty-free license for its own use but TI retains patent and licensing rights in the commercial arena."[71] An internal TI memo made clear that this was Deutch's baby: "When discussing TI's solar energy program with the press, please be sure and give John Duetsch [*sic*] the credit he deserves for creating a suitable contractual environment in DOE for the pursuit of this technology."[72]

In carrying out the contract, however, TI had to interact with ordinary DOE personnel who—as the scenario had forecast—did not necessarily agree with Deutch's support for TI. There were two main points of contact between TI and DOE: Hayes's SERI and JPL (which we saw in chapter 4 had a large DOE contract). Exchanges with these organizations did provide some benefits for TI. In particular, the TISES team welcomed the opportunity to keep tabs on developments across the industry: which types of systems (photovoltaic, photothermal, etc.) and which materials (cadmium sulfide, amorphous silicon, etc.) were in vogue and how far their competitors had progressed.[73] But the DOE contract also required TI to present results and cost calculations to SERI and JPL that had the potential to reveal the company's secrets; and DOE support came with occasional reviews of the project by department personnel who were wont to ask uncomfortable questions.

For instance, in 1979 Pete Johnson, the leader of the TISES group, attended a DOE briefing where "none of the participants attacked the concept or approach," yet where two rather stubborn objections were raised:

1. Paul Maycock questioned if the value of storage . . . for residential PV systems and if the "trouble and expense we go through with hydrogen storage and a fuel cell are worth the penalty we pay in efficiency and cost. . . ." He is focusing on the near term too much and the questioning doesn't represent his convictions as much as it does his personal need to raise a counterpoint. . . .
2. Satyen Deb (SERI) seemed to be particularly concerned about corrosion of the silicon exposed to the electrolyte on the front side of the array. Even after telling

him that the front metal protects the silicon and where it doesn't, the silicon self passivates and in our experience no further corrosion of the silicon takes place, he seemed to have a fixation on silicon corrosion.[74]

Neither of these is an outlandish question, yet in dismissing them TI again refused to countenance DOE's civilian solar power expertise.

With respect to Maycock's objection, fuel cells were—indeed, still are—a finicky technology that can be made to work in very controlled and expensive systems such as spacecraft but have largely proven uneconomical to mass produce. Yet the technology's problems are consistently obscured by a discourse that Matt Eisler dubs "fuel cell futurism," which is rampant in the military–industrial and petroleum sectors and which has had the effect (intended or not) of diverting resources away from less futuristic but more plausible energy technologies.[75] Another SERI reviewer pointed out some of the flaws in this way of thinking in May of 1978:

There is a serious question regarding fuel cell life. The more conventional H-O [hydrogen-oxygen] fuel cells have been made to last for five years, but these are very expensive devices. They [the TISES team] have made a very good start on developing an H-Br [hydrogen-bromine] fuel cell. It requires three membranes which causes it to be expensive. There is need for a sulfuric acid flush to get the bromine out of a part of the cell and it is very questionable in my mind that this device could be made with long life at low cost.[76]

The same reviewer also noted that

The metallized silicon balls implanted in glass have to withstand corrosive hydrobromic acid solution and while there is some evidence that this system might be stable in the short term, there is a serious question regarding a 30-year life which the analyses are based on. . . . [A] major concern is the safety of such a system since bromine gas [which is toxic and causes chemical burns] would be generated and required to be manifolded through the installation. This represents a serious threat. Also, if hydrogen is stored as a gas, it is a hazard and the ability to store hydrogen as a metal hydride has not been completely developed.[77]

In other words, DOE people were pointing out that TISES's underlying technologies, at least in their late 1970s form, (1) posed an environmental and health danger and (2) were too unproven and futuristic for civilian markets, even if their use in national security technologies such as spacecraft inspired confidence among military–industrial engineers.

These were just the kind of objections where TI's team thought it knew better. As Pete Johnson brusquely told reporters, "Corrosion is not a problem of note."[78] TI could have chosen to deal with these objections by redesigning the technology; but instead, it sought to go around the DOE's civilian staff and fall back on familiar military–industrial allies who were more likely to give TI the benefit of the doubt. An example of this that even Kilby found somewhat outlandish involved Dick Alberts, the Air Force procurement officer who had funded TI's initial foray into integrated circuit manufacturing in the early 1960s. In late 1978 Kilby

> met Dick Alberts in Washington. Alberts was with the Air Force in the early sixties, and was directly responsible for giving us the Air Force money for the early integrated circuits work. He is now with the Research Triangle Institute, but is being considered—or has been offered—the job of head of the photovoltaic section of DoE [i.e., the job then held by Paul Maycock]. If he takes the job he would be in a position to provide the other $35 million we will need to finish Project Illinois. To encourage him to take it I told him something of the project, and asked him to keep it in confidence. He was quite enthused, and will be of substantial assistance if he takes the job.[79]

Note the explicit assumption that the person "directly responsible for giving us the Air Force money for the early integrated circuits" would "be of substantial assistance" if he wrested the DOE photovoltaic program away from its current civilian leadership.

In the end, Alberts didn't take the job, but he did offer a suggestion for how to obtain Pentagon support for TISES:

> Sometime during 1980, the Illinois Project will need to begin a manufacturing methods program. My early guesses at the cost were about $35 million. While it is of course possible that this money could come from DoE, another possibility is becoming apparent. The MX [nuclear ballistic missile] program as it is presently configured will have 4600 firing sites, each of which will need about 100 Kw of electrical power. Although this could be supplied by individual diesel generators, there is a feeling that some form of solar power would be desirable. Conventional solar cells and storage batteries for this purpose would cost something over $2

billion. Because of the need for storage, Illinois type systems would be ideal for this application. I certainly realize that it would be too early to make any firm commitments for systems of this type, and also that it is unlikely that the MX will ever be deployed. Nevertheless, there is about $200 million for MX R&D in the current budget, and increases are planned for the future. I would therefore like to nose around SAMSO [Space and Missile Systems Organization] a bit and see what I can find out about the possibilities. An old friend of Dick Alberts now heads the group, and Alberts would be willing to set up the initial contacts.[80]

Kilby himself viewed the possibility of using TISES to power MX missile sites as "improbable" yet worth exploring precisely because "DoD is interested in the program to help offset the environmental opposition that they expect to encounter."[81] Once again, Kilby and his network could see their solar energy technology only in terms of *opposition* to and from the environmental movement, rather than as a point of cooperation with it.

SUNSET FOR TISES

The problem with the strategy of relying on trusted military–industrial networks (and rejecting civilian solarists) was that Kilby's military–industrial allies were only superficially, if at all, supportive of solar power. Most defense intellectuals and national security administrators believed that greater domestic oil production and a more robust military and diplomatic posture in the Middle East should be the United States' primary responses to the energy crisis. To the extent that they thought that the crisis necessitated development of non–fossil fuel energy sources, their preferred technology—by far—was nuclear power.

Thus, officials who had moved to the DOE from the Pentagon, NASA, or the AEC frequently disparaged solar power, often by making invidious comparisons to nuclear. Perhaps the clearest example came from Robert L. Hirsch, the assistant administrator for Solar, Geothermal, and Advanced Systems at ERDA. As reported in 1976,

In an extraordinary act, the federal official in charge of the development of solar energy has questioned the fundamental goals established for his office and the thrust of the programs under his direction. He also has drawn attention to the public expectations that have been generated by the solar program and the possibility that they cannot be fulfilled. . . . Hirsch confessed: "To a degree I get the feeling that people are reluctant to speak up, like the situation with 'the king that had no clothes.'"[82]

In *The Sun Betrayed: A Report on the Corporate Seizure of U.S. Solar Energy Development*, Ray Reece claims that Hirsch sought to slash SERI's funding and eliminate other solar programs out of fear that they would supplant nuclear energy.[83] That's perhaps speculative but certainly plausible. Most high-ranking officials at ERDA and the DOE came there from the *Atomic* Energy Commission, the world of nuclear wargaming (i.e., RAND, TEMPO, or the other think tanks we saw in chapter 2), or one of the nuclear services of the military (nuclear-armed missiles as well as nuclear-powered vehicles such as submarines). It would have been surprising if people with those backgrounds were *not* more friendly to nuclear power than solar energy.

It's unclear, then, why Kilby and TI would think that these people would help bring the TISES technology to market. One possibility is that they were misled by their own experience with integrated circuit technology. There, military customers such as Dick Alberts had been instrumental in helping TI get over the initial barriers to achieving the economies of scale needed to move to the much larger civilian market. Pat Haggerty, former president of TI, perhaps had that progression in mind when he projected in 1979 that "inherent in the work that Kilby has started is an industry that may be as large as the automotive industry and there certainly is a place for TI within it which will be the general size of General Motors. I mean it will have that sort of potential when it goes." The statement is an accurate description of what happened—via Pentagon assistance—with integrated circuits. In this instance, though, Haggerty was inaccurately forecasting the outcome of Kilby's solar energy work.[84]

Perhaps Kilby and TI knew that military–industrial support for solar power was thin but thought that they could convince their allies that TISES was no threat to fossil fuels and nuclear energy. As TI's 1978 proposal to the DOE put it,

It is now taken for granted that alternate energy sources must be developed if the future energy requirements of the world are to be met. It is also frequently assumed that solar energy must play an important part in the very long term. . . . All [alternate energy sources] will be marginal competitors to the conventional energy sources in the foreseeable future. No universal replacement of coal or nuclear energy is likely to be found in our lifetime or in that of our children. Although the costs of coal and nuclear energy will increase, they will do so at a fairly slow rate, and may actually decline in constant dollars. . . . Since no universal replacement for the present energy sources is likely to exist, each of the new technologies [such as TISES] must be tailored for a specific application. To be successful, a technology must serve at least one market that is large enough to justify the costs of its development.[85]

That is, solar power will be needed "in the very long term" but only to meet "future energy requirements," and *not* for environmental or ideological reasons. Thus, for the "foreseeable future," supporters of nuclear power and fossils fuels had nothing to worry about because systems like TISES would find only niche applications.

Perhaps Kilby thought that his military–industrial allies would help clear the solar field of countercultural solarists, leaving TI's system dominant. That had been one subtext of the Heilmeier–Deutch summit; a more brazen opportunity to cull Hayes and company came in 1980 with the election of Ronald Reagan. TI president Bucy was well connected to the Reagan campaign and was reportedly offered a job in the new administration. Thus, he asked Kilby for advice that he could funnel back to the presidential transition team. Kilby was probably a Reagan backer as well. At the least, his politics trended in the same direction as many of Reagan's voters, such that by the 1990s his friends could say that "like so many Texans, Jack didn't like [Bill] Clinton."[86] Certainly he was not averse to answering Bucy's request:

You asked about government officials who should be considered for replacement by the new administration. [Kilby named two Pentagon officials.] In the Department of Energy things are considerably more serious. All of the people who have been involved with the solar program should go. The list should start with Tom Stelson, Assistant Secretary for Conservation and Solar, and would include Fred

Morse, in charge of the residential applications, Paul Maycock head of the Photovoltaics Branch and Denis Hayes, Director of SERI. . . . In general it is my feeling that the level of competence of the people within DoE is a couple of notches below those in DoD.[87]

All the DOE solarists should go! Radical advice, yet in the end the Reagan administration's actions went substantially further than Kilby called for. We've already seen that JPL's federally funded solar programs were axed in 1982; even earlier, in June 1981, Hayes was fired and SERI's "budget [was] cut in half and its staff by a third."[88] That year the administration "proposed to cut the Federal budget for renewable energy sources from the 1981 fiscal year's $576 million to $193 million for 1982—a reduction of 67 percent. Another 50 percent cut is rumored to be in the works for the 1983 fiscal year."[89]

It's likely that such deep budgets cuts were not quite what Kilby had in mind. I surmise that he would have preferred that DOE continue to support solar power R&D but that Hayes and Maycock be replaced by military–industrial stalwarts such as Alberts to ensure that the DOE supported the "right" solar R&D. Indeed, the withdrawal of federal funding meant that the "development with government support" scenario that TI had been pursuing suddenly became unsustainable. This left Kilby and TI with the two other scenarios: go it alone or find a partner. Going alone, however, was even less feasible than a few years before thanks to TI's falling profits. In the late 1970s, TI management had thought—like many leading semiconductor companies—that it would be easy to make money embedding its chips into its own consumer products such as watches and calculators. And like its competitors, TI soon learned its error the hard way. Corporate profits were *halved* from 1980 to 1981, and the stock price nearly so.[90] TI's semiconductor division, in particular, went from a $170 million profit in 1980 to a $50 million *loss* the next year. And that was only the beginning: in 1983 "TI announced its first quarterly loss ever—$119 million, or $5 per share. Stock prices plummeted from $157 to $107 in a single day. Toward the end of 1983 TI pulled the plug on its disastrous home computer venture, swallowing a $145 million loss."[91]

These staggering and unexpected losses *could* have lent renewed urgency to TISES: as TI's consultants put it in 1982, "Increased competition and a slowdown in profit growth in traditional businesses has led TI to explore new large market opportunities."[92] But the double blow of huge losses in other divisions plus the withdrawal of federal support meant that TI *had* to find a new external partner if it wanted to continue with TISES. Yet TI's troubles and the federal retreat from solar power R&D would have made most potential partners wary. Thus, throughout 1982 Kilby and Pete Johnson fruitlessly pursued joint ventures with several foreign governments (Saudi Arabia, Japan, the United Kingdom, and West Germany) as well as companies from a variety of industries (oil, chemicals, utilities, venture capital). They also looked into a few outliers, such as Weyerhaeuser and Georgia Pacific (primarily known as wood and paper products companies) and Siemens and Hitachi (non-US electronics giants).[93] None of these entities, however, was interested—solar energy had lost its shine.

The standard explanation for the solar energy industry's collapse is that the price of oil sank to a point where TISES and similar systems were no longer competitive with fossil fuels.[94] It is true that in the mid-1980s the price of oil fell to (in real dollars) its 1973 value. It was as if the oil shocks of the long 1970s never happened—and therefore alternative energy technologies inspired by those shocks lost their justification so long as they were justified on the basis of price alone. Moreover, that decline in price was due in part to greater oil production in places like Alaska and the North Sea. That is, there seemed to be evidence of vast new oil and gas fields opening up, and therefore less urgency to solarists' claims that "alternate energy sources must be developed if the future energy requirements of the world are to be met."

When Kilby came up with the TISES idea in the early 1970s, a near future governed by resource scarcity seemed plausible in light of events such as the 1972 publication of the blockbuster *The Limits to Growth* report or the 1973 OAPEC embargo. By the early 1980s that vision had been replaced; now, the foreseeable future was one of resource abundance.

In 1983, TI's consultants explained the consequences of the switch from scarcity to abundance thus:

Rapid increases in energy costs and concerns about energy supply have let [*sic*] to projections of multibillion dollar markets for non-traditional energy technologies. Prompted by these market forecasts, government support and market place initiatives, many firms have undertaken efforts in PV, synfuels, storage and other new energy sources. However, more recent events have changed the energy situation, requiring a careful reassessment of energy technology programs

- Slower oil price increases in response to reduced usage—reducing the pressure for substitutes
- Slow progress in the development of alternative technologies—delaying the time of economic cross-over
- Reduced government support for energy technology development—increasing the cost to stay in the game at a time of economic pressure on most firms[95]

That, in turn, meant that the "many firms" that started solar and other alternative energy projects in the 1970s tried to sell off those projects in the early 1980s. The companies that TI sought to interest in the TISES system were probably looking at similar offers from other firms. An example quite similar to TI is the electronics giant RCA, which on the advice of the same consultants as TI's (the defense contractor Booz Allen Hamilton) sought offers from Siemens, Exxon, Shell, Texaco, Total, and Standard Oil of Ohio for its solar unit.[96] In the summer of 1983 Solarex, a photovoltaic company largely controlled by Amoco, purchased the RCA solar program.

In that kind of buyer's market, TI was unable to find anyone to give it satisfactory terms for a joint venture or even for outright purchase of TISES. Thus, in September of 1983 TI cancelled the TISES project and dispersed its approximately sixty employees to other parts of the company. This seems to have been a significant blow to Kilby. He immediately ended his leave of absence from TI and officially retired from the company on October 31, 1983. TISES was the last thing keeping Kilby invested in his formal connection to TI. Thus, with the project's cancellation he seems to have been anxious to end the TI chapter of his life.

In any case, soon after the cancellation Kilby and TI's lawyers drew up a new agreement because "due to the recent change in circumstances involving the TISES program, it is desirable to clarify our relationship."[97] Under the new terms, Kilby had two years to grant his own licenses to the TISES patents that TI had helped him obtain, but again that led to nothing. As he complained to his University of Minnesota colleague, Ray Warner,

> About a month ago TI decided to abandon the solar energy project there. I have told them that I would like to terminate my arrangements with them by the end of the month. If they are willing to release the patents on the project, I may see if I can do something else with the idea. If they are not, it will die. It may be just as well, as this is not a very good time to be peddling a solar project. We need another middle east crisis, I guess.[98]

Maybe so. That assessment certainly reflects how Kilby became interested in solar energy in the first place, and how he and TI framed the technology. But they could have framed TISES in other ways that did not rely on a Middle East crisis causing a national security panic and driving prices up.

In particular, they could have presented TISES as contributing to a more environmentally sustainable and more decentralized and democratic future. In this chapter I've offered a particular set of reasons why they didn't: namely, that those reformist solar futures were embedded in networks of countercultural, civilian, or charismatic solarists whom Kilby and TI spurned. Instead, Kilby and TI sought to advance their technology with the aid of military–industrial networks that they knew and trusted. They believed that they could effect a wholesale transfer of elements of the military–industrial complex over to civilian solar energy: personnel (e.g., Heilmeier and Alberts), technologies (e.g., fuel cells), management techniques (e.g., scenario planning), and legal frameworks (e.g., for intellectual property agreements with the Departments of Defense and Energy). Their experience with a similar civilianization of integrated circuits seemed to lend plausibility to their forecasts for solar energy. That plan failed, however, in part because their military–industrial allies were committed to fossil fuels and nuclear power and ultimately dismissive of solar power.

Kilby and countercultural solarists like Denis Hayes wanted many of the same things. But they pursued the common elements of their visions separately, or even antagonistically. Neither group was able to offer a vision for solar power that could withstand the decline in the price of oil and the ascendance of neoliberal policies. The two camps were never able to establish a sufficiently large common ground where the enthusiasm and inclusiveness of the one would complement the military–industrial savvy of the other. Perhaps even if that common ground had been opened up, the solar industry still would have dwindled in the 1980s. But we'll never know, because many square solarists like Kilby couldn't overcome their distrust and dislike of less-square solar energy promoters and the future social order that those people appealed to.

6

AN END TO EXCEPTIONALISM: PHILIPS, SIGNETICS, AND GLOBAL SILICON VALLEY

So far, this book has been an American story—at least on the surface. It would be possible to tell a more global version of the events I've discussed, but in most cases that version would deviate from American square scientists' and engineers' understanding of what was going on around them. The civil rights movement, for instance, had an important transnational dimension, which some New Left scientists attached themselves to—but most square technoscientists in the United States would likely have seen the global character of liberation movements either as unremarkable or as linking civil rights leaders to America's Cold War adversaries.[1] Squares would have been more attuned to the transnational character of other aspects of the long 1970s, but for the most part they would have focused on the American enactments of those events. For instance, many square technoscientists would likely have seen parallels between student protest movements in the United States and, say, France—though to someone like Calvin Quate it probably did not matter much that his counterparts in Paris were witnessing something similar to the laboratory occupations besetting Stanford. Obviously, many squares—such as Quate's friend and collaborator Rudi Kompfner—came to the United States from other parts of the world; but Kompfner was probably typical in maintaining an international research network while simultaneously viewing current events that affected science and science policy largely through a US lens.

A few of the background events that I have surveyed—particularly the moon landing and the war in Vietnam—were self-evidently linked to global affairs, however. Thus, America's place in the world obviously mattered to square technoscientists during the long 1970s, even if that context was only occasionally foregrounded for them. Yet, like most things, the occasions when that context was in the foreground inspired ambivalent reactions among the squares.

For instance, the Soviet Union's flagging ability to sustain techno-scientific competition with the United States was a cause for both self-congratulation and angst for squares whose livelihoods depended on that competition. Moreover, more reflective squares could see that the United States' "victory" in the space race and other superpower contests had done little to improve the nation's standing in the world. Indeed, regard for the United States declined in many countries in this period because of global opposition to the Vietnam War and US support for authoritarian governments such as the apartheid regime in South Africa.[2] Many squares had long served the national security state in one capacity or another, so they were implicated in the global criticism of America's military adventurism and antidemocratic foreign policy. As we've seen, a few squares were moved by such criticism to distance themselves from military patronage. Plenty of others, though, were more anxious about the *decline* in research funding from the military than they were about the *source* of that funding.

The superpower conflict wasn't the only source of squares' anxieties about the global situation, however. Indeed, the Cold War dominated the lives of American scientists and engineers less and less as national security R&D funding and military–industrial employment waned. Instead of the superpower conflict, squares started to worry more and more about global economic conditions: in particular, the sense, highlighted by the OAPEC embargo of 1973 and the deteriorating competitiveness of multiple American industries, that America's economic fortunes were slipping out of its hands.

Yet attitudes toward the precarity of American firms' positions in the global marketplace were highly selective; apparently, not all industries are created equal. The misfortunes of some inspired public lamentation but little federal effort to reverse course: for example, the furniture, textiles, and commodity chemicals industries.[3] The ritual gnashing of teeth was somewhat louder for, say, shipbuilding, consumer electronics, and steelmaking, but these were also allowed to wither. Such inaction contrasted sharply with the panicked scramble to resuscitate the US semiconductor industry from the mid-1970s onward, or the major foreign policy decisions guided by the needs of Big Oil throughout the postwar period.

In at least some of these cases, the industries that were allowed to decline were heavily unionized, and one might speculate that the union-unfriendly presidential administrations of the long 1970s lost little sleep over their plight. Indeed, when the federal government did intervene, it often did so in a way that led to the weakening of organized labor. Little federal help was available, for instance, to prevent auto industry jobs from flowing out of unionized factories in the Great Lakes region; but government support *was* available for building auto plants in union-hostile states in the South.[4] Similarly, the federal government stood by as coal mining employment shriveled in heavily unionized regions, even as it gave the oil industry a say in American diplomacy and military strategy. As Timothy Mitchell has argued, policies favoring oil over coal helped to shift political power from labor to management and capital.[5] In fact, the 1970s saw oil companies take over much of the US coal industry and move its operations to western states, specifically for the purpose of breaking the unions' power.[6]

Perhaps no industry illustrates this complicated interplay of global business, labor–management relations, and domestic politics better than the semiconductor industry—and so that is where this chapter will focus. Through the 1970s and into the 1980s, US semiconductor manufacturers lost market share to competitors, particularly those in Japan.[7] At the same time, the center of gravity of US semiconductor manufacturing shifted from the East Coast to the Southwest and West, and hence to regions with much lower rates of union membership.[8] At the same time, the "US" semiconductor industry became much more globalized in its operations, in part so that firms could avoid unionized workforces. Simultaneously, the "foreign" semiconductor industry became much more American in *its* operations. Though US semiconductor firms stoked and benefited from staunchly nationalist policies, in truth the industry's borders had been remarkably porous since the 1950s and became more so over time.

This chapter examines a firm that epitomizes that trend: Signetics. One of the iconic companies of Silicon Valley (*avant la lettre*) when it was founded in 1961, Signetics then became one of the leading firms in offshoring production to antiunion regimes. By the 1970s much of the semiconductor industry had caught up, and so Signetics lost one of its main competitive advantages. At the same time, Signetics was less innovative with respect to its core technology: it remained committed to military markets and therefore stuck with Pentagon-certified, ultrareliable bipolar transistors for too long, whereas its rivals moved to metal oxide semiconductor (MOS) transistors aimed at more profitable consumer markets. As a result of its technological lag and the loss of its manufacturing lead, Signetics neared bankruptcy (twice) and had to be rescued by the Dutch electronics giant, Philips, in 1975. One might imagine that the general ruction aimed at Japanese firms' *competing* with American ones would also apply to a Dutch firm actually *purchasing* one of the leaders in the American industry. But no—although a few industry observers complained that Signetics would become a "captive" supplier, the Philips takeover did not inspire calls to reorganize American R&D the way that competition from Japan did.

At the same time that it navigated economic perils, Signetics also drifted through the cultural and political dislocations of the long 1970s. The company didn't experience the protests that we saw pushing Stanford toward interdisciplinarity nor the political pressure that forced NASA to partner with the Department of Housing and Urban Development. But the era's cultural changes meant that both its management and its employees—as ordinary Americans—became quite different people over the course of the 1970s. Shifts in the US legislative and regulatory environment also pulled Signetics along with them (and both the company and individual employees played an important role in *creating* that regulatory environment). So did *global* events, such as the translation of youth protest movements into a wave of democratization in places like Portugal and South Korea.

To show how local, national, and global economic, political, and cultural changes converged, this chapter will take a "US and the world" perspective on Signetics and its network spanning three continents.[9] But this chapter also examines the coevolution of the United States and that most foreign of countries, Silicon Valley. That coevolution was, for the most part, rather quiet and incremental. Thus, the Signetics case is the squarest in this book: no parapsychology, no solar power, no groovy interdisciplinarity. Despite—or perhaps because of—that squareness, the changes experienced by Signetics were more durable (if still inadequate) than those I've described for UCSB, TI, Stanford, or NASA. Women moved up in the organization, and it became less acceptable to objectify them. It also became less acceptable for companies like Signetics to poison their workforces and the environment or to discriminate against people of color and people with disabilities. It became more common for American firms to be owned by foreign entities, and somewhat less common for them to cozy up to right-wing dictatorships. All those trends would continue—indeed, gain further momentum—in the 1980s; whereas parapsychology, solar power, and radical interdisciplinarity came to seem like outdated relics of the groovy era.

FAIRCHILDREN OF THE REVOLUTION

This version of Silicon Valley's history—square yet at the center of lasting change—both reinforces and undermines that region's mythos. Among some of the most prominent members of the Silicon Valley elite, the consensus has long been that the cultural experimentation of the long

1970s was a disaster with few long-term benefits—whereas, in contrast, the Valley's square *technology* has provided the only lasting good to come out of the groovy era. One can hear this version of history in Gordon Moore's oft-quoted 1973 claim that "we are really the revolutionaries in the world today—not the kids with the long hair and beards who were wrecking the schools a few years ago."[10] That perspective is also echoed by Stewart Brand:

JI: In hindsight, what's your assessment of 60s counterculture?
SB: Counterculture as a construct was a little strange to me at the time, and it's even stranger to me now. Free love went very badly. The drugs pretty much dead-ended. The Bucky Fuller domes leaked. The Communes failed. The music was good and the computer stuff was good.[11]

Whether or not it was good, the "computer stuff" has certainly stuck around. But so have a number of developments of the long 1970s that Brand and Moore conveniently forgot and for which they could take no credit: in particular, a regulatory state with some commitment (intermittent and inadequate as it was) to protecting workers and the environment and limiting discrimination on the basis of gender, race, or disability.

Examining Signetics allows us to get at both versions of the Silicon Valley story. We'll see that its technology *was* "revolutionary," not least in that it helped transform Silicon Valley—the physical region corresponding to the old Valley of Heart's Delight—into a global concept with no coherent geographic referent. But we'll also see that the regulatory state and the social movements that emboldened it were equally revolutionary in stimulating companies like Signetics to change. Both trends are illustrated well by Signetics; and Signetics nicely mirrors the evolution of Silicon Valley as a whole.

Indeed, even in its origins Signetics mirrored the standard template for Silicon Valley firms. One standard explanation for Silicon Valley's success is that it has long been a place with low barriers to entrepreneurship, where people want to, and can easily, form their own tech companies.[12] There are a number of factors that supposedly encouraged that culture, from the role of Stanford to the prevalence of garages, but one that both actors and academic analysts agree on is the presence of a single company where dozens of entrepreneurs gained technical skills, became disenchanted, and left to found businesses such as Signetics. As Stephen Klepper has shown, such incubating businesses can often be found at the core of industrial regions.[13] In the case of Silicon Valley generally—and Signetics in particular—that firm was Fairchild Semiconductor.

In a story that has been told many times, Gordon Moore, Robert Noyce, Eugene Kleiner, and five disgruntled colleagues founded Fairchild Semiconductor in 1957 thanks to the paranoid behavior of their former boss, William Shockley—a Nobel laureate for his coinvention of the transistor who later joined the Stanford faculty and became a notorious (and much-picketed) eugenicist.[14] Departure from Shockley's company was only possible, however, because Kleiner had connections to East Coast bankers who facilitated a start-up deal with Fairchild Camera and Instrument. That deal gave the eight cofounders large shares in Fairchild's new subsidiary, from which they soon became enormously wealthy. But the deal also left ultimate control of the company in the hands of an East Coast firm to which they had little loyalty. This created a dynamic in which the firm's founders—themselves renegades from an earlier company—had trouble inspiring loyalty from their own employees. Indeed, the way in which Fairchild Semiconductor came into being made its founders both wealthy and disloyal enough to invest in start-ups founded by renegades from Fairchild itself.

By the early 1960s, therefore, Fairchild had become a training ground where newcomers to the Valley could learn the industry, help develop a new product or piece of process equipment, and then leave to found their own company centered on a closely related product or tool. These "Fairchildren" are today revered for their contributions to the US semiconductor industry. In fact, Fairchildren came to dominate not just semiconductor manufacturing but also many of the Valley's auxiliary industries such as public relations and venture capital. These nonmanufacturing start-ups meant that Fairchild's culture of departure also assisted the emergence of the Bay Area's later high-tech industries: computing, biotech, and dot-com.

Signetics was among the very first of the Fairchildren. Indeed, it was the first company founded to manufacture Fairchild's signature invention—the integrated circuit. It may also have been where the term "integrated circuit" was popularized. As one of the founders put it, "At that time everyone referred to ICs as Solid Circuits (TI), Micrologic Circuits (Fairchild), molecular circuits, miniature circuits, and microelectronics by others. I believe Signetics was one of the first to refer to them as Integrated Circuits."[15]

Unsurprisingly, Signetics's founding strongly resembled that of Fairchild itself. Because of the departure of some of Fairchild's more popular cofounders in early 1961, a group of engineers in Fairchild's research lab became disenchanted with management and started plotting an escape. Two of them made contact with an East Coast bank despite not knowing the basics of entrepreneurship, such as what a business plan was (or that they needed one).[16] Yet the bank, Lehman Brothers, still agreed to invest—an indicator of the aura that Fairchild had already accrued. So in August of 1961, the four founders severed ties with Fairchild, purchased an 18,000 square foot building, and began setting up manufacturing facilities. None of them wanted to lead the new company; David James was chosen as president, largely because the others thought that his British accent made him a better ambassador. After initially calling themselves Dynalog, they settled on a portmanteau of "*sig*nals," "*net*works," and "electron*ics*" for their start-up's name.

Initially Signetics differed from Fairchild in not being a subsidiary of a larger East Coast firm, but in this respect too it quickly converged on the Fairchild model. The initial investment from Lehman Brothers was only enough to get Signetics through its first year, with the intention that Lehman would find a purchaser in that time. That buyer turned out to be Corning Glass Works, the well-known glass and ceramics manufacturer headquartered in upstate New York. Glass and microelectronics might seem like very different industries, but it is important to remember that

Silicon Valley's material basis is not just silicon, but also silicon dioxide—that is, the main constituent of ordinary glass. That's because circuits are composed of not just metals and semiconductors through which electrons move but also insulating materials that impede the flow of electrons in unwanted directions; and for that purpose glass and ceramics are well suited.

It's also important to remember that (as noted in chapter 5) many leading semiconductor firms started out in very different industries. In the early 1960s, born-silicon companies like Fairchild Semiconductor and Signetics were still the industry's underdogs; conglomerates such as TI, IBM, and General Electric, which had moved into semiconductors from other core businesses, were the big contenders. Corning executives seem to have wanted to follow those other big, established companies in leaping into microelectronics. In fact, when it bought Signetics, Corning already had an electronics plant in North Carolina that made diodes, capacitors, and other discrete passive electronic components. (*Passive* means that the components produced no amplification of an input signal; *discrete* means that they needed to be wired separately into a circuit.) The vice president in charge of that plant, Malcolm Hunt, was interested in expanding Corning's offerings to include integrated circuits with active components and saw the purchase of Signetics as one route to that goal.

Corning could have made a play for a number of Silicon Valley firms, but Signetics likely looked promising because its product line was well suited to the growing needs of the national security state. As one of the founders recalled later, "In '63 the DoD came out and emphasized that integrated circuits were the way to go and from that point on the military at least was pretty much sold on the idea. . . . [NASA too] were one of our better customers at the time."[17] Yet Signetics's success with NASA and the military only instigated Fairchild to undermine the renegades: the founders' former employer "copied our product exactly and had successfully done what we were doing. They cut the prices down to practically nothing. Bob Noyce admitted later 'we were bound to put you out of business.'"[18]

Thus, Signetics's position became precarious. In response, Corning asserted greater control by putting its own people in management positions and easing the founders out of upper management (and all but one of them out of the company). Soon, the expansion in military and NASA procurement meant that Signetics's revenues stabilized and then began growing rapidly. For example, in 1969 Signetics became "the one and only integrated circuit company whose parts . . . have been designated as standard parts for the Minuteman" nuclear ballistic missile.[19] The rapid increase in sales to the military meant that between 1964 and 1965 the number of Signetics employees more than doubled, and then nearly doubled again from 1965 to 1966.[20] Much of that increase came from the opening of new plants outside Silicon Valley: first in Provo, Utah (in a former bowling alley), and then in Seoul, South Korea. For the latter a new subsidiary was founded, Signetics Korea or SigKor—the first in a growing global archipelago of plants and subsidiaries. The good times that followed from the military's drive to close the missile gap lasted through the end of the 1960s, with the opening of another Utah plant and one each in Scotland and Germany and construction of a "fab" (silicon wafer fabrication facility) in Nebraska announced in 1969.

A VOLATILE INDUSTRY IN VOLATILE TIMES

Then the 1970s arrived with a bang. In 1969 to 1970, the US economy went into a brief recession fueled in part by contraction in military and space expenditures. The semiconductor industry was especially hard hit, though firms that had begun to move away from national security customers were somewhat insulated. As we saw in the previous chapter, Robert Noyce was an early skeptic of military markets, and therefore his new company, Intel (which he cofounded with Moore and Andy Grove in 1968), fared reasonably well. But Signetics's eggs were still largely in the national security basket. Construction of the Nebraska plant was cancelled, and in 1970 the company's revenues dropped by about 20 percent while its number of employees was nearly *halved.*

By 1971 revenues had recovered and the company began a round of expansion near its existing plants as well as starting a new plant in Setúbal, Portugal, and a new subsidiary, SigThai, based in Bangkok. Those extreme swings in revenue and employment demonstrate the volatility of the semiconductor business, which Charles Harwood, Signetics's president in the 1970s, explained thus:

> It was highly competitive between the companies, because we both made the same circuits that could go in the same socket. And when that happens, you get heavy price competition. If you wanted to take the model of a perfect [highly] competitive situation, that's what we had. And so it was vigorously waged. When times were good, we could raise prices, because there wasn't enough [product] to go around. As soon as times were bad, we wanted to fill the factories, so we cut our pricing below our full standard cost and this made it highly cyclical.[21]

Lessening dependence on the military offered a partial antidote to market volatility, and so the 1969 to 1970 recession stimulated Signetics to accelerate the diversification of its customer base. By 1973, only 15 percent of its business came from the military.[22] Even so, Corning's management grew tired of the semiconductor industry's extreme ups and downs and increasingly saw Signetics as a poor investment.

Both Corning and Signetics executives also seem to have been frustrated by their cultural differences. From the point of view of Signetics's founders, Corning was a "conservative, eastern compan[y]" that wanted to take charge even though it didn't understand how Silicon Valley worked.[23] From Corning's perspective, in the words of business historians Margaret Graham and Alec Shuldiner, "There was much that Corning might have taught the fledgling company about efficient manufacturing practices, statistical process control, and successful technology management, but Signetics employees expressed no interest in learning these things from its patron."[24] To get any traction at all with Signetics's people, Corning management had to offer stock options and profit-sharing programs, which were common in Silicon Valley but "unthinkable" among established East Coast firms.[25] They were also forced to bargain away Corning's own intellectual property to get favorable licenses for Signetics.[26]

Frictions arising from the companies' very different approaches were initially smoothed by Signetics's primary champion within Corning, Malcolm Hunt; but after he died in 1966, there was no one at the parent company who saw the value of accommodating to Silicon Valley's strange ways.

Thus, after the 1969 to 1970 recession, Corning management started to look for an exit strategy. As Harwood put it, "I think Corning pretty much fished in the icy waters, didn't end up with a major [semiconductor] corporation and got their money back."[27] The first step to recouping that investment was a public offering of Signetics's stock in 1973, though Corning still retained a majority share. That year a new recession triggered by the OAPEC embargo, however, again dented earnings. This time, Corning executives decided to find someone who would buy Signetics outright; by early 1974, they were in talks with Philips. A year later Corning sold Signetics to United States Philips Trust, the majority shareholder of Philips' North American concerns.

Signetics was somewhat less independent than Philips's other New World subsidiaries, however. North American Philips's other arms—such as Magnavox or Norelco—received corporate guidance from Andover, Massachusetts, but Signetics executives got their orders directly from the Netherlands.[28] Even so, Signetics's leadership, from Harwood on down, was initially left largely in place. Even some members of the management team who disagreed with Philips's strategy were allowed to continue; and a few employees from the Corning era remained with Signetics for years—some up to and even after it was finally subsumed by Philips in the 1990s. At least from that later era looking back, Signetics employees believed that Philips had never properly taken control or shown the leadership that they expected, leaving a "vacuum" in corporate strategy.[29]

Given Philips's hands-off approach, it's not entirely clear what it hoped to gain from purchasing Signetics. Philips's entry into Silicon Valley does not appear to have fit seamlessly into a comprehensive and integrated corporate strategy. To some degree, then, Philips's plans were probably quite similar to Corning's: that is, get a foot in the door of the most vibrant region of the US semiconductor industry and perhaps, with some luck, score big. Yet the Signetics that Philips bought in 1975 was not the company that Corning had purchased in 1962. One major difference was that by 1975 Signetics had become a global enterprise with plants on three continents. But even if we look primarily at its US operations we can see that Signetics, as a company made up of ordinary people, had changed significantly during the turbulent years of the late 1960s and early 1970s.

I've chosen Signetics as a case study because the ways in which it changed from 1962 to 1975 and beyond tell us something about the influence and attitudes of square scientists and engineers more generally. In the early 1960s, Signetics's corporate image—for both internal and external consumption—was very much set by the kinds of square engineers who crowd popular and to some extent academic histories of Silicon Valley: typically, young white men who both bonded and competed over common bad behavior. By the time Philips took over, though, those men had had to cede some of their authority to an increasingly diverse group: women "foremen," South Korean union organizers, Portuguese revolutionaries, and the like.

If we look at company newspapers from the time and oral history reminiscences from later, we can immediately see the cultural gulf between the early 1960s and the mid-1970s. In the earlier period, the integrated circuit business had only recently emerged, and so it displayed the anomie, youthful demographics, and hence immaturity that sometimes accompany brand new industries.[30] Relationships among the kinds of young, white, male, square Silicon Valley engineers who founded Signetics were densely networked—partly because so many of them had worked for Fairchild at some point, and because even those who left Fairchild remained concentrated in the Bay Area. That density of ties among firms fueled the cutthroat competition that Charles Harwood described earlier: companies were eager to obtain each other's trade secrets by whatever means necessary and so lubricated the flow of information with alcohol, sex, gossip, and espionage.[31] A not untypical episode recounted by a Signetics engineer involved all four (ellipses and brackets in original):

> I was visiting a customer in New Jersey. The sales people in the area I was visiting with . . . we had dinner and drinks in a bar [and most got "well oiled." I was the only sober one because I never touched alcohol. They also had arranged for a follow-on visit with some ladies of the night, but I refused to go and waited for them in the bar. After they came back, the drinking continued and one of our guys got rowdy and mistakenly crashed a removable section of the bar counter to the floor. The bouncer pounced on him and we all had to get in the way to prevent further mayhem. We were thrown out,] and then we left [for our motel at 3 AM. . . .] Somebody during that restaurant gathering had gone into my briefcase and, knowing that we had these good parts, stole them.[32]

Indeed, the term "Silicon Valley" was probably coined and certainly popularized by the industry's most prominent gossip columnist, Don Hoefler.[33] Hoefler was a former Fairchild publicist whose widely read weekly newsletters were filled with juicy items sourced to a network of moles and even, in one incident involving Signetics, bathroom graffiti.[34]

One strand of Silicon Valley culture, then, was forged through puerile bonding and competition among (mostly) young, white men over drinks at Trader Vic's or the Wagon Wheel. That atmosphere was never uncontested, but in recent years its descendant—the "brogrammer" world of Google and Facebook—has come under particularly heavy criticism for its exclusionary and often outright misogynistic, racist, and classist bent.[35] *That* Silicon Valley, however, has always relied on a manufacturing, maintenance, and administrative workforce that is and was much less white and male. Indeed, Signetics and other microelectronics firms often located their plants in places with a large number of women workers who were exiting a "sunset" industry—workers who by both their gender and their poor employment prospects could be paid low wages. In the former Valley of Heart's Delight, that sunset industry was fruit canning, but in other parts of the world microelectronics companies relied on former textile and agricultural workers who were more likely to be women and/or non-white than those firms' employees in sales, engineering, and management.[36]

These two Silicon Valleys interacted in complex but also historically evolving ways. For instance, Signetics's employee newspapers from the

1960s and early 1970s are filled with announcements of weddings and engagements between employees alongside annual swimsuit contests for young women employees and continual references to women as "girls," "gals," and, in the case of the Scottish plant, "lasses."[37] Of course, these forms of heteronormativity and objectification of women were hardly unique to Signetics or even to Silicon Valley. JPL, for instance, ran a "Queen of Outer Space" contest until the early 1970s, in which "photos of the girls are taken each year by William Figge, whose work appears in Playboy with enjoyable regularity."[38] As many scholars and activists have shown, objectification of women is endemic to technical organizations—but the forms that it takes are historically and culturally contingent.[39]

Among other things, the gender balance of Signetics's manufacturing workforce meant that the company's women employees were significantly more likely than men to face physical harm from the organic solvents, acids, and other chemicals used in making chips. As we'll see, those occupational hazards *became* an existential issue for Signetics in the late 1970s, but in the 1960s management largely laughed them off. For instance, when the new Utah plant opened in 1969, the site's corporate newspaper remarked:

> All the shower stalls without dressing rooms or doors were questioned by some of our group, but it seems they are for the treatment of acid splashes and you just walk right in and pull the ring. I wonder if they will have hairdo insurance in those departments. Can't you just hear the howl after a girl had spent four hours at the beauty shop and three minutes in the shower washing off acid and her new do?[40]

Managers' attitudes toward safety did change, but slowly—and only *after* the founding of the US Occupational Safety and Health Administration in 1970. We will return to that point later.

Attitudes toward, or at least opportunities for, women also shifted. From the late 1960s onward company newspapers often featured women being promoted to supervisory positions (sometimes as "foremen"), and the beauty contests seem to have ended in 1974.[41] A 1981 retrospective for the company's twentieth anniversary prominently featured a sidebar describing Opal Hern, who joined in 1965, reminiscing that "the most interesting change I've noted has been seeing the progress of women into a traditional male area of Supervisors."[42] We can't take Hern's statement as representative, of course; as Glenna Matthews has shown, other women had rather more negative recollections of the atmosphere at Signetics.[43] But the prominence given to Hern's perspective does offer evidence that by 1981 the company's leaders wished Signetics to be seen as having made progress in gender equity.

More generally, Signetics's corporate communications show that senior managers at the *end* of the long 1970s were more receptive to social change—or at least were aware of the legal system's and employees' growing expectations of such sensitivity—than Signetics executives at the *beginning* of that era. The president of Signetics from 1964 to 1970, James Riley, was as appalled by "the kids with the long hair and beards" as Gordon Moore and openly signaled that his employees should be as well. For instance, on the eve of the 1968 presidential election, he warned against the forces

> of disorder, riot and disrespect of the democratic order which America has built up over the last two hundred years. . . . This wonderful country has suffered indignity after indignity, witnessed by the world through the detailed coverage of our news media, at the hands of some of these groups. All this while the 95 per cent of the decent Americans go on working, paying taxes, serving in the draft, solving the problems of the day. Let's not let our beautiful system be fatally injured because the majority who are the backbone do nothing about the noisy minority.[44]

Reflecting in another column on the issues of the day, Riley presented equality of opportunity as one of the features of America's "beautiful system" but hinted that he viewed those fighting for it as part of the "noisy minority": "Equal Opportunity. . . . A term we're all accustomed to hearing. We see it in newspapers, on television, and hear it on newscasts. In fact it has become almost too common. . . . [W]e will not have achieved true equal opportunity until the day comes when we no longer hear the term."[45]

Yet by the time Riley was preparing to leave Signetics, the company newspaper was beginning to invite a wider range of views. The March

1970 issue, for instance, asked employees "How Will Historians View the Sixties?" All three of the answers it published are fascinating:

Hilda Young: The decade of the Sixties was a traumatic, bewildering and soul searching confrontation in all areas of human life. They called the youth of the Fifties the apathetic generation. So, of course, the youth of the Sixties would scare them. They expressed the right to be irreverent of "sacred cows," the right of dissent, the right to be different, the right to be wrong, and if one listened, the youth of the Sixties did have something to say. . . .
Rudy Griffin: The events of the Sixties produced a group of young people who are aware of the rights of the individual and not the state. This awareness will shape the Seventies.
Pat [from context, probably Patricia] Bruhn: Since historians inevitably name the periods they study, they may call the Sixties the "Age of Awareness." War, racism, poverty and even the "generation gap" have existed throughout human history. But sometimes men have been allowed the luxury of avoiding their existence. The combination of mass communications, the sword of nuclear technology, vastly greater numbers of educated, affluent young people and the emergence of formerly dominated cultures into world politics make this luxury impossible. During the Sixties this rug was lifted.[46]

This historian, at least, has no better diagnosis of the Sixties than Pat Bruhn's!

While Riley was at the helm, Signetics was the kind of company where employees were encouraged to take pride in "a big Signetics success in the Minuteman missile program."[47] Once Riley's tenure ended, though, other executives started to construct a more ambivalent corporate image in line with the employee views just quoted. As the manager of the Utah plant, Bill Gilchrist, put it in June of 1970 (the month Riley left to head a competing firm, Intersil),

The total market [for chips] is declining. . . . An important consideration in this matter is national priorities. In the 1960s it was aerospace, military efforts, moon shots, and like efforts strongly oriented to electronics. Now, the emphasis has shifted to stopping the war in Asia, slowing inflation, problems of over population, pollution, equal opportunities, and generally improving urban life. In a nutshell this has caused a slump in the electronics industry. Leaders in the field feel this is temporary. It takes time for industry to shift its emphasis, more time than it does national attitudes.[48]

Indeed, as Gilchrist predicted, over the next few years Signetics and its peers did shift their product lines to keep pace with changing national priorities.

That change in tone was at least correlated with, and perhaps accommodated by, Charles Harwood's arrival to replace Riley. Before Signetics, Harwood had been a loyal Corning man; but he ended up staying for fifteen years, through and well beyond the Philips takeover in 1975—so it was his management style that set the tone at Signetics for most of the long 1970s. And that management style was considerably squarer and more buttoned-down than Riley's. Harwood had come up through Corning's ranks and styled himself—in contrast to his Silicon Valley peers—as a staid and bland Yankee working for a small-town family company. In contrast,

> [Fairchild] was driven by scientists. And I'm not saying they weren't very capable as managers or in human resources, but their strength came from that. Charlie Sporck [of National Semiconductor] was an operations guy, manufacturing guy, came out of Fairchild. That was his strength. Charlie ran a very tight shop, very stingy, which I say in glowing terms, a very well-run, tight ship. Good folks. Jerry Sanders [of AMD] was a salesman. He was flamboyant. I'll give you a little story about Jerry. We went to call on one of our distributors shortly after Jerry had been there. They told me that they had lunch with Jerry after they'd just given AMD a big order and Jerry ordered a $1,600 dollar bottle of brandy for lunch. Now as a New Englander, that shocks me. But they'll never forget that act by Jerry. And that was part of his charm. . . . I was not a lightning rod like the other [semiconductor company] presidents, as I was told later. I think I was plain vanilla.[49]

Under Harwood, Signetics's public face became less jingoistic and paternalistic and more attuned to the mores of a changing nation—even, or perhaps especially, after Signetics passed into Dutch ownership.

But Harwood's arrival made less of a difference with respect to Signetics's other international entanglements. As mentioned earlier, in 1966 Signetics embarked on an aggressive expansion of operations both in the United States and worldwide. *Where* it expanded tells us a lot about Silicon Valley's attitudes toward labor, democracy, and globalization. In chronological order, Signetics's new facilities were established in Orem

(Utah), South Korea, Germany, Scotland, Portugal, Thailand, the Philippines, Albuquerque (New Mexico), and Sacramento (California). Utah, New Mexico, and California were US regions where Signetics could expect little union activity, whereas South Korea, Portugal, Thailand, and the Philippines were all, at the time, led by authoritarian, military-backed regimes willing to suppress union demands by force.

Signetics was not the first Silicon Valley firm to move its manufacturing to nondemocracies to avoid unions and US labor laws; that distinction apparently goes to Fairchild, which started a plant in Hong Kong in 1963.[50] But Signetics was one of the first to follow suit. Two reasons for offshoring that are apparent from Signetics executives' extant correspondence were to employ a low-wage workforce and for "gaining a foot-hold in the EEC" (European Economic Community), the European Free Trade Area, Japan, and to some extent India.[51] A third reason has been mentioned in the literature but is only implicit in Signetics correspondence: namely, these sites were conducive to an extreme form of labor discipline that was thought to be necessary to achieve the very tight tolerances required to make chips.[52]

A 1968 letter from Don Liddie, at the time Signetics's manager of international facilities, to his Corning bosses gives some of the flavor of his and his bosses' views:

> [A]s is the case with most everything in Korea, income tax payments have proven to be negotiable. . . . We, too, have found the labor situation is a lot more volatile than originally expected. . . . Unions have become much stronger in Korea by feeding on the new American companies. Their initial demands are likely to be in the area of 60–70% increases which generally boil down to norm of 20–25%. . . . In the final analysis we have found our Korean operation to be very profitable and the Korean government malleable although inexperienced.[53]

Similarly, Signetics's Scotland site was chosen specifically with the expectation that it would "have the opportunity to operate without a union."[54] It fulfilled that expectation for its first five years, while the operation was still small. By 1974, though, the plant had expanded to the point that the Amalgamated Union of Engineering Workers began organizing its employees.[55] Nothing was done immediately, but once Philips purchased Signetics, the Scottish site's activities could be farmed out to Philips's locations in the UK and Scandinavia. Thus the plant was closed rather abruptly in late 1976.[56]

Despite the "pluses" of operating in places where union activity could be avoided, Signetics executives were aware that there were drawbacks to relying on the military-backed regimes of South Korea, Thailand, Portugal, and the Philippines. Another 1968 letter outlining the "Portugal minuses" of locating a plant there listed:

1. Dictatorship.
2. 1967 budget indicates 40% for defense (primarily of African colony and Macao). [presumably a minus because too little was being spent on domestic infrastructure]
3. National expenses as a % of receipts was 128% in 1964. [a minus because the debt ratio was an indicator of future economic volatility]
4. Portugal not keeping pace with electric power needs.[57]

Despite these "minuses," Signetics still built its Portugal plant. Yet when the military junta there was overthrown in 1974, the transition to democracy apparently made the country less appealing. As Harwood noted later (and rather tendentiously), the "plant [was] closed by revolution" in 1976.[58] Signetics also closed its SigFil plant in 1985 in the middle of the Philippines' messy transition to democracy, though this may have been in planning for a longer time than the Setúbal closure.[59] In any case, the logic behind Signetics's overseas plants—access to cheap, unorganized labor—remained the same after Philips's takeover as before; and any plant that disobeyed that logic was slated for closure.

CLIMBING THE LEAGUE TABLES

Thus, by the time Philips took over in 1975, Signetics controlled a web of manufacturing sites with low union penetration, spread across three continents. The cost savings and market inroads provided by its Asian and European plants gave Signetics inherent advantages, yet those advantages remained underutilized during the Corning years because of Corning's lack of familiarity with how the semiconductor business was developing in the global region soon to be known as Silicon Valley. Common nationality aside, Signetics and Corning were in many ways more foreign to each other than Signetics and Philips were. Unlike their Corning peers, Philips's management team at least had extensive experience in semiconductor manufacturing. They didn't yet understand Silicon Valley's unusual approach to semiconductor manufacturing, but they clearly wanted to. Thus, Philips was much more willing than Corning to make capital and markets available to maintain Signetics's competitiveness. In return, Signetics offered Philips access to Silicon Valley networks and manufacturing know-how that the Dutch company—again, unlike Corning—was eager to acquire.

With Philips's backing, Signetics became a reasonably prominent firm—much more so than is reflected by its scant mentions in most histories of Silicon Valley.[60] Indeed, one of the most common talking points regarding the takeover was that it would boost both Signetics *and* Philips to the top tier in the global semiconductor industry rankings. To be sure, Philips was already a leader in some niches of that industry before the Signetics purchase. In Europe it was the "largest non-American supplier of solid-state components, [trailing] only Texas Instruments Inc. and Motorola Semiconductor"; it was also first in the world in discrete (i.e., nonintegrated) semiconductor components.[61] After the takeover, though, the combined firms occupied the more prestigious "No. 2 position after TI" in global integrated circuit sales.[62] Until the mid-1980s Signetics remained in the sixth or seventh spot among US firms, with Philips/Signetics together ranked second or third globally.[63]

Philips's move from number one in *non*integrated circuit components before buying Signetics to number two in *integrated* circuits afterward shows why the purchase made sense. As Mila Davids and Geert Verbong have argued, Philips had been quick to internalize discrete transistor technology in the 1950s but fell far behind once integrated circuits made their appearance.[64] That misjudgment was due both to technological inertia and to organizational overconfidence derived from Philips's history. Philips had started out in light bulbs and other electrical products in the 1890s; like its contemporary, General Electric, Philips later expanded from bulbs into the closely related technology of vacuum tubes—the forerunner of the transistor. When the transistor appeared in 1947, therefore, Philips was well placed to quickly assimilate solid-state electronics, with some assistance from its contacts with similar American conglomerates AT&T and RCA. Those contacts existed because Philips managers saw big East Coast conglomerates like GE, AT&T, and RCA as their company's peers, with whom they wanted to keep up. But Philips's leadership didn't see fly-by-night competitors that were located in the western and southwestern United States—such as TI, Motorola, and Fairchild—in the same light, and they were therefore unprepared for those companies to pull ahead of Philips. Davids and Verbong show that through the 1960s, Philips lost touch with the state of the art because it had far fewer contacts with the companies where innovations like the integrated circuit were being forwarded.

Signetics, in contrast, was closer to the state of the art in integrated circuit manufacturing, and its employees were plugged into the flow of information around Silicon Valley. Philips therefore bought Signetics largely to obtain a conduit to Silicon Valley's manufacturing know-how. Oddly, though, that explanation was not prominent in contemporary reports of the takeover. Instead, the most oft-mentioned justification was that Philips could use Signetics's chips in its consumer products. This possibility was sometimes obliquely referenced in the Dutch press: "The market for integrated circuits is surging because they are increasingly used in articles such as washing machines and watches," that is, goods that Philips did or might sell.[65] But it was more explicitly put forward by American observers with an edge of moral panic:

> Many in the U.S. electronics industry were also suspicious of the acquisition because all other major IC suppliers are U.S. owned. Some observers even speculated that Signetics would become a captive supplier to the huge multinational Dutch firm.[66]

Yet Signetics-to-Philips sales were relatively small. As a Signetics executive told reporters,

> The issue of Signetics as a captive source is ludicrous. . . . At our sales size, Philips could never buy everything we make. They represent a small percentage of our sales and won't even be our largest customer for a number of years.[67]

It is possible that Philips did want Signetics to sell a growing number of chips—at a discount—to the parent company.[68] But even if that were the case, it does not seem to have been a top priority.

In fact, to some extent sales from Philips *to* Signetics were almost as important as the other way around. By the late 1970s Philips had been manufacturing its own chips for many years, and some of the company's engineers were beginning to think about marketing some of the *equipment* that they had originally developed for use in Philips's fabrication line. Signetics offered a useful captive customer for that effort. For instance, when the Philips Electronic Instruments unit began selling an electron-beam writer (a tool for making the templates for chips) in 1979, "its first U.S. customer [wa]s Signetics"; its other early sales were also primarily to Philips subsidiaries.[69]

Similarly, when another unit, Philips S&I (Science and Industry), began exploring the market for "steppers" (a tool for projecting a circuit pattern onto a wafer multiple times) in 1978, its analysts concluded that suppliers such as Geophysical Corporation of American were too far ahead for Philips S&I to compete with directly at first. Instead, they decided to "concentrate in the next two years on [Philips's electronics subsidiaries] Elcoma, Signetics in the US, and Faselec in Switzerland."[70] By 1981, S&I could identify three big potential orders for its steppers: twenty-four to forty machines for Motorola, thirty to fifty between Elcoma and Signetics, and thirty to forty for IBM.[71] In the end, the S&I venture failed per se, though it later led to a very successful joint venture, ASM Lithography. Successful or not, though, these forays show that Philips and Signetics *both* offered internal markets for each other—the trade was decidedly two-way.

The same could be said for research. Like many Fairchildren, Signetics's research arm was kept small and closely integrated with manufacturing. This was partly pure necessity—the Fairchildren needed cash flow and hence had to ship product with little time or money for long-range research.[72] But it was also a matter of philosophy; Signetics's founders came out of Fairchild's large and independent research unit and saw that that organizational model did not work well in Silicon Valley. Philips, on the other hand, had a longstanding, highly independent lab in the Netherlands, the Philips Natuurkundig [Physics] Laboratorium (NatLab). The NatLab is justly famous both for its inventions (such as the compact disc) *and* for basic research superior to most universities—putting it in the same league as Bell Labs or IBM's flagship lab in Yorktown, New York.[73] In fact, Philips also operated a subsidiary lab in Briarcliff Manor, New York, just six miles from IBM Yorktown.

So Philips needed a "leg on the ground" in Silicon Valley in the same way that the Briarcliff Manor lab gave it access to the innovation region that was home to IBM, AT&T, RCA, and General Electric.[74] Here Philips was simply adopting a model established by East Coast conglomerates like IBM and Xerox, which already had longstanding labs in Silicon Valley. Philips was also following the lead of Japanese semiconductor and microelectronics firms such as Sony, Matsushita, and Fujitsu, which established R&D centers around the Bay Area in the late 1970s.[75] But the complicated relationship between Philips and Signetics meant that Philips's entry to the Silicon Valley R&D landscape had a rather different mandate than some of its peers: because Signetics and Philips *each* lagged behind the other in important respects, Philips's Bay Area lab had to establish a strong *two-way* knowledge transfer. Signetics needed the basic research capacity that Philips had built up over several decades as much as Philips needed Signetics's access to up-to-the-minute Silicon Valley inside information.

So in 1981 Philips opened a Silicon Valley facility that housed both an Advanced Technology Center (ATC), which was part of Signetics, *and* the Philips Research Laboratories of Sunnyvale, which was part of Philips's global chain of labs.[76] One stated aim of this colocation was for Philips's researchers to help their Signetics colleagues in moving from bipolar junction transistors to the newer MOS transistors. Most of the other Fairchildren had made this transition earlier, but Signetics was famously slow and began making MOS products profitably only after injections of capital and expertise from Philips—including a $250 million (roughly $700 million in 2019 dollars) "MOS offensive" announced in 1981.[77]

A further stated purpose of the lab was "to serve as a window through which Signetics can more readily access the knowledge and know-how that is being developed" at Philips, particularly "in the area of novel device concepts and device physics."[78] That is, Philips Sunnyvale would help Signetics not to miss out on future technological transitions as it had with the move from bipolar to MOS. In return, the ATC focused on Silicon Valley's strengths: improvements in process, rather than product—and above all, innovation in making manufacturing processes cheaper. ATC researchers seem to have had two main priorities for their process innovations: ensuring that Signetics and Philips kept pace with the industry in steadily moving to ever larger wafer sizes, and developing automation techniques to eliminate human workers on the fabrication line.

The ATC's focus on automation highlights one of the ironies of the Philips–Signetics relationship: namely, that Philips executives viewed the US labor environment in which Signetics operated in much the same way that Signetics executives viewed the labor environment for *their* offshore plants. Philips's executives had little to say about SigKor or SigThai, but they were vocal in praising America's "stimulating social-economic climate," which made it possible for them to operate more freely and cheaply "even with respect to staffing"—that is, they could hire and fire with many fewer constraints than in the Netherlands.[79] That liberty was perhaps exercised less with Signetics than, say, Magnavox—where Philips put almost 8 percent of the company on unpaid leave when it took over in 1974.[80] But Signetics's history of laying off almost half its employees in 1970 probably set an attractive precedent for Philips.

Unsurprisingly, this attitude provoked complaints from both Dutch and international unions and from the Dutch press that "Philips is moving work from Europe to the US."[81] Yet at exactly the same moment, advances in the automation of semiconductor manufacturing made it possible for Signetics managers to shrink their workforce in Southeast Asia without any compensatory increase in employment at its US locations. As Charles Harwood explained later,

> We closed down the plant in Manila because we had high speed equipment to put in for assembly. So after having had lots of people [doing hand assembly, we had fully mechanized factories]. . . . We never brought assembly back [to the US]. We put the high speed equipment in the overseas plants. [Which meant] we just didn't need as many plants. That's why we could shut some of them down.[82]

In other words, Philips wasn't so much moving work from Europe to the US as it was using its ownership of an American firm to help it streamline its semiconductor manufacturing workforce all over the world.

The ironies of the Europe–US–Asia semiconductor manufacturing dynamic (encapsulated by the relationships among Philips-Signetics-SigKor/SigFil/SigThai) can be seen clearly by comparing Dutch and US newspapers from the late 1970s. In those years Dutch papers were running headlines like "American chipmakers dominate" and articles skeptically detailing British and French attempts to recreate Silicon Valley in those countries—while at exactly the same time American papers were complaining that US-made chips were inferior in quality to ones made by Japanese firms and warning that South Korea, Taiwan, and "Europe chip away at U.S. high-technology lead."[83] In the long 1970s, everyone believed that others were doing better than they were.

MALAISE AND REFORM

That sense of falling behind presented itself in various national enactments such as "malaise" in the United States or the 1978 to 1979 "Winter of Discontent" in the UK. In many countries, the oil shock of 1979 undermined economies that had never really recovered from the oil shock of

1973. In the Netherlands, the unemployment rate quadrupled in the early 1970s (from less than 1 percent to approximately 4 percent) before stabilizing; then between 1980 and 1983 it skyrocketed again to more than 11 percent.[84] Dutch GDP actually shrank in 1981 and 1982. Unsurprisingly, then, Signetics seems to have performed rather better than the rest of the Philips organization until the mid-1980s—a source of steady though not stellar revenues while the rest of the company struggled against a weak economy.[85]

Malaise also had a very specific Silicon Valley enactment. Although the wider travails of the United States in the late 1970s are well known, both in a popular and in an academic sense, those of Silicon Valley are less well remembered.[86] A particularly nice illustration of that amnesia can be seen in the publication history of one of the most influential books about the Valley, AnnaLee Saxenian's *Regional Advantage: Culture and Competition in Silicon Valley and Route 128*.[87] There, Saxenian draws a fairly invidious comparison between Silicon Valley and the innovation region around Boston, with the latter pictured as stodgy and hierarchical and the former as innovative, mobile, and forward looking. That book, first published in 1994, gave an academic grounding to Valley elites' own belief that their aversion to tradition and credentialed authority explained their success. Curiously, though, some of Saxenian's earlier articles from the 1980s drew on the same empirical material as *Regional Advantage* but used that material to paint a more negative picture of a Silicon Valley that grew so quickly that it fell into a pitfall of existential success: overcrowded, unaffordable, polluted, politically immature, and economically threatened.[88] By the time the book version appeared, however, those problems were no longer seen—by Saxenian or by industry leaders—as particularly pressing.

Implicitly, the economic subject celebrated in Saxenian's book is the Foucauldian "neoliberal self as project"—a rootless individual who moves from company to company with no allegiances, seeking only to acquire skills, contacts, and reputation.[89] Thanks to network entrepreneurs like Stewart Brand, and even more so to the late 1990s dot-com boom and emergence of social media in the 2000s, this kind of self—what Gina Neff calls "venture labor"—has spread far beyond Silicon Valley and is increasingly the norm not just for employment but for all human activity.[90] We should, however, take the history of Signetics and its peers as a reminder that the period when that neoliberal subject was being formed was one of profound economic and cultural malaise. Champions of neoliberalism then and now would no doubt say that they provided emancipation from that malaise. But Signetics and its peers show us that at least in Silicon Valley, the late 1970s doldrums were ended in large part via good old-fashioned political organizing (by workers, employers, and community members) and through governmental intervention (at the local, state, and federal levels). Neoliberalism has prospered by taking credit for the achievements of older sociopolitical forms.

Nothing exemplifies Silicon Valley's late 1970s malaise more than the epidemic of theft that swept across the chip industry in this period. Both academic and popular histories have mostly neglected this story, but at the time it was difficult to avoid. Take, for instance, the October 1982 edition of *National Geographic*, almost half of which was devoted to "the chip: electronic mini-marvel that is changing your life" and "high tech, high risk, and high life in Silicon Valley."[91] Although much of the issue offered sunnily optimistic profiles of Valley legends like Jerry Sanders, Bob Noyce, and Steve Jobs, it also surveyed the negative aspects of the region's growth: rising inequality, pollution, housing shortages, and work pressure leading to drug abuse, "alcoholism, divorce, and depression." The prevalence of drug abuse in turn meant that Silicon Valley was home to a large number of addicts who were willing to steal or commit other crimes to pay for their habits—which, in turn, was pictured (literally, with a full-page photo opening the article on Silicon Valley) as feeding a black market for stolen chips and chip designs.

Nor was *National Geographic* exceptional in spotlighting theft and other problems afflicting Silicon Valley. Earlier in 1982 the *New York Times* profiled John Henry "One-Eyed Jack" Jackson, the alleged leader of a ring that stole 10,000 chips in one haul from Intel and about ten times that many chips from assorted other manufacturers.[92] If *National Geographic* presented Jackson and his ilk as a byproduct of the Valley's relentless work pressure, the *Times* described them as surfing on the industry's notorious up-and-down fortunes. That's because by the late 1970s an entire shadow industry—a "gray market" according to the *Times*—of electronics distributors had emerged to buy up large quantities of cheap chips at the bottom of the market and then sell them at inflated prices when demand rebounded. These gray marketeers were tolerated or even encouraged by mainstream manufacturers because the revenue that they supplied helped firms make it through tough periods, whereas during periods of high demand they helped manufacturers overcome shortfalls in supply. But distributors also operated through long chains of intermediaries, making it easy for stolen and counterfeit goods to be exchanged with few questions asked. During periods of the highest demand, some of those distributors and intermediaries found the opportunity to profit from stolen goods simply too good to pass up.

We can see here some parallels to critics who argue that today's social media giants similarly tolerate illegal or antisocial behavior because it makes them money and because they disdain even small steps to prevent it.[93] Susan Benner's eye-opening exposé, "Storm Clouds over Silicon Valley," offered a striking impression of the 1982 version of this attitude:

> Companies didn't report thefts, sometimes out of concern about security implications to customers and sometimes because they didn't even know the thefts had happened. . . . One large manufacturer, warned that a theft was planned for a specific date in the near future, prevented the theft but didn't fire the potential thief, who reportedly still works at the company and carries on a cocaine business. There was . . . an almost arrogant faith that a company's technological capabilities could overcome any other problems, that prosecuting thieves and tightening up security took time and resources away from the primary efforts—coming up with newer and better products and making money.[94]

Then, as now, the Valley's giants largely treated these crimes as victimless. Yet as Benner showed, the chip thieves and fences of the early 1980s were perfectly willing to maim or kill anyone who threatened their business.

Benner named Signetics as one of the firms involved in this illegal trade, and in fact the company claimed—with some justification—to have been one of the few to take some action against it. In April of 1982, the San Francisco public television station, KQED "aired a half hour documentary on chip thefts . . . [in which] Signetics and Intel were singled out and called Silicon Valley leaders in prosecuting chip thieves."[95] Two years later, Signetics initiated and cooperated with an undercover police sting operation that netted two arrests and $1 million in stolen chips.[96] Whether such actions had a deterrent effect is unclear. Certainly, the moral panic over chip thefts subsided after the early 1980s; but I think the question is still open for historical investigation whether Silicon Valley's gray and black markets truly disappeared.

Chip thefts were a striking symptom of cultural dysfunction, but not uniquely so in Silicon Valley's discourse about itself in the late 1970s and early 1980s: other matters of concern included hazardous chemicals used in chip fabrication and increasingly found in the water supply, leading to miscarriages, birth defects, and clusters of rare cancers; lax oversight of inventories, allowing chips to make their way to America's foes; and runaway growth leading to congestion, pollution, and housing shortages so severe that "in spite of the weather, the ocean, San Francisco, and the Valley's reputation as a hotbed of innovation, bright young engineers are turning down Silicon Valley offers and instead taking jobs" elsewhere.[97] Crucially, these problems were brought to Valley residents' attention not just by their daily experience but also through consciousness-raising efforts of journalists like Benner, politicians such as Sunnyvale mayor Dianne McKenna, and labor activists such as Joyce Maupin of Union WAGE (Women's Alliance to Gain Equality).[98]

Valley elites, meanwhile, were conspicuously ambivalent in their attitudes both toward their region's crises *and* toward efforts to overcome those crises through regulation and activism. On the one hand, some of the Valley's most oft-profiled giants such as Jerry Sanders used their soapbox to complain that government was willing to intervene to limit growth and pollution but not to fend off competition from Japan.[99] On the other hand, some industry leaders acknowledged that sprawl, pollution, and other issues could only hurt the Valley and should therefore be addressed. Indeed, a growing number of semiconductor company CEOs called for a more active role for the *state*, both in building much-needed infrastructure and in tighter regulation and oversight. Of course, they wanted the state to get involved on industry's terms—but what's unusual is that they identified the atomized, individualistic anomie of Valley culture as an impediment to the industry's attempts to organize to get government to listen to *them* instead of to labor unions and environmental activists.

The industry's counterorganizing and growing political maturity is covered well by Saxenian.[100] Signetics, largely because of its foreign ownership, was more a joiner than a leader in those efforts. It is true that Charles Harwood was among those complaining that US firms, particularly vendors of semiconductor manufacturing equipment, didn't have the cooperative culture that supposedly aided Japanese firms—a view that unsurprisingly met with pushback from the equipment manufacturers and their association, SEMI.[101] But Signetics lagged behind other semiconductor firms in forming a political action committee (PAC) so that it could lobby politicians.

Signetics's upper management did, indeed, want to form a PAC, but Philips blocked such a move. As one of the company's lawyers, William Dana, argued in 1981,

> Signetics is becoming more involved in concerns expressed on the local level about fire protection, safety, health, energy, sewage treatment, housing, transportation, and air pollution. There is a certain, vocal segment of the society who singles out industry as the source of these problems and seek to make industry bear the expense of implementing the programs that this group profess are the only answers. Absent an effective mechanism for establishing a dialogue between industry and the government, we could find that we were being railroaded to accept these solutions. While we are participating in numerous [trade] associations . . . it seems to me, we can increase our effectiveness within each of these organizations and in direct contact with government officials if Signetics were to establish its own Political Action Committee.[102]

Those plans were stymied, however, by a "*strict company policy* that neither North American Philips nor any of its activities or subsidiaries shall create or participate in a Political Action Committee in any form."[103]

Nevertheless, Signetics did participate in some of the cooperative industrial activities that marked Silicon Valley's turn away from the lawless free-for-all of the 1950s and 1960s. For instance, as Dana noted, Signetics was a member of organizations such as the Semiconductor Industry Association and the Santa Clara County Manufacturing Group. It did not, however, join the early semiconductor research consortia such as the Microelectronics and Computer Technology Corporation that was proposed in 1982. But it was one of the nineteen founding industrial affiliates of the Center for Integrated Systems (CIS), which John Linvill and James Meindl founded at Stanford in 1980; I have argued elsewhere that the CIS functioned as a kind of industrial research consortium.[104] It is perhaps notable in that regard that Meindl gave the keynote address at the opening of the Philips Research Laboratory Sunnyvale.

Signetics also participated in cooperative lobbying regarding state, local, and federal taxes and regulations. Unsurprisingly, the company—and the industry—mostly lobbied against government action. For example, Signetics executives were particularly incensed at the federal government's "using defense contracts to accomplish unrelated socioeconomic goals," particularly requirements for affirmative action hiring of African Americans.[105] They also praised the Reagan administration's suspension of regulations on hazardous chemicals in semiconductor manufacturing.

Yet Signetics and, as Saxenian argues, Silicon Valley manufacturers more generally did not uniformly oppose state intervention. For instance, Valley firms did indeed fight hard against municipalities' attempts to limit construction of new industrial facilities; but Saxenian shows that they came to recognize that to avoid such limits they needed to be open to higher public spending for new infrastructure. Signetics was no exception; for instance, in 1984 Harwood sent a memo to all Signetics employees urging them to vote for an increase in the Santa Clara County sales tax to fund upgrades in the county's highways to combat "increasingly bad traffic, wasted time and energy, and more air pollution."[106] The company also seems to have supported the Sunnyvale city government's efforts to expand its water treatment infrastructure—mostly because that would raise municipal limits on firms' use of city water in their manufacturing facilities.

So Signetics's political organizing was more ambivalent toward the role of the state than one might imagine. We can see that surprising ambivalence in the company's workplace reforms as well. For instance, with respect to exposure to toxic substances used in semiconductor manufacturing, Christophe Lécuyer has shown that Signetics was out in front of its peers in making its workplace safer.[107] Yet he also argues that Signetics's actions were largely the result of pressure from its employees and a network of journalists, lawyers, and labor organizers who supported those employees—pressure that Signetics's managers initially tried to resist but later tried to get ahead of. Indeed, eventually they reframed their grudging responses to that pressure as examples of Signetics's good corporate citizenship.

In particular, Signetics was forced into action by lawsuits and newspaper coverage related to the "Signetics Three" case. This concerned three employees—Marta Rojas, Cathi Hee, and Cathy Bauerle—who developed chemical sensitivities and health problems as a result of chronic exposure in a poorly ventilated laboratory.[108] Because the three women had asked the National Institute for Occupational Safety and Health (NIOSH) to investigate working conditions at Signetics, the company fired them in 1979. With help from labor organizers, they then sued Signetics, resulting in $25 million in damages and a spate of negative press reports, particularly by Susan Yoachum of the *San Jose Mercury News*.[109] As a result, Signetics adopted state-of-the-art industrial hygiene practices that slowly spread across the rest of the industry. These included steps that seem quite obvious today but that reflect the lawless conditions that pervaded the industry before the late 1970s: for example, installing chemical sensors in the workplace instead of relying on employees with an acute sense of smell.

Lécuyer argues that Signetics executives did not worry so much about the legal ramifications of workplace hazards because they had strategies for forestalling inspectors from agencies such as NIOSH—and, in any case, most multinational manufacturers can bear $25 million in damages relatively easily. What really worried Signetics executives was that workplace hazards would lead to a unionized workforce; improving industrial hygiene was therefore a way of showing workers that a union was unnecessary. My own dive into Signetics company newspapers lends credence to Lécuyer's argument. In its internal communications with employees, Signetics loudly and frequently praised a variety of workplace reforms. For instance, a 1982 company newspaper article announced that "IC industry gets good safety rating by state agency," and specifically "Signetics spends millions on health and safety . . . goal is to protect employees"; the article did admit that "Signetics finds traces of solvent in the ground" near a storage tank, but then outlined all the company's steps to clean up its mess.[110]

Similarly, in dealing with large plumes of polluted groundwater caused by chemicals leaking from their plants, Fairchild, Signetics, and other Valley firms seem to have been motivated more by fear of grassroots organizers—particularly a young mother named Lorraine Ross and an NGO, the Silicon Valley Toxics Coalition (SVTC)—than by oversight from the EPA.[111] To deal with the threat posed by Ross and SVTC, Signetics had to supplement its internal propaganda meant for employees' consumption with *external* appeals meant to burnish the company's image in the community. Thus, in the period when the company was most embroiled in controversies over occupational safety and public health, Signetics's charitable giving increased sharply. In 1978, Signetics budgeted just $16,500 in donations to organizations such as United Way and a local veterans' group.[112] For 1979, though, the budget rose to $22,000, and by 1981 charitable spending was budgeted at $100,000—nearly a sixfold increase in three years.[113] Much of that increase was targeted to highly visible outlets such as sponsorship of television shows on noncommercial Public Broadcasting Service channels, with the explicit intent of creating public goodwill.[114] In other words, Signetics executives went from not caring much what the public thought to caring quite a bit—all in a short period that coincided precisely with the Signetics Three case and other public relations calamities.

So Signetics changed a great deal during, and because of, the years of malaise. Of course, we should mostly credit pressure from activists and the state for those reforms, rather than internally generated reflection. After all, Silicon Valley's early years were marked by almost complete indifference toward pollution, discrimination, inequality, and the other ills that suddenly seized headlines in the late 1970s, so it's likely that executives at firms like Signetics would have maintained that indifference if left to their own devices.[115] But they *weren't* left to their own devices. Instead, they were stimulated to change by activist elements of civil society who pushed industry to reform and who pushed government to police that reform.

Notably, women such as Ross, Yoachum, and the Signetics Three were the most visible figures in efforts to reform Signetics and its peers. Signetics offers a particularly stark example of changes that probably would not have happened absent activist women applying pressure for a more responsive technoscience; but the other cases in this book similarly hint at the significant role that women played in creating conditions whereby square men either wanted to or believed they had to cooperate with reform.[116] That's not to say that women should have to bear the responsibility for making square men more responsible. Rather, I'm offering some empirical support for arguments that a more diverse and open technical workplace is somewhat more likely to be a responsible technical workplace.[117]

SIGNETICS NO MORE

Whatever drove the reforms of the long 1970s, by the end of that period they had become entrenched if still inadequate in many ways. A nice illustration is offered by the October 1984 Signetics company newsletter featuring three front-page articles: "Signetics blood drive"; "Signetics, two neighbors pursue water orders," detailing corporate efforts to "define and contain the extent of an underground plume of contaminants believed to have been caused by defective underground tanks that stored solvents"; and " [Hello] to Dave Herdrich," profiling a new deaf employee and praising his hearing colleagues who were learning to sign.[118] None of those three articles individually means much; but together they indicate that the company's leadership of the early 1980s wanted Signetics to be seen as a diverse, sensitive, responsible, generous corporate citizen in a way that their predecessors of the late 1960s did not.

What wasn't known at the time—though perhaps Harwood and other senior executives could foresee it—was that Signetics was nearing the end of its run. After several years as a decent source of revenues for Philips, in late 1984 Signetics entered a period of consistent red ink. This was something of a turnaround. It's important to remember that the Silicon Valley variant of malaise was driven in part by the region's existential *success,* whereas in the Netherlands and the rest of the United States, malaise reflected adverse economic conditions. Not that Signetics was immune to the era's national and global economic dislocations: for instance, both in 1973 and in 1979 its employees were encouraged to carpool and take other conservation measures, and the 1981 to 1982 recession forced "a 10-day, cost-reducing closing of its Sunnyvale operations . . . spread over six weekends during the summer months."[119] But with profits growing by 300 percent between 1980 and 1984, Signetics was floating higher in the water than its parent company.[120]

In the second half of the 1980s, though, Signetics steadily lost money. By 1987 the Philips–Signetics combination had dropped from second or third in the world semiconductor rankings to tenth, and the next year the combo dropped out of the top ten entirely.[121] At least part of the reason was that Signetics's hoped-for switch from bipolar transistors to MOS never worked out; as Signetics's final president, James Dykes, explained in 1990, Philips's light touch allowed Harwood and other Signetics executives to simply ignore the directive to move over to MOS transistors until it was too late.[122]

By that point, Philips itself was starting to unravel, with the result that the early 1990s recession forced it into a large-scale reorganization and "shock therapy" composed of mass lay-offs and disposal of unprofitable subsidiaries.[123] The parent company had already dissolved United States Philips Trust—Signetics's ostensible owner—and reasserted direct control over its US subsidiaries in 1987.[124] In 1990 a new unit, Philips Semiconductors, was set up to integrate Signetics and all Philips's other global microelectronics subsidiaries within a single organization.[125] In 1993 the Signetics name was effaced entirely, though the company's independent identity had largely disappeared well before then. The conventional wisdom that emerged in hindsight was that Signetics had always been a money pit and poor performer.[126] That seems a rather distorted view given the company's presence in the upper ranks of Silicon Valley until the mid-1980s—but it is perhaps a reason for Signetics's surprising absence in most histories of either Philips or Silicon Valley.

Signetics's disappearance is beyond the scope of this book. I doubt that following the Signetics story to its conclusion would add much to my core themes. But it is worth a very quick glance at what set Signetics on a path to poor revenues. One factor seems to have been the company's reluctance to update some of its facilities, particularly in the United States. In 1992, for instance, Philips Semiconductors closed the Orem, Utah, plant because it could no longer make state-of-the-art chips.[127] We can infer, then, that over the course of the 1980s, at least some of Signetics's manufacturing sites lost touch with the industry's forefront techniques—that is, by the late 1980s Signetics was missing exactly the kind of up-to-date know how that had motivated Philips's purchase of the company in the first place.

One reason that Signetics fell behind in manufacturing was that Harwood appears to have tried to substitute cheap management innovations for more expensive maintenance and replacement of hardware and facilities. Around 1980 he had become a convert to the management theories in Philip Crosby's hit book, *Quality Is Free*, which Harwood implemented at Signetics in the form of a fourteen-step, team-based quality control program. By 1984, "The successes of the quality movement of Signetics had been so great that . . . the prime movers spent time showing the mother company in Eindhoven how to get quality improvement going."[128] The business press enthusiastically celebrated Signetics's firmwide Zero Defects Recognition Day and the millions of dollars that Harwood's program was claimed to have saved.[129] Yet it was all something of a mirage; in late 1984 Signetics was forced to admit that it "did not fully comply with test specifications on a wide range of devices used in weapons and military equipment," resulting in a Pentagon ban on Signetics's chips.[130]

In the following months, Signetics began laying off some 6 percent of its US employees and close to 8 percent of its global workforce.[131] The company's troubles with the Pentagon, the downturn in profitability, and Harwood's perceived success as a quality-control innovator provided Harwood with an easy off-ramp to go found his own management consultancy in March of 1985. In the wake of his departure, almost all Signetics's senior managers also left, and about a third of the employees were laid off. The rump firm was divided between defenders of Harwood's program and those who thought that solving every problem using dedicated teams executing all fourteen steps was more administrative burden than a struggling company could bear. That tense atmosphere hobbled recovery and left the firm with little ability to contest absorption into Philips.

At the same time, Signetics and Philips made two bad bets on the future direction of the industry. One was that European firms and governments could cooperate in the 1980s as effectively as state and industry were reputed to have collaborated in Japan in the mid-1970s. This view led the Dutch and West German governments to join the multinational Joint European Submicron Silicon Initiative, through which they financed a third of the costs of a collaborative "MEGA Project" in which Philips (including Signetics) was supposed to develop static random-access memory (SRAM) and Siemens dynamic random-access memory (DRAM) extending to four million bits. This seems to have worked out reasonably well for Siemens, but by 1990 it was apparent that "Megaproject [is a] commercial flop for Philips."[132]

The other bad bet was that the Reagan administration's defense buildup would bring a boom in chips for the military services similar to that seen in the mid-1960s. Thus, in late 1981 Signetics opened a new plant in Sacramento, California, for its Military Products Division. As the plant manager put it, not very presciently: "Defense is the most exciting, the most dynamic field in electronics."[133] Yet despite the buildup, defense spending no longer meant as much to the semiconductor industry as it once did. Especially with the advent of the personal computer in the early 1980s, the semiconductor firms that prospered were the ones—such as Intel—that could get contracts to supply chips for *consumer* products such as personal computers. Obtaining military contracts, by contrast, was no longer a recipe for growth, especially for a company that no longer maintained the quality-control standards demanded by the military. Thus, after just eight years Signetics was forced to close its Sacramento plant and relocate most of its military division operations to the Utah location—which it also closed not long after.

The long 1970s didn't kill the Cold War entirely—the Reagan buildup and the apocalyptic rhetoric that accompanied it were real enough—but the long 1970s did mark the transition to a new era in which the military mattered much less to American scientists and engineers than it did before. Signetics was born and prospered in that earlier era of the Pentagon's preeminence on the technoscientific landscape. But it failed and disappeared in part because its managers—including, ironically enough, Philips's managers, with little loyalty to the Pentagon and no experience of its earlier influence—didn't recognize that the country and the world had moved on.

Yet the Signetics employees who had gone through the long 1970s—and participants in the global Silicon Valley more generally—were well aware that the world had changed a great deal in that time. Reading corporate newspapers from, and oral histories about, Signetics's early years is like watching an episode of *Mad Men*—except that its women employees in that era weren't just leered at and blocked from promotions but also poisoned by exposure to some of the most hazardous manufacturing chemicals imaginable. Inequities and occupational dangers certainly didn't disappear during the 1970s, but many Americans—including many square scientists and engineers—became much less accepting of them. The squares can't take much credit for that, however. Silicon Valley leaders like Gordon Moore may have been "revolutionaries" with respect to certain aspects of the technologies that today we take for granted. But with respect to cultural attitudes that many more Americans take for granted today than did before the long 1970s—well, there, the revolutionaries were people, famous and unknown, from all walks of life, living not just in the United States but around the world, who wanted a more democratic and inclusive society.

7

ENGINEERING AND APPLIED PHYSICS IN THE AGE OF FRACTURE

On January 1, 1984, a deal reached between AT&T and the US Department of Justice two years earlier went into effect, ending AT&T's regulated monopoly status while breaking its regional subsidiaries off as independent "Baby Bells." At the time, AT&T was one of the most influential companies in the world: employer of more than a million people at its height; crucible in which Unix, information theory, the transistor, computer music, and many other innovations were forged; operator of nuclear test sites and weapons labs; and of course the nearly unchallenged provider of telecommunications to most of a continent. Its breakup after ten years of antitrust lawsuits set American business, government, and technoscience on a new path of deregulated competition and eventual unfettered remonopolization that has lasted to the present. As such, the AT&T breakup signaled the end of the long 1970s as decisively as any other event.[1]

As in chapter 1, this chapter summarizes the arguments of the book and connects them to present-day ideas of "responsible research and innovation" through a somewhat abbreviated empirical case study. This time, my empirical examples are drawn from the career of a minor figure in AT&T's breakup, Arthur D. Hall III. From 1950 to 1966, Hall was an

AT&T employee, during which time he was particularly associated with the failed Picturephone project—a forerunner of today's Zoom, Skype, and FaceTime.[2] While at AT&T, Hall also wrote an influential book, *A Methodology for Systems Engineering*, and was a founding member of the Institute of Electrical and Electronics Engineers (IEEE) professional group on Systems Science and Cybernetics.

After 1966, Hall initially bounced around senior management positions at other companies while teaching part-time at the University of Pennsylvania. Then in 1970 he founded his own consulting firm, Arthur D. Hall, Inc. (later renamed Advanced Decision Handling, Inc., or ADH). Surprisingly, perhaps, some of AT&T's erstwhile competitors were among ADH's early clients. Through those clients Hall built a reputation as an expert witness who could persuasively undermine the legitimacy of AT&T's monopoly. As a result, the Department of Justice hired ADH to provide evidence for its antitrust suit—specifically, evidence that Bell's rates were unreasonably high and that its justifications for those rates (e.g., ensuring reliability and innovation) were unfounded.

By the mid-1970s, therefore, Hall was in much the same position as Jack Kilby in chapter 5: he had a steady revenue stream from his patents and consulting that enabled him to remain independent of a corporate world that he had come to hate. Like Kilby, Hall could turn his attention toward technological projects that he was passionate about and that held some promise of revolutionizing an entire industry. In Hall's case, those projects related to the farm that he had bought in the early 1960s and to which he applied his expertise in sensors, communications technology, and automation. Out of those efforts grew an idea for an automated agricultural *system* (in the systems engineering sense) called *Autofarm*. Having received a 102-page patent for Autofarm in 1977, Hall spent the rest of his career unsuccessfully seeking government and corporate investment to develop it for use in the United States, Canada, Israel, Saudi Arabia, Poland, and elsewhere.[3]

Like the other people and organizations presented in this book, Hall's case is not representative of American physical and engineering scientists, nor even of that community's square members. He ascended to relatively elite positions within the corporate world and in his professional society (the IEEE), and he had some influence in government, though not as much as he would have liked. His career path was probably messier than average for his time, though in that sense he was a harbinger for the mobile "venture labor" of the 1980s and beyond.[4] Although we've seen many instances of interdisciplinarity and field-hopping in this book, Hall was an extreme example: not only was his model of systems engineering highly interdisciplinary, but also his enactment of that model in his consultancy and especially in his attempt to revolutionize agriculture makes him an outlier.

So Hall is not representative, but he is illustrative. He took the trends examined in my other case studies to new extremes and thereby casts further light on them. In addition, his peripatetic movement through industry, academia, government, professional societies, and an independent consultancy allows us to reexamine all the domains covered in other chapters by tracing his complicated career trajectory. Moreover, the fact that Hall's papers are housed in a single collection (at the Hagley Museum and Library in Wilmington, DE) means that his individual outlook on all those domains and trends can be grasped synoptically.

Thus, examining Hall's career helps me to summarize points I've made in the previous chapters. His example also helps to fill in some of the gaps left by my other case studies. In particular, his case offers some perspective on institutions of the East Coast establishment: Ivy League universities, professional societies, and giant regulated monopolies such as AT&T, IBM, RCA, and General Electric. These institutions have not figured much in this book thus far (though I have written about them extensively in my previous monographs). In the long 1970s they still played an important role in American technoscience—not just through the numbers of scientists and engineers that they employed and the technologies that they were associated with but also through their outsize influence in political and business discourse. That influence waned sharply in the 1980s, but we need to understand what came before then. Hall gives us at least some clues to how people who were associated with these institutions experienced the long 1970s.[5]

Thus, examining Hall helps me expand the book's regional scope somewhat. Some readers will justifiably complain that this book's cases focus heavily on events in California and Texas. Hall, however, lived and/or worked in New Jersey, Pennsylvania, and Maryland, so he—along with the professional societies examined in chapter 1 and Corning Glass Works in chapter 6—brings the Mid-Atlantic states somewhat into the picture. With a little stretching that means the only regions missing from my study are the Southeast and the vast "flyover country" stretching from Ohio to Colorado. I have little confidence in any claims I could make about the former; the politics of race and postwar industrialization in the Southeast are not fully comparable to other regions, even the Texas of chapters 4 and 5. We need more work on the history of technoscience in the South during the groovy era, which this book cannot supply.[6]

I am more certain that one could find close Midwestern analogues to this book's case studies. To a small extent that's because a surprising number of figures in my other chapters hailed from the Midwest and, for the most part, received their initial technical training there: David Phillips, Virgil Elings, and Robert Noyce (Iowa); John Linvill (Missouri); William Magruder (Illinois); Jack Kilby and Glen Culler (Kansas); and Stanton Glantz (Ohio). There were enough continuities between the Midwest and the regions that are spotlighted in this book that someone like Kilby, for instance, was able to move effortlessly among Kansas, Illinois, Wisconsin, and Texas. More importantly, we have reasonably good evidence from other studies that at least some Midwestern research organizations made the same turn toward interdisciplinarity, inclusivity, and socially relevant applied research during the long 1970s as did their East and West Coast peers.[7]

Still, it's fair to critique this book's regional focus. We need to know much more about American science and engineering in the interior of the country—not just for the long 1970s but also more generally. For an exploratory study of square scientists and engineers in the 1970s, however,

California, Texas, and the Mid-Atlantic offer a reasonable microcosm. As noted already, the Mid-Atlantic was home to many of the Establishment institutions—such as AT&T and the APS—that saw their authority ebb over the course of the 1970s. California and Texas, meanwhile, benefited disproportionately from the early Cold War expansion of America's R&D capacity and therefore needed to make greater adjustments during the deflating bubble of the 1970s. We should thus expect that the contrast to the early Cold War was sharper in those states, though similar in form to that seen elsewhere.

California and Texas also represent opposite ends of a not wholly inaccurate spectrum in the American imaginary. That is, if I can show that squares were important actors even in supposedly groovy California, *and* that countercultural concerns were on the R&D agenda even in allegedly square Texas—well, then there's good reason to think that my findings will be robust for other regions too. That squares like Hall, moving among the Establishment institutions of the Mid-Atlantic, saw and did the same things as their peers in California and Texas lends further plausibility that my claims might be valid across other regions. Indeed, my main claim with respect to regional variation is that it didn't matter that much. Squares had to adapt to (and create!) local conditions, of course, but most of the situations that I've shown squares contending with were local manifestations of national or even global issues. Arthur Hall, working from his farm in Port Deposit, Maryland, was more tied to a specific place than the other actors in this book; but even when he tried to make very local interventions, he mostly had his eye on how his technical expertise could solve the problems of the nation or the world.

COMMON BEGINNINGS

The case of Arthur Hall helps me, then, to broaden the applicability of at least some aspects of my argument. Hall's main role in this chapter, though, is to embody almost all the trends I have surveyed in the other chapters. Despite his individual eccentricities, he stands in remarkably well for American square scientists and engineers in general. In what follows, therefore, I will gradually lay out different aspects of Hall's adaptation to the circumstances of the long 1970s while relating each of those aspects to examples from the other chapters.

Let us start, then, with the military–industrial "roots of the squares" that I examined in my introductory chapter. If we look across this book's case studies, one of the first things to notice is that nearly all the actors started from a similar place: a career immersed in the military–industrial complex. Some (e.g., James Meindl) were directly employed by the military at some point in their careers. Others (e.g., the NASA engineers in chapter 4) worked for government agencies that shared aims, technology, vendors, and personnel with the military. Others worked for those vendors, supplying hardware (e.g., Jack Kilby and TI) or knowledge (e.g., Philip Wyatt and Defense Research Corporation) to the military. Some (e.g., Calvin Quate) performed academic research supported by grants from military and other national security agencies. Most of the people profiled in this book moved among a variety of different ties to the military over the course of their early careers. Meindl, for instance, worked on nuclear submarine reactor controls for Westinghouse before joining the Army Research Lab at Fort Monmouth, then moved to Stanford, where his research was supported by the DOD.

Like Meindl, Hall moved from one form of entanglement with the national security state to another, both before and during the long 1970s. His most notable achievement—*A Methodology for Systems Engineering*—was written because "during the Korean War, the Bell Laboratories . . . due to its increasing military contracting and civilian business, experienced a severe shortage of systems engineers."[8] This led Hall's boss to ask him to figure out how to train systems engineers quickly, without the need for lengthy apprenticeships—and "thus was born the world's first formal course in systems engineering."[9] After leaving AT&T in 1966, he became vice president of two different companies (Jerrold Electronics and SCM) that made electronic equipment for the military. After forming ADH, he continued to serve military clients such as the Defense Communications

Agency (today known as the Defense Information Systems Agency) and give speeches for military organizations such as the Tactical Air Command.[10] He also envisioned adapting the Autofarm system for military applications such as "detection of chemical or biological agents"—an idea that he asked the DOD Small Business Advanced Technology Program to fund in 1981.[11]

Readers should note that direct and indirect interactions with the military–industrial complex such as Hall's didn't particularly set square physical and engineering scientists apart from their peers who were less square or who were affiliated more with the life or social sciences. Many American scientists and engineers found it hard to *avoid* working within the framework of the national security state in the 1950 and 1960s. Few did try to avoid it, at least not until the beginning of the long 1970s. Even those who were first to push *against* the national security state did so from a position of prior support *by* the national security state. For instance, Linus Pauling (a chemist) and Barry Commoner (a cell biologist) both became widely recognized public figures in the 1950s because of their opposition to nuclear testing, yet both had also conducted defense research during the war.[12] After the war, engineering and physical scientists probably were *somewhat* more beholden to the Pentagon than practitioners in the life and social sciences, but not by as much as one might think.[13] Even Noam Chomsky's linguistics research was supported by the Pentagon![14]

So the national security state hardly discriminated among science and engineering disciplines: it coopted them all. The Cold War may have had more of an asymmetrical effect, however, depending on the *kind* of science or engineering that one did. In particular, the national security state was possibly a more pervasive presence in the careers of scientists and engineers who conducted research that was intended for publication (in some form) than among those who didn't consider publishing. As the author of an influential textbook and founding editor of *IEEE Transactions on Systems Science and Cybernetics*, Hall fell quite easily into the former category, as do almost all the major figures in this book.

Even the industrial scientists and engineers that I've examined—people such as Jack Kilby who were more engaged in development rather than research—occasionally wrote technical articles and frequently published patents.

Admittedly, this is quite an imprecise criterion for demarcating the kinds of people who appear in my study from those who don't. Despite that imprecision, though, it is important to reflect on what we gain and lose by focusing on people who were more oriented to publication—as opposed to, say, scientists and engineers working in quality control laboratories or municipal water districts. In this study I look mainly at scientists and engineers whose work involved (semi)published research because those are the people who figure both in earlier historical studies of socially responsible research in the 1970s and in present-day enactments of responsible research and innovation—that is, by focusing on this class of people I can more directly intervene in both the historical and RRI literatures. Also, like most historians, I use documents to reconstruct the past, so my work is slanted toward historical actors who generated large volumes of relatively accessible documents.

Yet those reasons for preferring actors who publish over other kinds of technoscientists invite fair and serious criticism both of RRI and of my and other historians' studies. Published articles and patents—and the people credited as authors of such documents—represent only the smallest tip of the technoscientific iceberg.[15] The extent to which personnel involved in R&D who publish do or don't resemble their colleagues who don't publish should be an important question for future study. Nevertheless, we can learn a lot about the evolution of the American technoscientific enterprise even if we limit our scope to physical and engineering scientists who were doing research that could be published. Later in this chapter I will try to show what this group can tell us about present efforts in RRI. For now, though, let us stick with what this group of people can tell us about the long 1970s.

From about 1967 on, these people had to make hard choices about their relationship to the Cold War. The national security state's pockets were no longer deep enough to support all of them in the manner to which they were accustomed, and the politics of military support were becoming more complicated. As a result, some physical and engineering scientists moved toward positions that were highly critical of American militarism. Some of those critics, such as Chomsky or Carl Sagan, became well-known public figures about whom historians have written a great deal.[16] I've focused, however, on the much larger group of people who didn't seek or gain public prominence or whose feelings about the national security state were more mixed. As we saw in chapter 5, though, squares and vocal critics mutually regarded and constructed each other. It's difficult to understand the public roles of critics like Sagan if we don't also look at squarer, more reticent scientists' attitudes toward publicity seeking and avowedly activist research. At least some of the backlash against Sagan's support for the nuclear winter hypothesis, for instance, was framed as suspicion that Sagan was chasing a story that fit his very public politics but wasn't justified by the facts.[17]

Even so, public intellectuals like Sagan were not exactly alone. Many less visible but still influential figures also became more skeptical of the national security state in the 1970s. For instance, that era saw the formation of a cadre of physicists—including Hans Bethe, Richard Garwin, and Herbert York—who opposed antiballistic missile (ABM) defense despite their extensive ties to the Pentagon (which generally promoted ABM).[18] At least some of those who cautiously turned away from their earlier military patronage in this period linked their newfound criticisms of defense policy to their growing interest in "socially responsible" science. Bethe, for instance, was one of the relatively few physical scientists to take part in the National Teach-In to debate the Vietnam War in May of 1965; although Bethe was not "a supporter [of the war] but, at that time, he was not an active critic either," by the end of the long 1970s he was more open in his criticisms of defense policy.[19] Similarly, Garwin—one of the inventors of the hydrogen bomb—had, by the late 1970s, made such contributions to arms control that he received the APS's Leo Szilard Award and was made chair of its Panel on Public Affairs—although he was still such a square, behind-the-scenes player that the tag line of his biography is "the most influential scientist you've never heard."[20]

A similar example is Harvey Brooks, a semiconductor physicist at General Electric and then Harvard; a member of the President's Science Advisory Committee under Eisenhower, Kennedy, and Johnson; and a president of the American Academy of Arts and Sciences. Although publicly politically reticent all through the 1960s, by 1970 Brooks had reached his limit. As he wrote to Norman Hackerman, chair of the NSB,

> I have decided to associate myself publicly with two groups working against present administration policy in regard to the war in Southeast Asia. . . . I thought rather hard before taking this step. As you know, it has been my past policy not to take political stands, and I have always refused to associate myself with such groups as "Scientists and Engineers for Johnson," etc. I have always felt in the past that it was a mistake for scientists, as scientists, to identify themselves with political causes, especially those associated with particular political parties or candidates. But I feel the present situation is different and that the country is in such deep trouble that for the first time in my mind the costs of silence outweigh the costs of tarnishing the public image of science through political participation.[21]

At exactly the same time, Brooks was becoming one of the leading advocates of the new field of "technology assessment" (TA). Specifically, Brooks promoted a form of TA that would be inclusive of public opinion and reflexive about its own positioning and impact.[22] Reflexivity and inclusivity are two of the principles of modern RRI; as I'll discuss later, advocates of RRI today often cite TA as a predecessor. Thus, as Danielle Shanley has shown, Brooks should be seen as one of the forebears of RRI; and he came to that status for reasons that are difficult to separate from his rejection of the Vietnam War and of his earlier political quietism.[23]

MIXED MESSAGES

If the Vietnam War pushed some people to the (broadly defined) left, however, it also pushed some to the right. Naomi Oreskes and Erik Conway's study of a small but influential ring of climate denialist physicists, for instance, identifies that group's sense that colleagues betrayed them over

Vietnam as a starting point for the denialists' long drift toward political lobbying and lucrative corporate contracts for research undermining the scientific consensus on a range of issues.[24] Interestingly, one of the denialists' main opponents in the debate over whether cigarettes cause cancer, the antismoking activist Stanton Glantz, appears in this book as a Stanford aeronautical engineering student and advocate of socially responsible engineering. Between them, Glantz and his denialist foes illustrate American scientists' and engineers' divergent responses to the war.

We know a *little* about why denialists like Fred Seitz moved right and idealistic students like Glantz moved left—but we don't know nearly enough. Simply invoking "Vietnam" is only a partial explanation, because people reacted to the Vietnam War in many different ways, and because attitudes toward the war were entangled with other technoscientific questions of the day like ABM defense and the supersonic transport.[25] We need studies that parse the motivations of people like Glantz and Seitz as well as even more radical figures such as Charles Schwartz; we also need studies of the interactions *among* those strands of American science and engineering during and after the long 1970s. This book isn't that study, though. What this book does instead is examine just the people who were ambivalent or publicly quiet about their positions on the war and the national security state more generally—that is, I exclude actors on the ideological flanks, such as Seitz and Schwartz, and concentrate on the more reticent or hesitant group running from Glantz on the left to Jack Kilby on the right. This group's attention and goodwill were resources that were fought over by left and right, hawk and dove, square and groovy.

True, at times, some of this book's actors' views were—or appeared to be—discernible to a wider public. For instance, Stanford administrators and prowar faculty members regarded Glantz as a troublesome peacenik because of his campaign to compile statistics on Pentagon support for campus research. A closer look at Glantz's writings, though, shows that he wanted to mediate between Stanford's warring factions, rather than declare one side right and go home. He also opposed calls to ban research support by the military, preferring instead that Stanford's engineers simply diversify their sources of funding to avoid relying disproportionately on the national security state. That is, whatever opinions were imputed to him, Glantz was an ambivalent square, not an absolutist of either left or right.

The same can be said of many of the figures in this study. Virgil Elings, for instance, signed a couple antiwar petitions, whereas James Meindl's crew cut was perceived by coworkers as declaring that he supported the war—yet both men led ideologically diverse research and teaching programs.[26] That is, even though they held broadly discernible views, their practice was accommodating to other square scientists and engineers, even those whose broadly discernible views pointed in the other direction. In this respect, Arthur Hall was once again fairly typical. Even though he left behind a large volume of published and unpublished records, I can't find any clear documentation of his views on the war. Circumstantially it seems likely that he supported it, though I also suspect that he had little trouble assimilating the US military's defeat into his critique of large government and corporate bureaucracies. In any case, Hall seems to have drawn ideas from people who were both pro- and antiwar. In his proposals for Autofarm, for instance, he approvingly cited Barry Commoner—a strident opponent of the war—as an authority on the costs of pollution and the need to replace fossil fuels with photovoltaic energy.[27]

The flip side of squares' ambivalence was their flexibility in accommodating to the times—because their commitments were light, most squares didn't strongly oppose either the things that were changing or the things that stayed the same. Sometimes cheerfully, sometimes grumpily, the people profiled in this book moved beyond military patronage and enlarged their outlook in the 1970s. The course that Hall taught at the University of Pennsylvania on Communications Technology and Public Policy gives some of the flavor of that accommodation: his final lecture was originally entitled "Communications and Military Interaction," but in revising the course in 1973 Hall amended the syllabus by simply crossing out "military" and replacing it with "social."[28] Similarly, he changed the line "Implications and applications for the military in education, identification, etc." to read "Implications and applications for the military, education, law enforcement, trade, etc. Public issues of monopoly, privacy, financing public uses, etc."[29] If military applications were out of favor and "social" ones were in, well, Hall could adapt—no problem.

As the latter syllabus revision shows, ambivalence did not equate to indifference or ignorance. Squares were keenly aware of where the public and their patrons wanted them to focus. Note the expansive list of topics covered elsewhere in Hall's revised syllabus: "techno-economic factors encouraging and limiting the use of communications in the military, in law, education, medicine, and international development, in order and justice, urban decay and rebuilding, pollution control, privacy, nation building, in peace, and in peace-keeping."[30] That list of topics overlaps considerably with Bill Gilchrist's list in chapter 6 or Rudi Kompfner and John Baldeschwieler's in chapter 3 or with the rosters of early 1970s projects undertaken at UCSB in chapter 2 or at NASA in chapter 4. There was near-unanimity among square technoscientists as to the list of societal problems that they were being asked to lend their expertise to in this period.

Yet when historians of American science and engineering in the 1970s have looked at the topics included on that list—mass transit, public housing, disability technologies, environmental monitoring and remediation, education, health care—they have usually treated them in isolation, rather than as part of a broader movement. Perhaps more contentiously, I would also argue that histories of this period have exaggerated the prevalence in the 1970s of topics that were *not* on this list—such as parapsychology research or the construction of high energy particle accelerators—because those other topics were sexier or more scientifically prestigious than the ones that *were* on the list. Parapsychology and particle accelerators certainly merit historical attention, but not in isolation from attention to the socially relevant research topics of the day.[31] Indeed, we can't really understand parapsychology or high-energy physics of the 1970s if we think that their practitioners separated themselves from socially responsible research. True, some parapsychologists were obsessed with the spooky and exotic to the exclusion of more mundane topics; but at least some—for example, Elizabeth Rauscher or David Phillips—moved freely between parapsychology and topics on the "societal relevance" list. Similarly, some high-energy physicists thought that solving nature's mysteries was sufficient contribution to society; but others, such as Donald Glaser or Virgil Elings, left the field to pursue projects of more immediate and obvious societal value.[32]

One reason that we should treat the topics on the consensus list collectively rather than separately is that there was an accompanying consensus that these topics should be tackled in similar ways. In particular, almost everyone agreed that all these problems required interdisciplinary approaches. Yet there was far less unanimity as to what interdisciplinarity meant or how it should be organized. Again, Arthur Hall's Communications Technology and Public Policy course lends some perspective. The course was offered within the Systems Engineering department of the University of Pennsylvania's Moore School of Engineering; the chair of that department, Kenneth Fegley, hoped that the course could help him forge interdisciplinary links to other parts of campus. Thus, in 1973 Fegley wrote to George Gerbner, dean of Penn's Annenberg School of Communications, to say that "we are eager to have students from the Annenberg School of Communications who wish to obtain competence in the telecommunications area take graduate courses offered by our department."[33]

This seems to have been a sincere invitation, since Hall's course covered topics such as "cable television systems and the 'global village'," and "the graphic arts." Fegley probably believed that Hall would be open to students of diverse backgrounds, because Hall's own systems methodology combined engineering with approaches "more in the realm of the behavioral sciences, insofar as they deal with psychological and sociological aspects of individuals and of groups."[34] Hall, however, objected to having nonengineers in his classroom. Thus, Fegley was forced to tell Gerbner that although he had

> anticipated that Arthur Hall would be able to accommodate students with diverse backgrounds. . . . He has advised me, however, that he believes it is necessary for

> students taking this course to hold a B.S. in science or engineering or the equivalent. To admit students with a B.A. in the humanities would, he states, be an experiment at the expense of the student.[35]

Despite that check on his hopes for interdisciplinary education, Fegley promised that "we will continue to work toward bringing students with various backgrounds but common interests together in telecommunications courses. We believe that the value of many courses to students whose background is engineering will be enhanced by interaction with students whose academic home is the Annenberg School."[36]

The tensions between squareness and interdisciplinarity can be glimpsed across the other chapters of this book as well. Take, for instance, one of the highly interdisciplinary clusters we encountered at Stanford in chapter 3: there, we met John Chowning (Stanford's Music Department) and Calvin Quate and Rudi Kompfner (Stanford's Electrical Engineering Department), as well as John Pierce (Bell Labs and later Caltech). Pierce, Quate, and Kompfner all knew each other from Bell Labs; indeed, Pierce and Kompfner were very close. But Pierce was also quite fond of Chowning, who had visited Bell Labs to learn computer music techniques from Pierce and Max Mathews.[37]

Thus, Pierce—who wrote science fiction under the penname J. J. Coupling—enthusiastically supported the groovy and interdisciplinary Chowning in his bid to obtain tenure at Stanford. As Pierce wrote to the Stanford Music Department, "I hotly (rather than warmly) endorse John Chowning's abilities and attainments. Chowning is one of the very few men (and fewer musicians) who has used the computer wisely and originally in producing music. I would put him among [the top] three."[38] In contrast, the much squarer Quate was, as an associate dean, apparently one of the people who blocked Chowning's initial attempt to secure tenure.[39] Kompfner, meanwhile, had been the manager who gave permission for Billy Klüver, a Bell Labs engineer, to collaborate with artists such as Jean Tinguely, Robert Rauschenberg, and John Cage.[40] Yet Kompfner had drawn the line at Cage and Klüver's request to use Bell Labs's horn antenna to pipe in "white noise from space [rather] than just placing a noise generator" on stage. It's not hard to imagine Kompfner's exasperation with Cage and Klüver's retort (paraphrased by Douglas Kahn) that "he did not understand that a 'sssssss' is not just any 'sssssss.'"[41]

And yet! Quate himself considered patenting a device that adapted James Meindl and John Linvill's Optacon technology so that it could convert printed music directly into digital sound.[42] Kompfner, meanwhile, was the son of a pop music composer in Vienna; he himself was an amateur artist and former professional architect. Those interests don't seem to have figured much in his work at Stanford or Bell Labs, but in his papers is preserved an achingly earnest semiautobiographical essay about music, technology, and cognition that he sent to Pierce.[43] That is, unlike cultural entrepreneurs such as Pierce, the squarer Quate and Kompfner were grumpy about hip nonengineers like Chowning and Cage elbowing their way into the traditional technological jurisdiction of engineering. Nevertheless, Quate and Kompfner didn't foreclose the possibility of crossover between engineering and the humanities, medicine, and other fields. Indeed, they sometimes facilitated or even participated in collaborations aimed at realizing those crossovers. It's just that for Quate and Kompfner, as for Arthur Hall, interdisciplinarity was best achieved by the scientist and engineer contributing to other fields—and *not* by practitioners of those other fields appropriating the resources or credit that they thought belonged to science and engineering.

COLLIDING WITH REALITY

So squares understood the way in which the world was changing around them, and they responded by gravitating to a widely agreed-upon list of topics of growing public interest, yet they were decidedly ambivalent about those topics and the nontechnoscientists associated with them. Almost the only thing that squares weren't ambivalent about was the relevance of their own expertise in solving the problems of civil society. They were confident that they could help and that they therefore deserved a seat at the table.

That belief stemmed largely from the squares' earlier success in national security contexts, which convinced them that the combination of Cold War technologies and methodologies could solve *any* problem. We can see that confidence, for example, in Hall's listing of all the problems of modern agriculture that Autofarm would solve at one go:

> [Autofarm's] objective is to bring a systems approach to the farm, together with technologies that have not heretofore had major or revolutionary affects [*sic*] on the farm, including electronics, communications, and space technology. The end results are to drastically increase the productivity per acre of farm land, and to . . . reduce soil, water, noise, and air pollution created by current agricultural technologies. Finally it will reduce, or eliminate, farms as consumers of fossil energy and will eliminate the farm as a dangerous place to work.[44]

Moreover, squares like Hall believed that along with Cold War "electronics, communications, and space" technology, they could and should also adapt the *institutional* apparatus of national security R&D that had made those technologies possible in the first place: institutions such as funding and oversight mechanisms, intellectual property protections, and revolving-door career trajectories in which personnel moved easily among industry, government, and academia. Hall, for instance, was a prolific patenter and proposal writer, and he vigorously circulated his patents and proposals through the network of colleagues whom he knew (or at least was connected to) through the IEEE, AT&T, Penn, the DOD, aerospace manufacturers, and other corners of the military–industrial–academic complex.

To be sure, sometimes the squares were justified in the belief that they could adapt Cold War technologies and institutions to solve civilian problems. We saw, in particular, that microelectronics technology successfully crossed over from the military to civilian realms, for example, in the scientific instruments that Virgil Elings and his students invented in chapter 2 or the medical technologies that James Meindl, James Angell, and others developed in chapter 3. Indeed, we saw in chapter 6 that microelectronics companies that *didn't* sufficiently transition from military to civilian markets were outcompeted by ones that did.

Hall's career contained some similar successes in bringing Cold War technology, expertise, and institutions to bear on civilian problems. His consulting work in the AT&T antitrust case, for instance, began with the production of two scenarios anticipating how the Bell System would function with and without being broken up.[45] As we've seen, scenario building was a technique that originated in the nuclear wargaming world of think tanks like RAND. At AT&T, Hall had incorporated RAND's simulation and game theory techniques into his groundbreaking systems engineering methodology. That combination of scenario building and systems engineering was apparently then the selling point of Hall's consulting firm, ADH: scenarios backed by systems expertise were what ADH offered to clients such as the City of New York, the National Basketball Association, and the NSF.[46]

But there were also many cases where square technoscientists' optimism was unfounded.[47] Hall's Autofarm system exemplifies that overreach spectacularly, so it's worth turning our attention to it briefly before coming back to the larger context in which it was embedded. As figure 7.1 shows, Autofarm consisted of a network of sensors monitoring "all important parameters in the homogeneous agricultural production area" (soil moisture content, presence of weeds or pests, weather, readiness for harvesting, etc.); an automated decision tree for responding to sensor data; and a physical apparatus that could be directed by that automated agent to go into the field or orchard and apply fluids (water, pesticides, herbicides, etc.), harvest and sort crops, prune trees, and so forth. The sensors, the automation, and the use of solar panels and fuel cells for powering it all—all these, as Hall acknowledged, drew on technologies and expertise developed for or by the space program and the military. Hall also borrowed technologies—such as Grumman's Sailwing Wind Generator and General Electric's Geniponics system—from military–industrial companies that were similarly adapting aerospace products for civilian markets in the 1970s.[48] Much like the NASA engineers in chapter 4, then, Hall was confident that space-age technology originally developed for a narrow purpose could be applied to complex and open-ended domains on earth.

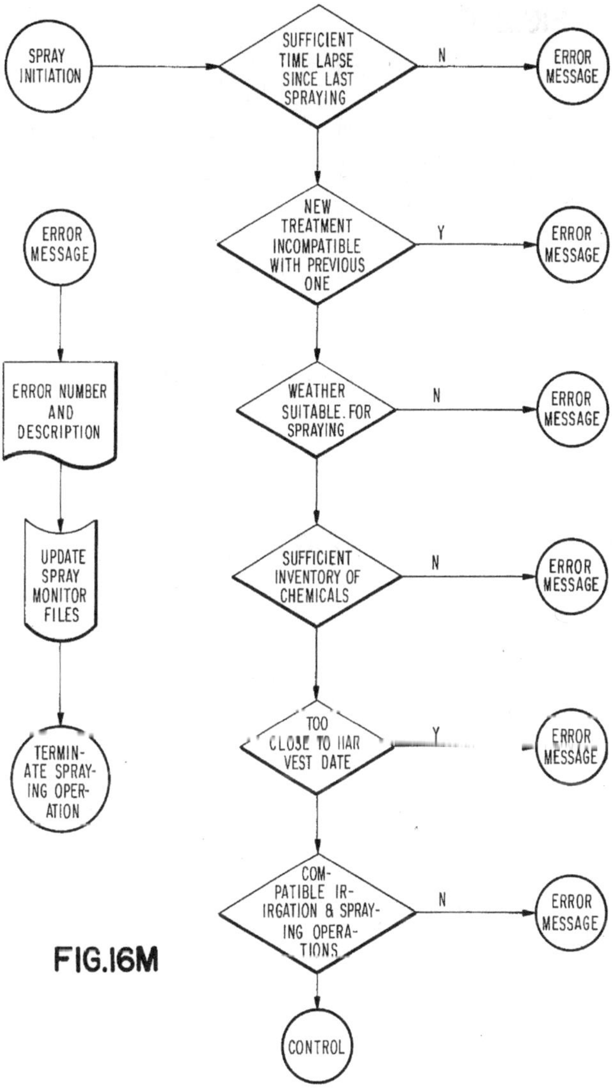

Figure 7.1

Three of the figures from Arthur D. Hall's Autofarm patent, Highly Automated Agricultural Production System, US Patent #4,015,366, issued April 5, 1977. The figure on the *left* shows a decision tree governing spraying of herbicides, pesticides, and so forth. The figures on the *right* show the Autofarm machinery configured to prune (*top*) and harvest (*bottom*) fruit trees.

FIG.29C

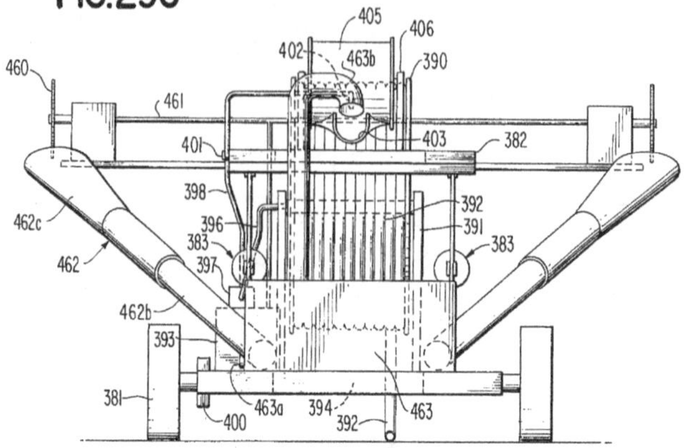

FIG.30A

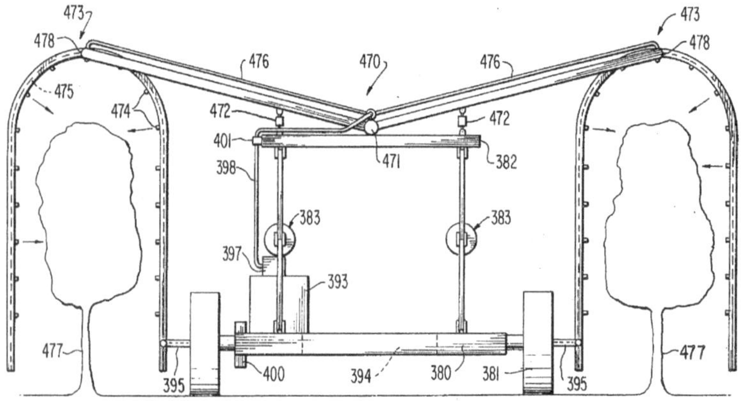

Figure 7.1 (continued)

Remember, though, that the engineers in chapter 4 abandoned their attempts to apply space-age expertise to terrestrial problems after just a few years. Likewise, GE abandoned Geniponics and Grumman the Sailwing Wind Generator; and most other aerospace companies that ventured into mass transit, alternative energy, public housing, and environmental remediation in this period did the same. Hall, however, never did abandon Autofarm—he continued trying to interest investors until the 1990s. The difference between him and NASA's, GE's, or Grumman's engineers, I think, was that Hall was a free agent—he was self-employed and therefore did not have a manager trying to redirect his efforts, and he could brush off external critics who thought that his ideas were unworkable. Hall's vision for Autofarm therefore shows us what the square technoscientific imagination could run to in the absence of constraints.

As it turns out, Hall's imagination could run very far indeed. At a minimum, he foresaw Autofarm replacing humans for a variety of agricultural tasks: monitoring crops and environmental conditions, treating crops in various ways, harvesting them, and preparing them for the market. Even today, only a few of these tasks have been automated to nearly the degree that Hall thought was already possible in the 1970s. But more than that, Hall believed that the components of Autofarm could be generalized to solve a wide variety of problems outside, or at least not normally associated with, agriculture. In a retrospective account he offered a list of those problem areas (here quoted in part):

AUTOFARM Spin-offs and other uses for subsystems. . . .

B. Automatic remote spectrometry, for water, soil, and air—pollution sensing, battlefield surveillance for poison gases, mineshaft sensing, etc.
C. Audience response systems in theaters, schools, prisons.
D. Water filtration and purification by automated biomass production. . . .
G. Aquaculture: joint fish and biomass production for food and energy
H. Automated gasifier using wood/coal/corn stover for [propulsion of] trucks, ships, and autos.
I. Automated silviculture for renewable fuel production.
J. Total environmental control for space stations, phytotrons, and greenhouses for food production, air and waste recycling.[49]

Hall never achieved *any* of these applications, at least not on the scale that he had imagined. Some of them were possibly within reach at the time, but others were (and are still) quite speculative. His confidence that military–industrial technologies of automation, systems engineering, power, and sensing could provide medium-term solutions to every civilian problem was, thus, profoundly misplaced.

By stepping through some of the items on Hall's list we can start to grasp the immense scope of his vision. Of course, for a few of these items, time has shown that scope to be justified. With item C, for instance, Hall had in mind a large distributed cybernetic feedback network whereby audiences (say, theatergoers at a play or students in a classroom) could shape the performances that they were viewing, or whereby system controllers (say, prison guards or teachers) could monitor the collective murmurings and nonverbal reactions of individual agents. Hall had proposed a theater-based experimental setup along these lines as early as 1968; that version was intended to help psychology and communications researchers (perhaps with applications such as advertising and commercial film and television in mind) to subject an audience to various stimuli and correlate those stimuli with the intensity of responses such as anger and joy.[50] Versions of this idea are today fairly common, for both liberatory and oppressive applications—though Hall seems not to have made such a distinction.

Most of the other applications on this list, however, are unrealistic even today. Items H and I, for instance, envisioned automated cultivation and harvesting of vast forests of monocultured hybrid poplars to be used as a replacement for fossil fuels. Indeed, in at least one proposal Hall painted a picture of autarkic farm families growing vegetable crops in greenhouses and poplars in timber lots, converting the poplars to biogas to power their living quarters and greenhouses, and skimming excess carbon dioxide from the gasification process to stimulate growth of the greenhouse crops.[51] He also foresaw growing potatoes in the greenhouses, and then converting their starch to ethanol to power the farm family's vehicles!

Yet even that highly speculative vision of autarky was far from the outer edge of Hall's imagination. As item J in his list indicates, he also thought that Autofarm could be used in fully closed systems—particularly on board permanently inhabited space stations—to endlessly recycle human waste back into food again. David Munns and Kärin Nickelsen have shown that this was a longtime (but unsuccessful) locus of military-funded square scientists' efforts in the early 1960s—efforts that never led to implementation in any actual space habitation.[52] The most extensive attempt at this kind of high-tech total autarky has been the earthbound Biosphere 2; and the failure of that habitat's closed-loop system resulted in starving, oxygen-deprived "bionauts" descending into paranoid anarchy.[53]

Finally, item D on Hall's list particularly exemplifies the strange politics of the squares' optimism that Cold War technologies could be applied to the problems of civil society. For a more detailed description of item D, let me turn to a letter that Hall wrote in 1981 to the chair of Maryland's Hazardous Waste Facilities Siting Board:

> Now, all these pleasant pursuits [on Hall's farm] have been disturbed by rumblings that this County may become the depository of toxic waste which nobody seems to want and everyone fears. My first inclination to the citizens was sympathy. My more considered response is that this County should actively promote its becoming the central depository for as wide a collecting area as is economically feasible, providing that it can assimilate and employ certain new technologies which I believe can in effect convert an unmitigated disaster into a blessing. In the past three years I . . . became a student of some brilliant research done almost exclusively by NASA, which established that certain subtropical water plants have the properties of being able to concentrate and store absolutely astounding quantities of heavy metals, phenols, and other toxic chemicals, while producing equally astounding quantities of cellulose biomass, its energy in turn being recoverable. . . . These studies led to my conviction that the prospective location of hazardous waste dumps in our County need not be a disaster as nearly all of our citizens are portraying it to be, but in fact can be an enormous industrial boon to economic development.[54]

That is, despite Hall's fairly conservative politics, he didn't follow the denialist path taken by other conservative scientists. Hall acknowledged, here and in many of his other writings, that pollution and global warming were real problems that citizens should be concerned about.[55] It's just that he thought Autofarm and other technologies inspired by the Cold War could solve those problems—indeed, could solve them so completely that citizens should *welcome* toxic waste dumps in their backyards![56]

MUTUAL AMBIVALENCE SOCIETY

To observers looking back from the twenty-first century, Hall's stance seem highly ambivalent if not contradictory. Hall's politics were not the clear-cut politics of left or right, but rather the ambiguous politics of the cybernetic system—with himself, or some other square scientist or engineer, imagined as the rightful system designer and controller. He was deeply opposed both to "authoritarian powers [such] as business corporations" and to a government and political system that (apart from the "folk heroes"—like himself—who broke up AT&T) was unwilling to curb the power of big business.[57] Yet at the same time, Hall foresaw the need for some system-level entity to "manage the environment, restore it, and live within its limits," if necessary by enforced "birth control and planned parenthood as one way to limit consumption of earth's resources. In work on this problem, systems engineers and methodologists can be the most valuable members of the team."[58]

Hall's views complicate the conventional American political spectrum—both now and in his own day. His politics were profoundly ambivalent about government, business, democracy, environment, and so on: he was promarket but antibusiness; progovernment but antibureaucracy; in favor of "freedom" but also in favor of coercive population control. That ambivalent mix of views is part of why Hall merits a chapter of his own in this study. Throughout this book, I've taken political ambivalence and/or reticence as *the* defining characteristics of the squares. Indeed, I hope that studying the squares can lead to a more nuanced understanding of ambivalence and the ambivalent in general. Most histories of science—maybe even most histories full stop—revolve around figures with pronounced views; we therefore don't yet have a good vocabulary for talking about those whose views were indistinct or muddled, even though such people often form a very large majority in any given setting.

To give a small example of what such a vocabulary might include, I would note that Hall's views offer an opportunity to distinguish between two different forms of ambivalence: ambivalence in the moment versus retrospective ambivalence. The former applies to squares who were ambivalent about the politics *of the time*—Harvey Brooks's reluctance to be associated with the antiwar movement or Bill Rambo's waffling about the value of interdisciplinarity are examples. The latter, meanwhile, applies to squares who appear ambivalent only in hindsight, simply because the apparent contradictions between their various views were only established later. For instance, Arthur Hall and Jack Kilby were enthusiastic supporters of alternative energy but almost certainly also backed Ronald Reagan's election as president in 1980—a position that would not necessarily have appeared as ambivalent until a few years later, once Reagan and the movement that he led had more clearly repudiated alternative energy. Both forms of ambivalence are important in understanding American science and engineering in the 1970s. The ubiquity of ambivalence in the moment shows us the lasting legacy of attempts to bind American science and engineering to an ideology of "apolitical" quietism; whereas retrospectively ambivalent figures such as Hall and Kilby show us how confusing the political landscape of the 1970s was to actors at the time (and is to its chroniclers now).

Both these forms of ambivalence characterize square scientists' and engineers' views of the world around them. But if we reverse our perspective and ask how the world saw the squares, then we find an almost equal measure of ambivalence. On the one hand, American technoscientists had many admirers; I've shown in my other case studies, particularly in chapter 4, that many organizations and social movements were eager to enroll square physical and engineering scientists in their projects. On the other hand, we've also seen that that cooptation often failed because the squares wouldn't or couldn't offer what their partners wanted. More generally, as we saw most clearly in chapter 3, Americans saw science and engineering as both the cause of and solution to many of society's problems. I've also occasionally noted instances of Americans' concerns that scientists and engineers were too narrowly specialized or too confident in their own abilities, and therefore that they could cause further problems if their technoscientific hubris were left unchecked.

The reactions of various observers to Hall's Autofarm system display the full range of these ambivalent attitudes. There were, indeed, a few who lauded it, at least on initial inspection. One US Department of Agriculture (USDA) official, for instance, called it an "exciting perspective of the agriculture we will see in the not too distant future."[59] A USDA economist, Leroy Quance, positively reviewed Hall's manuscript of a book about Autofarm for John Wiley & Sons, saying

> there are new unprecedented technologies on the horizon such as photosynthesis enhancement and bioregulators in crop production. Our analysis indicates that these new technologies will not have much of an impact before the year 2000. . . . But the computerization of agricultural production as exemplified by *AUTOFARM is the sleeper*. This technology already exists and only has to be put to use to . . . usher in, at a much earlier date than expected, such . . . technologies as photosynthesis enhancement and bioregulators in crop production.[60]

Hall seems to have been especially grateful for Quance's appraisal of Autofarm as a "sleeper," since he invoked it frequently in letters seeking investors for Autofarm.

Yet the response to those dozens of letters to companies and government agencies was almost uniformly negative. Most of the organizations that Hall approached simply replied that they didn't have the money to pursue his ideas or that they weren't in the business of automated agriculture—hardly surprising because, as we've seen in other chapters, the 1970s were a period of strained budgets and numerous failed crossovers from one industry to another. Some of these organizations, though, explained their rejection of Hall's requests with reference to negative evaluations from in-house experts in the fields that would've been overtaken by Autofarm. Almost none of those experts were impressed. For instance, FMC Corporation, a chemical and agricultural technologies company, cited this assessment in declining to invest in Autofarm:

> The material furnished me regarding Mr. Hall's ideas in regard to his "automatic orchard" was very interesting, but I am somewhat concerned as to whether or not such far-reaching changes can be implemented in our lifetime. . . . I believe that Mr. Hall's greatest contribution will be that he has certainly let himself think boldly and without limitation, and as a result we must admit that there is some merit to some of his suggestions, although I cannot believe that the implementation of any great number of these items would be economically feasible in the foreseeable future. Although I am intrigued with [Hall's] overall proposal, I must return to the cold, hard facts of life and advise you that I can see no practical manner in which we can implement his suggestions into marketable products. Many of his suggestions have already been tried and have not met with market acceptance.[61]

As we've seen over and over in this book, square physical and engineering scientists' optimism crashed on the reality that "problems are much more complicated than meets the eye." Often, other people had studied those problems in depth before the squares got there, had better claims on those problems than the squares did, and therefore rejected the squares' interventions.

Some squares took complication and rejection well, but many became frustrated, leading to anger and/or disaffection. Some were irritated from the start but accommodated nonetheless—we saw that in the contrast between William Rambo's public and private comments on interdisciplinarity in chapter 3. Others gave reform a chance but came to believe that it had been a mismanaged debacle. Again from chapter 3, Calvin Quate's praise for the NSF's cancellation of its RANN program and his retreat from attempts at collaboration with biomedical researchers show where such frustrations could lead. Others didn't become angry so much as they lost interest in socially relevant projects, or they gave up such projects once they no longer thought that there was any pressure or incentive to do them. From chapter 4, the disappearance of NASA projects that were oriented to terrestrial social problems largely falls into this category.

With Hall, frustration was often tinged with anger. For instance, on hearing that his proposal to the RANN program had been denied because of budget cuts, Hall expressed his irritation to NSF thus:

> I am particularly disappointed at this outcome since last year the preliminary proposal . . . and this year the formal proposal [were] solicited. . . . and since the preparation and the printing of this proposal has been very expensive. . . . [NSF] could have told me frankly at the outset that there was no money for the proposal even if it should prove directly responsive to . . . national goals. . . . I must say that I do not think NSF has given proper consideration to this proposal of its merit by people who are professionally qualified in the field of agriculture. So that it is a matter of record that some individuals who are professionally qualified in the field of agriculture have seen the proposal and think it is sound, attached herewith [are letters of support from the Agricultural Research Service].[62]

Elsewhere, Hall showed a certain self-awareness that his irritation might be counterproductive. In his answer to a rejection letter from a subsidiary of the Atlantic Richfield oil company, for instance, he acknowledged that he was responding partly "to unload some of my frustrations" and to "show I'm at least not some overalled farmer nut, but instead am a practicing engineer-scientist with a few practical results to his credit."[63]

Rejection is never nice, of course; but the particularities of Hall's and other squares' responses to rejection tell us something important about how the American science and engineering community changed during, and as a result of, the long 1970s. On the one hand, people like Hall, Quate, and Rambo did have *something* to legitimately complain about. The reforms of the long 1970s were often thrown together hastily, and cut back or cancelled with just as little warning. The Nixon administration's short-lived New Technology Opportunities (NTO) program and the NSF's somewhat more enduring but extremely fickle RANN effort both illustrate the disorienting effect of rolling out new, experimental initiatives and then unexpectedly rolling them back up again.

On the other hand, some squares' complaints reflected an unjustified attitude of entitlement. For this group, American scientists' and engineers' contributions in World War II and the Cold War had made R&D funding a right—a right that the reforms of the 1970s took away. Not all squares adopted this view, of course. Some took the reforms of the 1970s in stride: they recognized that the existential need for scientists and engineers to help America win the Cold War had passed, so resources for R&D would necessarily be thinner in the future. For this latter, more fatalist, group, either you could accept that you had to do more with less or you could reinvent yourself in a new line of work; one hallmark of the era was the sizable number of people who left science and engineering, either temporarily or permanently. But there were plenty who, like Hall, took a less magnanimous position: that the country owed scientists and engineers *more*, not less.

Thus, Hall's response to rejection at home was to inflate his ambitions even further; and when he couldn't find anyone at home to support those inflated ambitions, then he started to look abroad for patrons who would. If the US government and American firms didn't want to fund him because their experts didn't think that he could squeeze a little extra productivity out of wheat fields and apple orchards, well, then the governments of Canada, Egypt, Israel, Kuwait, and Saudi Arabia would beg him to turn tundra and desert into productive farmland![64] And if no one wanted Autofarm for growing wheat or apples, well, then he would find someone who needed it for disposing of toxic waste and replacing coal and oil in the energy supply!

That kind of inflation often succeeded in the early Cold War: if no one wanted your nuclear airplane, well, offer them a nuclear spaceship instead; if no one wanted your analog flight simulator, well, convince them that they really needed a general-purpose real-time digital computer; if no one wanted your ground-based ABM system, spin a vision of space-based missile defense replete with X-ray lasers and "brilliant pebbles."[65] But in the long 1970s that kind of Cold War gigantism hit its limits in North America and Western Europe.[66] In Saudi Arabia, though, one could still talk of towing icebergs to the Middle East and turning the desert into a breadbasket—which is why people like Hall and Kilby sought Saudi help when rejected at home.[67]

Admittedly, Cold War gigantism did make a partial comeback in the United States with the election of Ronald Reagan in 1980. Like many voters, therefore, Hall believed that Reagan would restore his and his peers' supposedly forfeited rights and allow the nation to get beyond the failed experiments of the 1970s. As he wrote in late 1980 to a conservative Congressman close to the Reagan transition team,

> Because of my conviction that our government needs all the competence it can get, and because I know that, at least in the areas of my expertise, it has suffered from incompetence, it has seemed to me that I should offer my talents to the new administration.[68]

In an attachment, Hall noted his political ambivalence and reticence: "I have not been a political person, although I have generally voted and provided financial support for Republicans."[69] But now, "concerned about the future of our country," he asked that Reagan's transition team consider him for eighteen different jobs, including Director of ARPA and Secretary of the Army!

Needless to say, despite his optimism about Reagan's election, Hall was not appointed Secretary of the Army—or to any other position in the Reagan administration. We saw something similar in chapter 5, where Jack Kilby was hopeful that Reagan's people would cleanse the DOE of its hippies and incompetents, leaving the field clear for the government to support TI's solar technology—but in the end the new administration just abandoned solar power entirely. Likewise, Signetics opened a new plant for military products in 1981, in anticipation of Reagan's defense buildup—only to find that military contracts weren't as lucrative as they had been in the 1960s. In other words, for some squares who had prospered in the early Cold War and who had become frustrated with the experiments of the 1970s, Reagan's gauzy nostalgia and promises to boost defense spending must have seemed like a lifeline—yet in the end the new administration disappointed them by instituting a new neoliberal order rather than a return to the early Cold War. For both the squares who voted for him and the ones who didn't, Reagan therefore offered yet another reason for frustration and ambivalence.

RRI'S ANTECEDENTS IN THE LONG 1970S

With Reagan's election we come to the end of the long 1970s. We now have everything we need to survey square scientists' and engineers' trajectories through that decade; and from there, we can start to relate the 1970s to today. In particular, we can use the 1970s to evaluate current efforts to convince square scientists and engineers to adopt a more "responsible" form of research and innovation.

If we step back and look at the careers of the squares who've appeared in this book, we see that "responsible" research and innovation—in at least a broad sense—flickered in *and out* of square technoscientists' lives. Most of the people I've profiled in this book traveled a roughly similar arc with respect to social responsibility: they started the 1960s in (and many maintained long-term loyalty to) the military–industrial complex; in the early 1970s they made forays into interdisciplinary applied research directed to civilian social problems; and by the 1980s most of them had become frustrated and lost interest in—or were channeled away from—those problems. Surprisingly, perhaps, with the end of the Cold War, many squares returned to the kinds of projects that they had pursued in the 1970s. We can see that arc in the rise, fall, and return of interest in biomedical applications among Stanford's electrical engineers. We can see it in Virgil Elings's drift away from AEC funding and toward environmental and biomedical research in the early 1970s; his turn toward entrepreneurship and selling tech to the semiconductor industry in the 1980s; and his final transformation into a philanthropist (focusing in particular on educational innovation) in the 2000s. And we can see it in NASA's meandering path from a tool of Cold War prestige to a pawn of détente, back to an auxiliary of the national security state under Reagan, and once again a facilitator of peace and post–Cold War cooperation in the 1990s.

In other words, square technoscientists' relationship to "societal relevance" and "responsible research and innovation" has been highly nonlinear. Every era in which the talk and practice of "social responsibility" has gained favor has given way to an era in which it has been rejected—and vice versa. This is perhaps not so surprising, but it does run counter to the discourse of the present-day field of responsible research and innovation. Practitioners of RRI seek to make technoscience *more* responsible; some even propose that by exposing scientists to RRI principles early and often enough they could cause a permanent shift toward more responsible behavior.[70] That possibility should of course be explored. But the possibility that scientists and engineers might choose to become *less* responsible is hardly ever mentioned in RRI studies. Yet scientists and engineers do regularly make that choice, and RRI practitioners therefore need to understand what makes scientists and engineers (and their organizations) more *and less* responsible. What are the conditions of possibility of, and the incentives for, adopting, ignoring, *and discarding* the kinds of principles and practices advocated within RRI?

One obstacle to answering this question is RRI's impoverished sense of its own history—RRI practitioners simply aren't aware of the several decades of moves toward *and away from* something like responsible R&D. Admittedly, many programmatic overviews of RRI do contain a brief potted history of the field, often starting with the postwar "social contract for science" and then hopping from the antinuclear movement to the Human Genome Project, mad cow disease to nanotechnology.[71] And those potted histories are correct as far as they go—but they don't fully acknowledge that previous attempts at something like RRI have failed, or at least have given way to periods that were much less encouraging of responsible research and innovation. These potted histories therefore avoid the inconvenient fact that the discourse of responsibility is only intermittently persuasive to scientists and engineers, and therefore that discourses of *irresponsibility* are persuasive the rest of the time.

Folk histories of RRI also often put forward a skewed vision of what the field's predecessors were actually up to and the context in which those forerunners operated. Take, for instance, a recent debate over whether RRI represents an extension, a critique, or a "travesty" of the field of technology assessment.[72] The answer to that question hinges in part on whether technology assessment was or was not exemplified by the US Congress's Office of Technology Assessment (OTA), which supposedly "did not include an emphasis on public or stakeholder participation."[73] Yet that characterization of the OTA is only partially accurate. True, as the OTA institutionalized over time, it did indeed become more technocratic and less public facing.[74] But when the former Congressman Emilio

Daddario founded the OTA in 1972, he imbued it with an ideal of public participation—that is, at its founding the OTA was decidedly *not* a place where "it was generally assumed that experts' knowledge dedicated to the study of intended and unintended impacts of new technologies would be sufficient to provide political representatives with useful tools to govern science and technology."[75]

Moreover, at the time the OTA interacted with a number of other institutional experiments—such as the RANN program and the NTO initiative—that were similarly dedicated to supporting more responsive R&D. RANN's links to the OTA, in particular, were quite close: while he was a Congressman, Daddario had offered the amendment that led to the formation of RANN's predecessor, the IRRPOS program.[76] These connections were well understood at the time; in a speech in early 1972, for instance, Arthur Hall lumped NTO, OTA, and RANN together as common symptoms of "a great concern, that has been building over the past few years, for the need to forecast and to assess the threats and opportunities of new technologies, and to sense and measure emerging societal needs before they reach crisis proportions."[77]

Yet the variety of institutional manifestations of that concern are not widely known today. Experiments like RANN and NTO were more short lived than the OTA and therefore have mostly disappeared from RRI's historical memory. Nor have historians of science and technology done much to keep the memory of RANN and NTO fresh. Thus, RRI practitioners are not very cognizant of their field's *failed* forerunners and are even less aware of the reasons for their failure. Few contributions to the RRI literature, for instance, draw lessons from the fact that scientists and engineers such as Calvin Quate came to loathe RANN and rejoiced at its demise.

The OTA, RANN, NTO, and other programs—and the field of technology assessment more generally—also interacted at the time with a number of other scientific and intellectual movements that have generally been excluded from RRI's genealogy despite their relevance to any discussion of responsible technoscience.[78] As Jessica Smith has shown, for instance, the Corporate Social Responsibility (CSR) movement was forged by mining engineers who absorbed positive lessons from their engagements with the public in the 1970s.[79] Likewise, the appropriate technology movement has been written out of the history of RRI, even though AT directly influenced national laws (in the United States and elsewhere) and supranational policies (of the United Nations as well as nongovernmental organizations) requiring technoscientific projects to account for their social and environmental impacts in a manner not dissimilar to RRI today.[80] Appropriate Technology was also widely adopted as a label for funding streams that supported the same kinds of projects that RANN and NTO fostered at the time and that RRI might encourage today. For instance, the DOE had an Appropriate Technology Small Grants Program that Arthur D. Hall applied to in 1981 to support development of Autofarm.[81]

Early advocates of TA such as Brooks and Daddario frequently appeared on the AT conference circuit and adopted some AT ideas. So if TA can be taken as an antecedent of RRI, then AT surely should as well—because early on there was no clear line between the two movements. And as Danielle Shanley argues, if we look at the shared ideals of *early* TA and AT, then we can see that both movements *did* "include an emphasis on public or stakeholder participation" that some RRI practitioners believe is original to their own program.[82] Appropriate technology in particular had a widespread grassroots appeal—indeed, it attained a degree of grassroots participation that RRI advocates today can only dream of. Appropriate technology wasn't just open to the public—it was *of* the public.

To take a parochial example, my hometown of Lawrence, Kansas, was also home to an Appropriate Technology Collective that among other things pressured the city and county government to adopt recycling and energy conservation measures, and joined other groups in opposing construction of the nearby Wolf Creek nuclear power plant.[83] As their newsletters and meeting notes show, the Douglas County AT collective also made common cause with Indigenous activists, church groups, astrologers, small farmers experimenting with alternative energy technologies—you name it. In putting into practice an ideal that it is "important to involve key groups in community and make it a democratic process," AT set a standard for public participation that RRI has yet to catch up to.[84]

Unfortunately, groups like the Douglas County AT collective had little success in bringing square scientists and engineers into their democratic process. Such groups formed part of the environment in which the squares in this book operated, but none of the archival collections that I have examined contain significant examples of practicing scientists and engineers who collaborated with grassroots AT organizations. No doubt there are examples out there, but they can't have been common. What I have run across, however, are examples of square scientists and engineers who rejected collaboration with such organizations and who were disdainful of open and idealistic forms of innovation. Squares were willing to adapt to the times and converse with other square scientists and engineers of (moderately) opposing views; but they demanded deference to their technical expertise and were often suspicious of potential collaborators who did not share the squares' backgrounds in Cold War modes of R&D.

Part of the tragedy of chapter 5, in particular, is that TI and Jack Kilby declined to seek allies among, say, the readers of the *Small Farm Energy* newsletter (an appropriate technology-oriented monthly to which the Douglas County AT collective subscribed)—preferring, disastrously, to look to the national security state and wealthy homeowners instead. Of course, even if Kilby and TI had wanted to engage with *Small Farm Energy's* readership, it's hard to see how that newsletter's definition of "appropriate" technology would have afforded engagement with a giant, profit-seeking military–industrial company: "Appropriate alternative energy is characterized by the Energy Project as being home built, easy to manage and maintain, built with recycled and locally available materials, low-cost, and cost-effective."[85] The middle, as throughout this book, was largely excluded.

Not entirely excluded, though. Arthur D. Hall, for one, was willing to reach out to people who had similar goals but different approaches. For instance, in 1981 Hall wrote to Robert Rodale, an influential publisher of books and magazines related to health, wellness, and organic farming, to highlight both the differences between their visions and the potential common ground on which they could agree:

> I fully agree about the need for another agricultural revolution comparable to that by Jethro Tull. Because of our different backgrounds, we differ on the choice of specific technologies, government interventions, and other means of bringing it about. For instance, I think we cannot do without modern chemical technologies for pesticides, fertilizers, growth enhancers, etc. However, by specific sensing of plant needs, I think we can use only a fraction of the chemicals we do now. . . . Doing this calls for using information and computer science which heretofore have been used almost not at all in agriculture. . . . [R]eal (peaceful) revolutions are generally not wanted, even if they are impecably [*sic*] sound technically and economically. The . . . USDA rejected my revolution on the mistaken and foolish grounds that it would save labor, and they were enforcing a policy of not encouraging any labor-saving technology.[86]

Yet even if Hall could recognize the common ground between himself and Rodale, he couldn't imagine building anything there. Instead, Hall's natural allies were the same as Kilby's. Like Kilby, Hall was a product of the national security state and of big business, and it was among those institutions that both of them felt comfortable even after they became disillusioned and left big business to become their own bosses. Hall therefore sought support for Autofarm from national security pillars like the DOE, the Army, and the Battelle Memorial Institute (one of the main contract operators of the DOE's National Labs), as well as corporations connected to the national security state such as IBM and Control Data Corporation. He also approached food conglomerates like Campbell Soup Company and Green Giant.[87] And, like Kilby, he tried to interest the oil industry, which was then using its large cash reserves to diversify into seeds, animal feed, and other agricultural markets. Yet Hall didn't, as far as I can tell, seek support from grassroots groups like the Douglas County AT collective.

PRINCIPLED DISAGREEMENT

So RRI owes some kind of historical debt to scientific, intellectual, and social movements of the 1970s such as technology assessment, appropriate technology, and CSR. And practitioners of RRI owe it to themselves to learn more about those predecessor movements' failed rapprochement with square technoscientists like Hall and Kilby—if only to better understand the scale of the challenge today. RRI practitioners could also learn a great deal from the institutional enactments of those predecessor movements, such as the OTA, RANN, and NTO—if only to understand that most such experiments burn out quickly with few enduring achievements. The setbacks of the present might be avoided, and will certainly be easier to endure, if we better acknowledge the setbacks of the past.

Of course, even if RRI owes something to TA, AT, and CSR, well, times have moved on. We now live in a neoliberal world where experiments like OTA and RANN would be unlikely to get off the ground. The technoscientific reformers of the long 1970s had somewhat different aims, and operated in a much different environment, than RRI's promotors do now. As noted in chapter 1, the phrase "responsible innovation" was temporarily in vogue in the 1970s; but it didn't mean exactly what is meant by that phrase today. There is no universally accepted definition of the current version of responsible innovation, but there is *some* agreement that it rests on four principles: inclusivity (more than just scientists and engineers must be involved); anticipation (some thought must be given to future consequences and circumstances); reflexivity (participants should reflect on how their background and positioning influences their views); and responsiveness (RRI activities should lead to actions that respond to the concerns raised in the course of deliberations).[88] Those four principles are a twenty-first century invention; they simply weren't how anyone explicitly defined socially responsible science in the 1970s.

And yet, *versions* of all four principles can be glimpsed in the cases presented in this book. The long 1970s was a festive time of experimentation in how to organize science and engineering; and many of those experiments in some way included practices of inclusivity, responsiveness, anticipation, or reflexivity. That said, the episodes in this book show that none of RRI's four principles straightforwardly gave rise to a more responsible mode of technoscience. Indeed, on some occasions those principles facilitated *less* responsible technoscience—that is, a technoscience that was less humane, less democratic, or less sustainable.

That point is perhaps easiest to see with the principle of responsiveness. American technoscientists of the long 1970s had little choice but to be responsive to any number of voices: student protestors, the Nixon administration, Congress, local communities, activists representing various social movements, and others. Some squares responded more quickly than others, and some were happier to do so than others; but they all responded in some way to shifting budgetary priorities, campus takeovers, long lines at the gasoline pump, and so on. As we've seen in most of the chapters of this book, every time a new national "crisis" (urban, environmental, energy, economic) came along, square scientists and engineers responded.

But responding didn't make them *responsible* in any meaningful way. Their responses were problematic in a number of ways and were for the most part rejected. Critics of the era's technoscientists—such as the STS scholar Langdon Winner—argued that the squares' responses were too top-down, too high modernist, too confident.[89] Perhaps the technoscientists whom I've profiled would object that they were never given the latitude to adequately respond before the priorities that the country set for them were once again upended: by the time they had solutions to the urban crisis, they were being told to respond to the environmental crisis instead; by the time they had solutions to the environmental crisis, they were forced to respond to the energy crisis instead. When scientists and engineers did persevere until they could offer a mature response to an earlier crisis, the country had moved on and didn't want what they offered. At least that's what Jack Kilby thought about the rejection of his solar energy system and Arthur Hall believed about the rejection of Autofarm. Constant responsiveness didn't lead to a technoscience that anyone believed actually addressed society's problems.

Similarly, the ambivalence of RRI's "anticipatory" criterion should be apparent from this book's cases. The tools of anticipation—scenario planning, technological forecasting, and other methods such as those that Hall described as "time series-extrapolation for projecting trends, morphological analysis for discovering discontinuities in the future, the delphi technique for organizing the opinions of experts"—originated in the world of nuclear wargaming before moving into the oil and aerospace industries.[90] As Hall noted in 1972, "There is an emerging methodology called *technological assessment and forecasting*. This methodology has its roots and many of its techniques and attitudes in older disciplines called systems engineering and operations research."[91] By the time that these techniques were applied to socially responsible research and innovation, they had already absorbed the imprint of the national security state and big business.

Thus, agencies like NASA—and its vendors such as Boeing—had extensive experience in crafting scenarios and were able to deploy them for socially relevant projects such as designing integrated utilities for public housing. But they also deployed exactly the same anticipatory techniques in aid of environment- and economy-destroying boondoggles like the solar power satellite. Indeed, NASA and Boeing used scenarios to argue that the SPS was preferable to alternatives such as energy conservation and terrestrial solar or wind power.[92] Similarly, companies like TI used scenarios as a guide to action—but their scenarios only guided them toward collaboration with the national security state, Saudi princes, and oil companies. Hall's scenarios led him to predict that the break-up of AT&T would "be yeast for the boom in telecommunications that AT&T has retarded for so long . . . [and] greater freedom" for users, shareholders, managers, and workers.[93] Maybe so. But Hall certainly didn't foresee that the companies that succeeded AT&T—such as Google and Facebook—would give rise to an anomic panopticon of all networked to all.

If we turn to the RRI principle of reflexivity, we again see the same ambivalence. Now, I have to admit that some of the clearest examples of reflexive practice that I've come across were associated with the actors in this study with whom I have the most sympathy: for example, Rudi Kompfner's heartfelt thoughts on music or David Phillips's reflections on carving a middle path for parapsychology research between skepticism and credulity. But reflexivity alone wasn't enough: self-awareness didn't necessarily lead to a more responsible technoscience. Phillips, for instance, imbued his teaching in the Master of Scientific Instrumentation program with incredible idealism; but by the 1980s he was working on projects funded by the Army and trying to use parapsychology to anticipate the movements of the stock market. Similarly, Hall sometimes wrote quite sympathetically about his circumstances and the origins of his views. More often, though, he used self-reflection merely to (sometimes literally) litigate grievances: over credit for various inventions, or the rejection of Picturephone and Autofarm, or how AT&T had supposedly treated him.

Some of the choicest examples of reflexive technoscience that I've come across were contained in the era's new semiunderground publications, such as *Grindstone: A Forum for Controversial Issues of Special Interest to the [Stanford] Engineering Community*, a newsletter to which Stanton Glantz was a prominent and very earnest contributor. Glantz also used the Stanford Workshop on Political and Social Issues (SWOPSI), which he ran in 1971, to interview faculty members and put them in a reflective mood. Take, for instance, this interview quote—probably from Glantz's adviser Holt Ashley, a Stanford aeronautical engineer:

> Q. Why do you want to move into the areas that you don't seem to be able to move into?
>
> A. There are several reasons. First, I think it is possible to get stale by devoting too much of one's "creative life" to a relatively narrow specialty as I sometimes feel I have been doing. Second, I think there is unquestionably a fairly direct relationship between a great deal of my students' and my research and military technology applications. Up until 5 or 6 years ago that fact didn't bother me very much and I was satisfied that the problem seemed to me to have an inherent interest. Today I'm certainly not as happy as I was then about many aspects of U.S. Foreign Policy and the domination of the military in the national scene, so it follows that I would like to find some new things for myself and my students to do that don't seem so closely coupled with that one portion of our national activity. I would like to find research activity which seems to me to have a really constructive social potential.[94]

This kind of reflection could indeed undergird a more responsible approach to R&D. But Glantz was an anomalously sincere and energetic actor (as was Ashley), and SWOPSI and *Grindstone* were exceptional institutions—we can't assume that they will have counterparts today. Perhaps one lesson of the 1970s is that RRI practitioners today should put more energy into building up institutions such as SWOPSI and *Grindstone* and treating them as critical infrastructures for sustaining reflexivity. A great deal of effort has gone into constructing infrastructures for the other principles of RRI—particularly inclusivity and anticipation—but less effort has gone into creating infrastructures for reflection.

Indeed, at least with respect to the principle of inclusivity, the infrastructures that RRI practitioners rely on today—science museums, consensus conferences, citizens' juries, public "open days" for academic laboratories—are in fact products of the long 1970s. Take science museums, for instance: such institutions do, of course, have quite a long pedigree, but their earlier incarnations tended to be quite hierarchical and paternalistic. As Rebecca Onion has shown, the long 1970s ushered in a new kind of science museum—exemplified by Frank Oppenheimer's Exploratorium in San Francisco—which was supposed to be open to all, with no distinction drawn between "experts" and "members of the public."[95] No doubt some laboratories had offered open days before the late 1960s, but it was only in the 1970s that institutions such as the CCRMA made the public's access to their sites a selling point in gaining federal funding.[96]

These efforts at inclusion were motivated in part by R&D organizations' growing belief that they could no longer take local communities for granted—or that if they did, the consequence could be picketing and sit-ins or (as in chapter 6) shocking newspaper articles that could inspire new local taxes and regulations. Firms like Signetics therefore began investing more of their money and their employees' time in civic activities, and they recruited a (somewhat) more inclusive local workforce. Organizations such as the UCSB physics department or NASA's Johnson Space Center encouraged their people to conduct research projects jointly with members of local communities. Yet despite the era's incentives for more inclusive R&D, many square scientists and engineers either weren't interested in interacting with local publics or simply didn't know how. Thus, as we saw in chapter 2, inclusive activities were often led by people in marginal positions within the sponsoring organization rather than the organization's more established members.

Established technoscientists were reluctant to adopt more inclusive practices in part because they believed that the costs of those practices outweighed the benefits. We've seen a number of cases in which that belief was clearly mistaken: the inclusive and egalitarian outlook of the UCSB MSI program, for instance, helped it recruit and teach students and led to jobs for the students and start-up companies for the teachers. And we've seen cases—such as in chapter 5—in which the lack of inclusivity had something to do with a project's failure. But we've also seen a number of cases in which frustrations with inclusivity led technoscientists to retreat from reform; Calvin Quate's exasperation with biomedical researchers' disinterest in acoustic microscopy is the clearest example. And we've also seen cases in which inclusivity led to outcomes that benefited third parties more than the R&D organization or the "included" societal partners. NASA's Meal System for the Elderly, for instance, was supposed to aid poor and elderly people with low mobility—but in the end affluent able-bodied camping enthusiasts and the companies that marketed to them probably saw the most return on that research. In other words, inclusivity—just like the other three RRI principles—was a decidedly ambivalent indicator of "responsible" outcomes. Inclusivity helped some technoscientific projects to become more socially responsible but contributed to other projects' failure or their capture for unjust or exclusionary ends over the longer term.[97]

MORE OR LESS RESPONSIBLE

So the technoscience of the 1970s certainly featured numerous experiments with all four principles of today's RRI (inclusivity, anticipation, reflexivity, responsiveness). Yet that experimentation was characterized by ambivalence and ambiguity: sometimes giving rise to more socially responsible outcomes, sometimes decidedly not. The four principles of RRI would therefore seem to be equally amenable to societal relevance and irrelevance, responsibility and irresponsibility.

One could argue that it's not the four principles in isolation that are necessary for more responsible R&D, but rather their synergistic combination—and that that synergistic combination of all four is what sets modern RRI apart from its predecessors *avant la lettre*. Fair enough. It is admittedly difficult to identify an R&D organization in the long 1970s that synergistically enacted all four principles at once. But my sense is that there is too much interpretive flexibility in the enactment of the four principles to use them as a litmus: too often they contribute to outcomes and objectives that don't seem very socially responsible.

So I would argue that the four principles don't help us demarcate socially responsible research and innovation from its opposite. Certainly, the four principles aren't timeless demarcations of responsibility that can be straightforwardly applied to the past. They may not even have much applicability beyond the near future, because RRI is already—as of 2019—being mockingly "buried"; whatever movement succeeds RRI will offer some other set of demarcations. In this book I've therefore tried to avoid timeless criteria for identifying "responsible" R&D; instead, my test for responsibility consists of questions that are keyed to the specific conditions of American physical and engineering science in the 1970s, though they could be adapted for today. For instance, was this person or organization becoming more oriented to taking life or to fostering the good life; to helping the environment or harming it; to serving the vulnerable or the powerful; to chasing profit for its own sake or to pursuing some broader set of social goods (potentially including but not limited to profit)?

By that standard, we can see that—in general—America's square scientists and engineers and their R&D organizations did practice a more responsible technoscience in the long 1970s than they had in the 1960s or they would in the 1980s. My case studies provide reasonably good evidence that even the most grudging squares were at least as successful in enacting a reformist technoscience as their more enthusiastic countercultural peers. Over the course of the 1970s, however, many squares abandoned their experiments with reform or decided that profitability was an adequate measure of social responsibility. No doubt some squares had equated social responsibility with profitability all along. At least at the beginning of the long 1970s, though, profit was more often framed as a *means* for achieving social responsibility rather than as a metric of responsibility or as an end in itself. As Arthur Hall told Emerson Electronics executives in 1972, efforts such as RANN and NTO "reflect some assessment of social needs and an attempt to match them with technologies that might satisfy the needs, and convert them to new markets."[98]

Of course, squares weren't alone in drifting toward an exclusive focus on profitability and disregard for other measures of the public interest. In the 1980s, many countercultural reformers became good Reaganite market worshippers as well. It was hard not to! And those few reformers who persevered with the experiments of the 1970s (such as the New Alchemists or Science for the People) usually ran out of steam by the early 1990s at the latest. Curiously, personal politics or lifestyle didn't seem to matter that much in determining whether someone moved toward or away from socially responsible R&D; whether square or groovy, almost everyone moved toward such projects in the 1970s and away from them in the 1980s. In other words, structural incentives for or against reform were far more determinative than personal politics or lifestyle. So if we want a more socially responsible research system today, we should put at least as much energy into devising new structural incentives as we do into reforming scientists' and engineers' personal views.

But first, we need to know a lot more about *which* structural conditions incentivize more responsible research and which ones encourage less responsible research. My case studies don't settle that question, but I think they do point in some fairly clear directions. Let's start with the optimistic "more responsible" side of the equation. Optimistic, if rather tawdry, that is: for one of the basic findings of this study is that with great funding comes great responsibility. Scientists and engineers of (nearly) all political stripes responded to new funding streams that were dedicated to solving societal problems. In the long 1970s such funding took a variety of different forms: existing organizations such as NASA diverted existing funding in new directions; other existing organizations formed entirely new funding streams (e.g., RANN at the NSF); and a few new organizations (such as the EPA) entered the scene. Funding for socially responsible R&D was still hard to get; but given that defense R&D budgets were shrinking rapidly, funding for R&D oriented to public housing, mass transit, disability technologies, alternative energy, environmental sensing and remediation, and other areas suddenly looked very attractive indeed.

Square scientists and engineers showed significant creativity in looking for money in unexpected places and in revising their portfolios to include the societally relevant research that funders wanted. Sometimes their efforts were quixotic: for example, Hall's and Kilby's soliciting of Saudi money or David Phillips's proposals for parapsychology funding from local peace activists. But sometimes they succeeded: for example, Calvin Quate's grant from the Hartford Foundation for acoustic microscope research, or Virgil Elings's grant from the American Heart Association for development of a new blood flowmeter. And the funders themselves were similarly creative and willing to run risks: witness the joint DARPA–National Bureau of Standards grant that supported Quate and other researchers in aid of the US semiconductor industry; or the joint NSF–National Endowment for the Arts scheme that funded Stanford's CCRMA.

If money was an effective carrot for encouraging more responsible R&D, then there were also several similarly effective sticks: investigative journalism, probing questions from Congress, and laboratory sit-ins. Many squares were particularly bothered by the latter, bemoaning direct-action protests as an end to freedom of inquiry. Yet even those who objected ended up doing more or less what the protestors wanted. As a result, reform was almost certainly in its fullest swing while conscription was in effect and young people were therefore easy to mobilize for protest. The end of the draft in late 1972 (along with the oil shock the next year) concluded the first phase of the long 1970s, with a steady decrease in protests and hence a decline in reformist activities after that point.

Not that I am arguing that a return to conscription would make American technoscience more responsible today! Indeed, one of the clearest signals from my cases is that square technoscientists became ever more responsible while the national security state was shrinking and (mostly) retreated from reforms once the Cold War heated back up. That observation shouldn't be taken as a blanket condemnation of defense research in general. My view—which I think most squares shared—is that *some* kinds of defense research are in *some* circumstances justified and should even count as socially responsible. How much of what kinds of defense research and under what circumstances—those questions are where the disagreements lie. Clearly, the military's enormous R&D budgets in the early Cold War contributed to some socially progressive outcomes; John Linvill and James Meindl's Optacon is as good an example as any. But so long as the national security state was the black hole at the center of American R&D, the considerations of other parts of society were secondary. When the national security state became merely one among many players, the research system became more inclusive and responsive. The lessons of the cases in this study are perhaps most compatible with Stanton "Glantz' goal of reducing DOD funding to 'say 10–20% of the research'" in places like Stanford's School of Engineering.[99]

While the Pentagon was retreating, the civilian state was advancing—and that, too, was closely associated with more responsible technoscience. R&D funding from civilian federal agencies facilitated reform; but in some respects the more durable changes were wrought by regulatory and administrative changes, not just from the federal government but also at the state, county, and municipal level. Environmental regulations made companies like Signetics less lax in disposing of chemicals used in manufacturing and made researchers at places like NASA or Science

Spectrum more interested in designing instruments to detect pollution. Equal opportunity and antidiscrimination laws effected lasting if inadequate changes in the workforces of almost all R&D organizations.

The observation that a shrinking defense establishment and a growing regulatory state foster responsible innovation runs counter to some prominent RRI voices. Particularly in the United States, some leading RRI proponents speak glowingly of the Pentagon's "can-do approach" and ability "to catalyze rapid innovation" while excoriating "fifty years of growing public investments" on the civilian side that supposedly generate nothing but "contradiction, controversy, and confusion."[100] This view is also linked (at least institutionally, but I think also ideologically) to a view that RRI allows "governance from within" which avoids the dysfunctions of "regulations, which often are directives from upstream or downstream, are largely static, can become outdated, and can fail to apply clearly to dynamic and changing R&D processes and contexts."[101]

I find both these views problematic—indeed, problematic *with respect to the aims of RRI*. The view that the Pentagon offers the right model for responsible R&D, in particular, utterly disregards the calamity and irresponsibility of the Cold War. The superpower confrontation was, of course, not cold at all; it stimulated fifty-plus years of armed conflict all over the world. And even in its coldest corner—the stockpiling of nuclear weapons for an Armageddon that never came—the Cold War cost untold lives: from the mining, processing, and disposal of uranium and other incredibly hazardous materials; from the construction, storage, transport, and maintenance of and training for the use of nuclear weapons; and especially from the hundreds of thousands of premature deaths traceable to nuclear testing.[102] Some squares were drawn toward socially relevant research because they grew to be appalled by these horrors. Others adopted socially relevant research but still continued to believe that confronting communism was worth the costs. Both those views can be defended, but to argue in good faith that the Pentagon's "can-do" attitude provides a good model for responsible R&D, one must do a full accounting of the costs of that attitude.

Meanwhile, any form of RRI that dismisses the regulatory state risks discarding the most versatile and effective means of encouraging scientists and engineers to practice a more democratically responsive form of R&D. The Clean Water Act of 1972 and other regulations that were instituted in the 1970s were a direct expression of popular social movements such as environmentalism and appropriate technology. Nationally influential individuals like Rachel Carson and Barry Commoner and locally influential ones such as Susan Yoachum and Lorraine Ross did much to build public support for their cause—but that broad appeal was largely translated into effective outcomes by the actions of a democratically responsive regulatory state. Similarly, participants in the feminist and civil rights movements worked hard to stigmatize discrimination; but they worked just as hard to get laws passed to make discrimination legally impermissible too. Why, then, would RRI's promotors strive to make technoscience more inclusive while simultaneously rejecting one of the most effective tools—the state, and specifically the regulatory state—for reaching that end?

Still, it's probably true that funding, regulation, and civilianization alone aren't enough to sustain reform. Those approaches need to be complemented by other avenues to more responsible R&D. We could, for instance, try to recruit and retain more scientists and engineers who are already more socially aware, especially those who are minoritized in some way. After all, we have some evidence that minoritized scientists and engineers differ from their peers in, as Joanna Weidler-Lewis puts it, "the belief that 'good' engineers use engineering skills to create a better world for all."[103]

We can also try to ensure that the environments in which square scientists and engineers operate are richer in cues that they should reform their practice. Admittedly, in the 1970s such cues often emanated from sit-ins and investigative journalism—communicative genres that often caused squares to harden their attitudes. Yet direct action of that sort at least required squares to account for themselves. And when the senders and receivers of such cues could meet on common ground—as at the postprotest discussions at Stanford's Systems Techniques Laboratory in 1968—then attitudes sometimes softened. Cues that originated from scientists' and engineers' families, friends, students, and close personal networks, however, were probably even more persuasive. That, at least, is one plausible abductive leap from the examples of Virgil Elings and David Phillips in chapter 2.

Durable change probably also depends on finding a mode of reformist technoscience that square scientists and engineers *enjoy*. Evidence for this in my case studies is murky, but I do have the sense that many squares were affectively drawn to and energized by the new societally relevant topics that they adopted in the late 1960s and early 1970s. They were fascinated by the technical challenges and heartened to do something that they thought would directly and obviously benefit others (in contrast to the indirect or inscrutable benefits derived from basic or defense-oriented research). Interactions with people outside their usual circles also sometimes sparked joy, at least early on before the frustrations of miscommunication set in. Hall, for instance, could be quite sour in his correspondence with other "professional scientists, engineers, and managers" in the United States, but when he got to meet similar experts in Poland in 1974, he described it as "one of the high points in my entire professional career" for the opportunity it provided to meet "as we say colloquially, 'the salt of the earth,' which means the kind of people that we like to know as personal friends."[104]

Some of the sense of fun and social worth that squares experienced in the early 1970s seems to have been associated with a spirit of experimentation. Organizations that were under stress experimented with lots of different ways to relieve that stress: new interdisciplinary centers, new degree programs, new collaborations, and so forth. Previous studies have shown how groovy counterculturalists took the era's organizational stressors as an opportunity to experiment with a more fulfilling and enjoyable kind of science.[105] But I've tried to show that at least some squares took the same license and went in not dissimilar directions—for example, Linvill, Meindl, and Bliss's Optacon (and the start-up they founded to sell it) or James Angell's pursuit of electronic music and biomedical electronics. One particular kind of bureaucratic stress—existential success—beset a number of organizations in this study and occasioned a variety of experiments with socially responsible science and engineering. That could indicate that RRI practitioners should think about how to leverage organizational stagnation, stress, and renewal. As Thomas Heinze and Richard Münch argue, organizational renewal *is* research; thus, we should view responsible organizational renewal *as* responsible research.[106]

Throughout this book, I've offered quotes that give some sense of the positive affect that some squares associated with organizational experimentation. Of course, I've also offered quotes that evince the negative emotions that other squares associated with the changes going on around them. And here we can start to talk about the things that made American science and engineering *less* responsible. Clearly, when influential people like Bill Rambo groused—even in private—about reforms being imposed on the physical and engineering sciences, their complaints inhibited reform. But I think a more important factor in reversing the trend toward socially responsible research was simple bureaucratic attrition. Organizations tried out lots of new initiatives in the early 1970s, but by the end of the decade they were ready to cull those experiments and select only the ones that promoted current organizational objectives. And as the 1970s turned into the 1980s, those objectives grew more oriented to economic competitiveness and the resurgent superpower confrontation than to societal relevance.

Many squares offered little resistance to that culling because by that point they thought that socially responsible R&D just wasn't as rewarding as it had once been. Some of that weariness reflected a growing realization that the technical challenges that they had set out to solve weren't *just* technical and possibly weren't soluble at all, and that any solutions they offered would likely be ignored or rejected. Those kinds of puzzles are, well, not very fun. We can hardly be surprised that under those conditions, many scientists and engineers chose to turn away from socially responsible R&D and to seek new projects where they could be more certain of success and approbation.

Of course, square scientists and engineers also shouldn't have been surprised that their efforts didn't meet with instant approval. They *should* have known that the public acceptance of technoscientific expertise must be earned through the hard work of persuasion, and not just through the hard work of calculation and design. Yet when squares tried to understand why their ideas weren't taken up and why the joy had been drained out of their experiments with socially responsible R&D, they almost never assigned any blame to themselves. The hippies at SERI were too technically incompetent to appreciate TI's solar energy system; the bureaucrats at the NSF were too inept to organize funding for Autofarm; the bean counters at HUD were too cheap to pay for space-age public housing; biomedical researchers were too conservative to understand what the acoustic microscope offered them—these were the kinds of explanations that came more readily to the squares.

Such explanations are products of the most intractable factor militating against socially responsible science and engineering: namely, the enormous value placed on (certain kinds of) technical merit, and the disregard for those deemed not to have (those kinds of) merit. Not all, but most of, the scientists and engineers whom I've profiled here seemed to agree with the physicist Alvin Weinberg's view: that is, social problems can be solved with a "technological fix," but only so long as those who don't understand technology listen to those who do.[107] We can see this stance in Arthur Hall's refusal to allow communications students into his classroom, or in his desire to be seen as "a practicing engineer-scientist with a few practical results to his credit" rather than an "overalled farmer nut." More perniciously, we can see it in his claim that systems engineers should oversee a program of "birth control and planned parenthood as one way to limit consumption of earth's resources."

To be sure, technical merit shouldn't be disregarded. The classic retort—that the critic, too, would rather fly in an airplane designed by a competent engineer than by an incompetent one—isn't wrong. But it isn't right, either. As scholars such as Amy Slaton have pointed out, scientists' and engineers' ascriptions of competence and incompetence too easily line up with their own interests in terms of class, race, and gender.[108] So long as they don't question that alignment, technoscientists aren't showing the hard-nosed skepticism that they apply to everyone else.

That is, ascriptions of incompetence should always be two-sided. Maybe, for instance, Jack Kilby was right that "all of" the DOE's solar energy specialists were technically inept and should be fired; but Kilby, too, got some important things wrong. He failed to persuade others of the merits of the TISES system. Nor did he and his colleagues design TISES in a way that would have enduring appeal even after the price of oil dropped; he put his faith in untrustworthy allies and he failed to enroll other potential allies who might have kept TISES going. Similarly, maybe Calvin Quate was right that biomedical researchers were too hidebound and incurious to recognize the advantages of the acoustic microscope. But Quate, too, was counterproductively conservative. He failed to educate himself in how biomedical researchers make judgments, and he failed to immerse himself in their world enough to bring them around. Instead, he turned back toward audiences in semiconductor manufacturing with whom he was more comfortable and who didn't cast any doubt on his expertise.

In other words, we have to see "good" science and engineering as relatable to some kind of shared good, rather than evaluating what counts as good in solely technical terms. Research and development are not machines for generating countable metrics of technical merit: *X* many truths, *Y* many bombs, *Z* many dollars. Rather, science and engineering are social activities that create much that is "good" that cannot be measured by technical standards: meaning, solidarity, opportunity, material advancement, spiritual enlightenment, mystery, fascination, and so on. Science and engineering necessarily draw on the society of which they are part and parcel for the valorization of these more-than-technical goods that are generated out of technical activity. It is these more-than-technical aspects of science and engineering that offer some possibility of greater common ground with the rest of society—some hope that the middle need not always be excluded.

Of course, like all human endeavors, science and engineering have a noxious side too: they can bring discrimination, pollution, violence, anomie, boredom, and so on. The long 1970s saw all these aplenty. Technical competence—at least as usually understood—does not seem to safeguard against these more negative aspects of technoscience. Quite the opposite, in fact. So, on their own, technical measures of good science and engineering do not seem capable of delivering good science and engineering. If we want good technoscience, then the kinds of questions we have to ask are not purely technical—at least not in the usual sense. Rather, we have to ask what and who is technoscience for? Why do we do it this way and not another way? What kind of world do we imagine technically competent science and engineering will lead to?

To make this point more eloquently, I'd like to give the final word to one of this story's nontechnoscientists: Richard Lyman, historian and president of Stanford for much of the long 1970s. Not that historians have all the answers, but Lyman was for the most part a perceptive and sympathetic observer of American science and engineering in this period. He understood that science and engineering are heterogeneous activities, not homogeneous machines—and that therefore technically competent science and engineering cannot, on their own, get us to a society that anyone would want to live in. He didn't shy from pointing out that achieving more responsible science and engineering takes money and patience—you can't get there overnight or on the cheap. Most of all, Lyman saw that creating a technoscience that is responsible to its time can only be accomplished by overcoming the excluded middle and encouraging mutual comprehension and not just mutual ambivalence. Or, as he put it in 1970,

> If we are in difficulties partly because our functions are many, and our focus can therefore never be single, it will do us no good to try to return to some simpler day. . . . Instead we ought to glory in the fact that some people are learning to appreciate Keats in one part of the campus, while others are solving problems of

linear programming in another. Glory in it, and make a towering virtue of necessity by exposing the one group to the other, and each to a thousand further groups, at every available opportunity. It is an arrogant assumption of some humanists that no computer man reads Keats, and no electronics buff can dig Scarlatti. It is an arrogant counter-assumption of some technologists that no humanist has anything important to contribute to life in the technitronic age of the future. The university exists in part to attack such arrogant parochialisms, and with the help of donors of courage, steadfastness, and a capacity to see things in perspective, the attack will in the long run succeed.[109]

NOTES

Chapter 1

1. Charles Schwartz, "I pledge that I will not participate in war research or weapons production; I further pledge to counsel my students and urge my colleagues to do the same," *Bulletin of the American Physical Society* 15.1 (1970). 28.

2. Letter from William L. Lehmann to W. W. Havens, November 26, 1969, and letter from W. W. Havens to Charles Schwartz, December 16, 1969, both Folder 11, Box 7, American Physical Society Records, 1899–1989, Niels Bohr Library, American Institute of Physics (henceforth NBL).

3. Charles Schwartz, letter to the APS Council, January 25, 1970, Folder 11, Box 7, American Physical Society Records, 1899–1989, NBL.

4. Charles Schwartz, "For the Schwartz Amendment," *Physics Today* 21.1 (January 1968): 9–11. Schwartz's was the first in a set of letters to the editor on the question "Should the APS Discuss Public Issues?" The other letters offer an enlightening cross-section of the voices both for and against Schwartz's 1968 proposed amendment.

5. Charles Schwartz, "Petition to the President of the American Physical Society," circa March 17, 1970, Folder 11, Box 7, American Physical Society Records, 1899–1989, NBL.

6. Ad Hoc Committee on Schwartz Proposal, "First Draft of Report to Professor C. S. Wu," July 28, 1975, Folder 13, Box 7, American Physical Society Records, 1899–1989, NBL. This amendment was apparently related to Schwartz's accusation that a Berkeley colleague, Thomas Pigford, and the university administration had covered up Pigford's involvement with the nuclear industry to make Pigford's pronouncements on the safety of nuclear power seem disinterested and objective. See Marcy Kates, "Controversy over an Alleged Consulting Conflict at UC," *Berkeley Gazette*, June 11, 1976: 1.

7. Sarah Bridger, *Scientists at War: The Ethics of Cold War Research* (Cambridge, MA: Harvard University Press, 2015), 198–205.

8. Interview of Charles Schwartz by Patrick Catt on July 19, 1995, Niels Bohr Library & Archives, American Institute of Physics, College Park, MD, www.aip.org/history-programs/niels-bohr-library/oral-histories/5913.

9. Jay Orear, letter to John Bardeen, October 29, 1968 (and other items in this folder), Folder 9, Box 7, American Physical Society Records, 1899–1989, NBL.

10. For the purposes of this study, the names of individuals who wrote to the APS are immaterial; here, and in similar instances, I will simply cite correspondents by their initials. RFT, letter to W. W. Havens, January 12, 1970, Folder 11, Box 7, American Physical Society Records, 1899–1989, NBL.

11. SGK letter to Luis W. Alvarez, January 9, 1970, Folder 11, Box 7, American Physical Society Records, 1899–1989, NBL.

12. DP, KHW, SCP, and DLH letter to Samuel Goudsmit, January 16, 1969, Folder 11, Box 7, American Physical Society Records, 1899–1989, NBL.

13. OL, letter to W. W. Havens, December 2, 1969, Folder 11, Box 7, American Physical Society Records, 1899–1989, NBL.

14. Luis W. Alvarez letter to SGK, January 21, 1970, Folder 11, Box 7, American Physical Society Records, 1899–1989, NBL.

15. RAB letter to the APS Council, April 6, 1969, Folder 10, Box 7, American Physical Society Records, 1899–1989, NBL.

16. HW letter to R. Hobart Ellis Jr. (editor of *Physics Today*), March 11, 1969, Folder 10, Box 7, American Physical Society Records, 1899–1989, NBL.

17. Bridger, *Scientists at War*; Kelly Moore, *Disrupting Science: Social Movements, American Scientists, and the Politics of the Military, 1945–1975* (Princeton: Princeton

University Press, 2013); Sigrid Schmalzer, Daniel S. Chard, and Alyssa Botelho (eds.), *Science for the People: Documents from America's Movement of Radical Scientists* (Amherst: University of Massachusetts Press, 2018); Alice Bell, "The Scientific Revolution That Wasn't: The British Society for Social Responsibility in Science," *Radical History Review* 2017.127 (2017): 149–172; Simone Turchetti, "Looking for the Bad Teachers: The Radical Science Movement and Its Transnational History," in *Science Studies during the Cold War and beyond: Paradigms Defected*, ed. Elena Aronova and Simone Turchetti (New York: Palgrave Macmillan, 2016): 77–102.

18. Naomi Oreskes and Erik Conway, *Merchants of Doubt: How a Handful of Scientists Obscured the Truth on Issues from Tobacco Smoke to Global Warming* (New York: Bloomsbury Press, 2010); Lawrence Badash, *A Nuclear Winter's Tale: Science and Politics in the 1980s* (Cambridge, MA: MIT Press, 2009); Jacob Darwin Hamblin, *Arming Mother Nature: The Birth of Catastrophic Environmentalism* (New York: Oxford University Press, 2013).

19. David Kaiser, *How the Hippies Saved Physics: Science, Counterculture, and the Quantum Revival* (New York: W. W. Norton, 2012); W. Patrick McCray, *The Visioneers: How a Group of Elite Scientists Pursued Space Colonies, Nanotechnologies, and a Limitless Future* (Princeton: Princeton University Press, 2013); David Kaiser and W. Patrick McCray (eds.), *Groovy Science: Knowledge, Innovation, and American Counterculture* (Chicago: University of Chicago Press, 2016); Eden Medina, *Cybernetic Revolutionaries: Technology and Politics in Allende's Chile* (Cambridge, MA: MIT Press, 2011); Fred Turner, *From Counterculture to Cyberculture: Stewart Brand, the Whole Earth Network, and the Rise of Digital Utopianism* (Chicago: University of Chicago Press, 2006); Andrew Kirk, *Counterculture Green: The Whole Earth Catalog and American Environmentalism* (Lawrence: University Press of Kansas, 2007); Peter Sachs Collopy, "The revolution will be videotaped: Making a technology of consciousness in the long 1960s," PhD dissertation, University of Pennsylvania, 2015.

20. The phrase is used by Richard Lyman, president of Stanford, in "Revolt against Reason Can Have 'Ominous Consequences'," *Campus Report* January 27, 1971, page 1.

21. I'm thinking in particular of Jack Sarfatti in *How the Hippies Saved Physics* and Fernando Flores in *Cybernetic Revolutionaries*. Stewart Brand, the central actor of Turner and Kirk's monographs, is more of a political enigma, but he did become a venture capitalist and prominent supporter of military–industrial technologies such as the ARPANET, space shuttle, and nuclear power. Many of the figures in

McCray's *Visioneers*, as well, moved effortlessly from a countercultural presentation of self to ardent support of Ronald Reagan and military–industrial technologies such as the Strategic Defense Initiative.

22. On McCarthy, see for instance Noam Cohen, *The Know-It-Alls: The Rise of Silicon Valley as a Political Powerhouse and Social Wrecking Ball* (New York: The New Press, 2017). Earnest's quote comes from his website: https://web.stanford.edu/~learnest, December 28, 2017, accessed June 6, 2019.

23. Bridger, *Scientists at War*.

24. I don't claim any originality in using the term *square scientist*, though I don't think that I borrow it directly from anyone else. However, related terms have circulated before and probably influenced my usage. For example, Erinn McComb, "'It's Hip to Be Square': Individual Control, Masculinity, and Engineering within NASA, 1958–72" (Cleveland: Society for the History of Technology, November, 2011); and David H. Onkst, "The triumph and decline of the 'squares': Grumman Aerospace engineers and production workers in the Apollo era, 1957–1973," PhD dissertation, American University, 2011.

25. Peter Galison, *Image & Logic: A Material Culture of Microphysics* (Chicago: University of Chicago Press, 1997), 396–400, 403–411, offers a good overview of Alvarez's square philosophy of enthusiastically working on defense projects while switching between his physics and "engineering hat[s]" as need demanded. On the mass extinction hypothesis, see William Glen (ed.), *The Mass-Extinction Debates: How Science Works in a Crisis* (Stanford: Stanford University Press, 1994). For an examination of Alvarez's rejection of a few different varieties of nonsquareness (e.g., his own Hispanic identity and the pseudoscientific "nuts" who often wrote to him), see Jesús Rubén Martinez, "The adventures of Luis Alvarez: Identity politics in the making of an American science," PhD dissertation, University of Texas at Austin, 2011.

26. Lillian Hoddeson and Vicki Daitch, *True Genius: The Life and Science of John Bardeen, the Only Winner of Two Nobel Prizes in Physics* (Washington, DC: Joseph Henry Press, 2002).

27. Matthew Wisnioski, *Engineers for Change: Competing Visions of Technology in 1960s America* (Cambridge, MA: MIT Press, 2012). Perhaps closer to my focus is Matt Wisnioski, "Inside 'the System': Engineers, Scientists, and the Boundaries of Social Protest in the Long 1960s," *History and Technology* 19.4 (2003): 313–333.

28. Eric J. Vettel, *Biotech: The Countercultural Origins of an Industry* (Philadelphia: University of Pennsylvania Press, 2006).

29. Nicolas Rasmussen, *Gene Jockeys: Life Science and the Rise of Biotech Enterprise* (Baltimore: Johns Hopkins University Press, 2014).

30. David Kaiser, "The Postwar Suburbanization of American Physics," *American Quarterly* 56.4 (2004): 851–888.

31. Committee on Women in Physics, "Women in Physics," January 30, 1972, Folder 11, Box 24; and Warren E. Henry (chair of APS Committee on Minorities in Physics) letter to Dr. S. H. Koenig, November 7, 1972, Folder 15, Box 19; both in American Physical Society Records, 1899–1989, NBL.

32. Gene Bylinsky, "How Intel Won Its Bet on Memory Chips," *Fortune* (November 1973): 143–144. The quote pops up several places in the literature, but I first ran across it in Ross Knox Bassett, *To the Digital Age* (Baltimore: Johns Hopkins University Press, 2002). Arnold Thackray, David Brock, and Rachel Jones, *Moore's Law: The Life of Gordon Moore, Silicon Valley's Quiet Revolutionary* (New York: Basic Books, 2015) paints a picture of a very square individual, at least in the sense used in this book. Politically, Moore was a "'middle of the road' Republican by temperament and conviction." However, his defining characteristic, and one of the main themes of the biography, was his profound reticence. In that sense the "kids with the long hair" quote was something of an exception.

33. Edwin Layton, *The Revolt of the Engineers: Social Responsibility and the American Engineering Profession* (Cleveland: Press of Case Western Reserve University, 1971), 9.

34. In other countries, engineers did form labor unions; there was nothing natural about the American situation. See Peter Meiksins and Chris Smith (eds.), *Engineering Labour: Technical Workers in Comparative Perspective* (London: Verso, 1996).

35. Jon Gertner, *The Idea Factory: Bell Labs and the Great Age of American Innovation* (New York: Penguin, 2012).

36. David Kaiser, "The Atomic Secret in Red Hands? American Suspicions of Theoretical Physicists during the Early Cold War," *Representations* 90.1 (2005): 28–60.

37. Jessica Wang, *American Science in an Age of Anxiety: Scientists, Anticommunism, and the Cold War* (Chapel Hill: University of North Carolina Press, 1999).

38. Audra J. Wolfe, *Freedom's Laboratory: The Cold War Struggle for the Soul of Science* (Baltimore: Johns Hopkins University Press, 2018).

39. Kaiser, *How the Hippies Saved Physics.*

40. Donna Riley, Amy E. Slaton, and Alice L. Pawley, "Social Justice and Inclusion: Women and Minorities in Engineering," in *Cambridge Handbook of Engineering Education Research*, ed. Aditya Johri and Barbara M. Olds (New York: Cambridge University Press, 2014): 335–356.

41. This turn is too broad to summarize fairly, but notable examples from histories of science and technology include Margaret W. Rossiter, *Women Scientists in America, v. 2 Before Affirmative Action* (Baltimore: The Johns Hopkins University Press, 1995); Marie Hicks, *Programmed Inequality: How Britain Discarded Women Technologists and Lost Its Edge in Computing* (Cambridge, MA: MIT Press, 2017); Londa Schiebinger, *Plants and Empire: Colonial Bioprospecting in the Atlantic World* (Cambridge, MA: Harvard University Press, 2004); Alondra Nelson, *Body and Soul: The Black Panther Party and the Fight against Medical Discrimination* (Minneapolis: University of Minnesota Press, 2011); Warwick Anderson, *Colonial Pathologies: American Tropical Medicine, Race, and Hygiene in the Philippines* (Durham: Duke University Press, 2006); Clapperton Chakanetsa Mavhunga, "A Plundering Tiger with Its Deadly Cubs? The USSR and China as Weapons in the Engineering of a 'Zimbabwean Nation,' 1945–2009," in *Entangled Geographies: Empire and Technopolitics in the Global Cold War*, ed. Gabrielle Hecht (Cambridge, MA: MIT Press, 2011): 231–266; Linda Nash, "The Agency of Nature and the Nature of Agency," *Environmental History* 10 (2005): 67–69; Chris Pearson, "History and Animal Agencies," in *The Oxford Handbook of Animal Studies*, ed. Linda Kalof (New York: Oxford University Press, 2017): 240–257; Bruno Latour, *Aramis, or, the Love of Technology* (Cambridge, MA: Harvard University Press, 1996).

42. Jennifer Schuessler, "In History Class, Capitalism Sees Its Stock Soar," *New York Times*, April 7, 2013: A1.

43. Hub Zwart, Laurens Landeweerd, and Arjan van Rooij, "Adapt or Perish? Assessing the Recent Shift in the European Research Funding Arena from 'ELSA' to 'RRI'," *Life Sciences, Society and Policy* 10 (2014): article 11.

44. Danielle Shanley, "Reimagining the Future through Revisiting the Past: The Value of ANTi-History in Thinking about RRI's Possible Future(s)," *Journal of Responsible Innovation* (2020), https://www.tandfonline.com/doi/full/10.1080/23299460.2021.1882748.

45. Rene von Schomburg, "A Vision of Responsible Innovation," in *Responsible Innovation*, ed. Richard Owen, John Bessant, and Maggy Heintz (New York: John Wiley & Sons, 2013): 51–74.

46. David H. Guston, Erik Fisher, Armin Grunwald, Richard Owen, Tsjalling Swierstra, and Simone van der Burg, "Responsible Innovation: Motivations for a New Journal," *Journal of Responsible Innovation* 1.1 (2014): 1–8. The original quote contains a reference that I have left out: Erik Fisher, Roop L. Mahajan, and Carl Mitcham, "Midstream Modulation of Technology: Governance from Within," *Bulletin of Science, Technology & Society* 26.6 (2006): 485–496.

47. Cyrus C. M. Mody, *Instrumental Community: Probe Microscopy and the Path to Nanotechnology* (Cambridge, MA: MIT Press, 2011) and *The Long Arm of Moore's Law: Microelectronics and American Science* (Cambridge, MA: MIT Press, 2017).

48. Shanley, "Reimagining the Future through Revisiting the Past."

49. Rebecca Press Schwartz, "The making of the history of the atomic bomb: Henry Dewolf Smyth and the historiography of the Manhattan Project," PhD dissertation, Princeton University, 2008; Michael A. Dennis, "Echoes of the Past. Henry Guerlac and Radar's Historiographic Problem," in *Tracking the History of Radar*, ed. Oskar Blumtritt, Hartmut Petzold, and William Aspray (New Brunswick: IEEE–Rutgers Center for the History of Electrical Engineering, 1994), 285–298.

50. For example, Stuart W. Leslie, *The Cold War and American Science: The Military-Industrial-Academic Complex at MIT and Stanford* (New York: Columbia University Press, 1993); Rebecca Lowen, *Creating the Cold War University: The Transformation of Stanford* (Berkeley: University of California Press, 1997); G. Pascal Zachary, *Endless Frontier: Vannevar Bush, Engineer of the American Century* (Cambridge, MA: MIT Press, 1999); Peter J. Westwick, *The National Labs: Science in an American System, 1947–1974* (Cambridge, MA: Harvard University Press, 2003).

51. Works by nonhistorians on the history of commercialized technoscience include David C. Mowery, Richard R. Nelson, Bhaven Sampat, and Arvids A. Ziedonis, *Ivory Tower and Industrial Innovation: University-Industry Technology Transfer before and after the Bayh-Dole Act in the United States* (Stanford: Stanford Business Books, 2004); Walter W. Powell and Jason Owen-Smith, "Knowledge Networks as Channels and Conduits: The Effects of Spillovers in the Boston Biotechnology Community," *Organization Science* 15 (2001): 5–21; Paul Rabinow, *Making PCR: A Story of Biotechnology* (Chicago: University of Chicago Press, 1996); Fiona Murray, "The Role of Academic Inventors in Entrepreneurial Firms: Sharing the Laboratory Life," *Research Policy* 33 (2004): 643–659.

52. Among works that depict the long 1970s as a pivot from the early Cold War to the present era, see Elizabeth Popp Berman, *Creating the Market University: How*

Academic Science Became an Economic Engine (Princeton: Princeton University Press, 2012); Lillian Hoddeson, Adrienne W. Kolb, and Catherine Westfall, *Fermilab: Physics, the Frontier, and Megascience* (Chicago: University of Chicago Press, 2008); Stephen J. Macekura, *Of Limits and Growth: The Rise of Global Sustainable Development in the Twentieth Century* (New York: Cambridge University Press, 2015); McCray, *The Visioneers*; Matthew Wisnioski, "How the Industrial Scientist Got His Groove: Entrepreneurial Journalism and the Fashioning of Technoscientific Innovators," in *Groovy Science*, ed. David Kaiser and W. Patrick McCray (Chicago: University of Chicago Press, 2016): 337–365. Although he locates his analysis a decade earlier, I find the trends outlined in Jon Agar, "What Happened in the Sixties?" *British Journal for the History of Science* 41.4 (2008): 567–600 highly complementary with my study.

53. David Kaiser, "Cold War Requisitions, Scientific Manpower, and the Production of American Physicists after World War II," *Historical Studies in the Physical and Biological Sciences* 33.1 (2002): 131–159; Vettel, *Biotech*, 22-29.

54. John Krige, *American Hegemony and the Postwar Reconstruction of Science in Europe* (Cambridge, MA: MIT Press, 2006); Atsushi Akera, *Calculating a Natural World: Scientists, Engineers, and Computers during the Rise of U.S. Cold War Research* (Cambridge, MA: MIT Press, 2007); Philip Mirowski and Esther-Mirjam Sent, "The Commercialization of Science and the Response of STS," in *The Handbook of Science and Technology Studies*, ed. Edward J. Hackett, Olga Amsterdamska, Michael Lynch, and Judy Wajcman (Cambridge, MA: MIT Press, 2008); 635–690.

55. Carol Gruber, "The Overhead System in Government-Sponsored Academic Science: Origins and Early Development," *Historical Studies in the Physical and Biological Sciences* 25.2 (1995): 241–268.

56. David Hounshell, "The Cold War, RAND, and the Generation of Knowledge, 1946–1962," *Historical Studies in the Physical and Biological Sciences* 27.2 (1997): 237–267.

57. Scott G. Knowles and Stuart W. Leslie, "'Industrial Versailles': Eero Saarinen's Corporate Campuses for GM, IBM, and AT&T," *Isis*, 92.1 (2001): 1–33.

58. David E. H. Edgerton, "The Linear Model Did Not Exist: Reflections on the History and Historiography of Science and Research in Industry in the Twentieth Century," in *The Science-Industry Nexus*, ed. Karl Grandin, Nina Wormbs, and Sven Widmalm (Sagamore Beach, MA: Science History Publications, 2004): 31–57; Benoît Godin, *Models of Innovation: The History of an Idea* (Cambridge, MA: MIT Press, 2017).

59. Harvey Brooks, "Applied Science and Technological Progress," *Science* 156.3783 (June 30, 1967): 1706–1712.

60. Walter Sullivan, "Science," *New York Times*, February 1, 1970: E7.

61. Dian Olson Belanger, *Enabling American Innovation: Engineering and the National Science Foundation* (West Lafayette: Purdue University Press, 1998).

62. Rebecca Slayton, "From a 'Dead Albatross to Lincoln Labs: Applied Research and the Making of a Normal Cold War University," *Historical Studies in the Natural Sciences* 42.4 (2012): 255–282.

63. Cyrus C. M. Mody and Hyungsub Choi, "From Materials Science to Nanotechnology: Interdisciplinary Center Programs at Cornell University, 1960–2000," *Historical Studies in the Natural Sciences* 43.2 (2012): 121–161.

64. Bassett, *To the Digital Age*.

65. Jongmin Lee, "Engineering the environment: Regulatory engineering at the U.S. Environmental Protection Agency, 1970–1980," PhD dissertation, Virginia Polytechnic Institute and State University, 2013. As Lee argues, the EPA contained many engineers who regarded themselves as "moderate environmentalists"—that is, squares who implemented the reforms imagined by more vocal activists.

66. Belanger, *Enabling American Innovation*; Juan C. Lucena, *Defending the Nation: U.S. Policymaking to Create Scientists and Engineers from Sputnik to the 'War against Terrorism,'* (Lanham, MD: University Press of America, 2005).

67. An example that appears in chapter 3 is the Semiconductor Technology Program of the mid-1970s, which was jointly sponsored by the (civilian) National Bureau of Standards and the Defense Advanced Research Project Agency (with supplemental funding from the Defense Nuclear Agency, the Air Force, and the Navy). This program sponsored academic and corporate development of measurement technologies for use by the US semiconductor industry. W. Murray Bullis (ed.), *Semiconductor Measurement Technology Progress Report, January 1 to June 30, 1976* (Washington, DC: Department of Commerce, 1977).

68. Paul Forman, "The Primacy of Science in Modernity, of Technology in Postmodernity, and of Ideology in the History of Technology," *History and Technology* 23 (2007): 1–152; Alfred Nordmann, Hans Radder, and Gregor Schiemann (eds.), *Science Transformed? Debating Claims of an Epochal Break* (Pittsburgh: University of Pittsburgh Press, 2011).

69. Berman, *Creating the Market University.*

70. Josh Whitford and Andrew Schrank, "The Paradox of the Weak State Revisited: Industrial Policy, Network Governance, and Political Decentralization," in *State of Innovation: The US Government's Role in Technology Development*, ed. Fred Block and Matthew R. Keller (Boulder: Paradigm Publishers, 2011): 261–281.

71. Daniel J. Kevles, "Big Science and Big Politics in the United States: Reflections on the Death of the SSC and the Life of the Human Genome Project," *Historical Studies in the Physical and Biological Sciences* 27 (1997): 269–297.

72. Martin Kenney, *Biotechnology: The University-Industrial Complex* (New Haven: Yale University Press, 1986); Maureen McKelvey, *Evolutionary Innovations: The Business of Biotechnology* (New York: Oxford University Press, 1996).

73. Aant Elzinga, "Features of the Current Science Policy Regime: Viewed in Historical Perspective," *Science and Public Policy* 39.4 (2012): 416–428.

74. Mowery *et al.*, *Ivory Tower and Industrial Innovation*; Paula Stephan, *How Economics Shapes Science* (Cambridge, MA: Harvard University Pres, 2012).

75. There are many studies of 1968, both in particular countries and in transnational perspective. I am particularly indebted to Elaine Carey, *Protests in the Streets: 1968 across the Globe* (Indianapolis: Hackett Publishing, 2016).

76. Philip Jenkins, *Decade of Nightmares: The End of the Sixties and the Making of Eighties America* (New York: Oxford University Press, 2006).

77. Carl York, draft report "Scientific and Engineering Manpower," Folder "White House—Manpower [1970–1972]," Box 12, White House Central Files, Staff Member and Offices Files, Office of Science and Technology, Edward E. David Jr., Richard M. Nixon Presidential Library (henceforth David papers).

78. Norman Hackerman, statement before the Subcommittee on HUD–Independent Agencies of the Committee on Appropriations, US House of Representatives, March 2, 1976, Folder 4, Box 3, Norman Hackerman personal and professional papers, 1956–1986, MS 407, Woodson Research Center, Fondren Library, Rice University.

79. D. L. Brooks, "Subjects of Major Interest to NSB during 1971," November 17, 1971, Folder 4, Box 3, Norman Hackerman personal and professional papers, 1956–1986, MS 407, Woodson Research Center, Fondren Library, Rice University.

80. On the less violent end, see Leslie, *Cold War and American Science*, chapter 9. On the more violent end, see Tom Bates, *Rads: The 1970 Bombing of the Army*

Math Research Center at the University of Wisconsin and Its Aftermath (New York: HarperCollins, 1992).

81. Kaiser, "Cold War Requisitions"; Vettel, *Biotech.*

82. See the essays and the introduction to Kaiser and McCray, *Groovy Science.*

83. G. J. Schroepfer letter to Norman Hackerman, June 10, 1975, Folder 12, Box 3, Norman Hackerman personal and professional papers, 1956–1986, MS 407, Woodson Research Center, Fondren Library, Rice University.

84. We need a dedicated study of the Mansfield Amendment, but sketches of it can be found in, for instance, David H. Guston, *Between Politics and Science: Assuring the Integrity and Productivity of Research* (New York: Cambridge University Press, 2000), 78-80.

85. Peter Carpenter (assistant director CMR), "The 1970 Military Authorization and Appropriations Bills and the Fulbright (Mansfield) Amendment," circa January 14, 1970. In Folder "Center for Materials Research," Box 14, Stanford Sponsored Projects Office Collection SC 344 ACCN 1986–200, Department of Special Collections and University Archives, Stanford University Libraries (hereafter SUA).

86. Memo to file from William Massy (Stanford Vice Provost for Research) and Pat Devaney, June 19, 1974, re: Meeting with Paul Shewmon (Director of the Division of Materials Research, NSF), re Materials Research Laboratories. In Folder "National Science Foundation," Box 125, Richard Lyman Presidential Papers (hereafter Lyman Papers) Series 1, SC 215, SUA.

87. First quote: memo to the NSF file from William Massy, February 3, 1975, re: Meeting with Tom Meloy, Division Director for Engineering, January 30, 1975. Second quote: Massy and Devaney memo to file, June 19, 1974. Both from Box 125, Folder "National Science Foundation," Lyman Papers.

88. Massy to file, February 3, 1975.

89. Ibid.

90. Ibid.

91. Massy and Devaney, memo to file, June 19, 1974.

92. William Massy, letter to F. P. Thieme (Chair, Committee on the Eighth National Science Board report, National Science Foundation), June 30, 1975. In Folder "National Science Foundation," Box 125, Lyman Papers.

93. "Talking Business," *New York Times* April 19, 1979: D2.

94. Jenkins, *Decade of Nightmares*, 188–194. To some extent, Jenkins buys into (or at least doesn't critique) the unrealistic rhetoric surrounding high tech, especially biotechnology: biotech's boosters in the 1970s presented it as the key to a limitless future even though the industry was several decades from making a profit.

95. Jacqueline Dowd Hall, "The Long Civil Rights Movement and the Political Uses of the Past," *Journal of American History* 91.4 (2005): 1233–1263.

96. Nelson, *Body and Soul*; Norman Dain, "Critics and dissenters: reflections on 'anti-psychiatry' in the United States," *Journal of the History of the Behavioral Sciences* 25.1 (1989): 3–25; David Miguel Molina and P. J. Blount, "Bringing the Moon to Mankind: The Civil Rights Narrative and the Space Age," in *NASA and the Long Civil Rights Movement*, ed. Brian C. Odom and Stephen P. Waring (Gainesville: University Press of Florida, 2019): 44–60.

97. Daniel T. Rodgers, *Age of Fracture* (Cambridge, MA: Belknap, 2011).

98. In making this claim I expand on the arguments in Elizabeth Popp Berman, "Not Just Neoliberalism: Economization in US Science and Technology Policy," *Science, Technology & Human Values* 39.3 (2014): 397–431.

99. What follows takes some inspiration from the work of Nikolas Rose. In particular, the three arms of neoliberalism that I outline somewhat resemble Rose's three interlinked dimensions of the "enterprising self" (political, institutional, and ethical) in "Governing the Enterprising Self," in *The Values of the Enterprise Culture: The Moral Debate*, ed. Paul Heelas and Paul Morris (Milton Park, UK: Routledge, 1992): 141–164.

100. Michel Foucault, *The Birth of Biopolitics: Lectures at the Collège de France, 1978–1979*, trans. Graham Burcell (New York: Palgrave Macmillan, 2008); Matthew T. Huber, *Lifeblood: Oil, Freedom, and the Forces of Capital* (Minneapolis: University of Minnesota Press, 2013).

101. Jacob William Ward, "Information and control: Inventing the communications revolution in post-war Britain," PhD dissertation, University College London, 2018.

102. Philip Mirowski, *Science-Mart: Privatizing American Science* (Cambridge, MA: Harvard University Press, 2011).

103. Peter Galison and Barton Bernstein, "In Any Light: Scientists and the Decision to Build the Superbomb, 1952–1954," *Historical Studies in the Physical and Biological Sciences* 19.2 (1989): 267–347.

104. Rebecca Slayton, *Arguments That Count: Physics, Computing, and Missile Defense, 1949–2012* (Cambridge, MA: MIT Press, 2013); Donald MacKenzie, *Inventing Accuracy: A Historical Sociology of Nuclear Missile Guidance* (Cambridge, MA: MIT Press, 1993); Scott Kirsch, *Proving Grounds: Project Plowshare and the Unrealized Dream of Nuclear Earthmoving* (New Brunswick: Rutgers University Press, 2005).

105. Hugh Gusterson, "A Pedagogy of Diminishing Returns: Scientific Involution across Three Generations of Nuclear Weapons Science," in *Pedagogy and the Practice of Science: Historical and Contemporary Perspectives*, ed. David Kaiser (Cambridge, MA: MIT Press, 2005): 75–107.

106. On the original reorientation of the National Labs in the 1970s, see Peter J. Westwick, *The National Labs: Science in an American System, 1947–1974* (Cambridge, MA: Harvard University Press, 2003). In the 1980s the National Labs largely withdrew from their forays into civilian-oriented research. However, after the end of the Cold War some of those 1970s efforts returned, with a more commercial spin. See Andrew Schrank, "Green Capitalists in a Purple State: Sandia National Laboratories and the Renewable Energy Industry in New Mexico," in *State of Innovation: The U.S. Government's Role in Technology Development*, ed. Fred Block and Matthew R. Keller (Milton Park, UK: Routledge, 2011): 96–108.

107. Matthew Jones, "Through the List-Serv, Darkly: User Groups and the Emergence of Data Mining in the 1990s," *Society for the History of Technology meeting* (St. Louis: October 13, 2018).

108. Thomas F. Gieryn, "Boundary-Work and the Demarcation of Science from Non-Science: Strains and Interests in Professional Ideologies of Scientists," *American Sociological Review* 48.6 (1983): 781–795; Andrew Abbott, *The System of Professions: An Essay on the Division of Expert Labor* (Chicago: University of Chicago Press, 1988).

109. Daniel J. Kevles, *The Physicists: The History of a Scientific Community in Modern America* (Cambridge, MA: Harvard University Press, 1997 [1971]); Joseph D. Martin, *Solid State Insurrection: How the Science of Substance Made American Physics Matter* (Pittsburgh: University of Pittsburgh Press, 2018).

110. RSC, letter to Joseph E. Mayer (president of the APS), February 28, 1973, Folder 12, Box 26, American Physical Society Records, 1899–1989, NBL.

111. Martin, *Solid State Insurrection.*

112. New England Combined Chapter of the AVS, workshop on "Optical Thin Films, Their Properties and Application to Solar Energy Conversion," June 5, 1974, in Folder 55, Box 14, American Vacuum Society Records, 1953–2007, NBL.

113. Oreskes and Conway, *Merchants of Doubt*; S. Fred Singer, "Report to the Council of the American Geophysical Union from the Committee on Environmental Quality," April 19, 1970, Folder 21, Box 24, American Geophysical Union Records, 1919–2012, NBL.

114. Ad Hoc Committee on Population Exposure, "Radiation Dose to the General Population," March 1971, Folder 16, Box 20, American Association of Physicists in Medicine Records, 1896–2007, NBL.

115. Long range planning committee memo to Board of Directors of AVS, re: Preliminary Thoughts and Comments on the Course of AVS, appendix to April 10, 1969 board meeting minutes, Folder "Board Minutes 1969," Box 1, American Vacuum Society Records, 1953–2007, NBL.

116. Minutes, Board of Directors meeting, December 8, 1983, in Folder "Board Minutes, 1983," Box 1, American Vacuum Society Records, 1953–2007, NBL.

117. J. R. Arthur, "AVS President's Report: Fourth Quarter 1983," *Newsletter of the AVS* (November 1983), in Folder "Newsletters, 1983–1986" Box 4, American Vacuum Society Records, 1953–2007, NBL.

118. Jeanne Baker (member of Zehe's defense team) letter to Edward N. Sickafus (president of the AVS), November 13, 1984, Folder 59, Box 2, American Vacuum Society Records, 1953–2007, NBL. Kristie Macrakis, *Seduced by Secrets: Inside the Stasi's Spy-Tech World* (Cambridge, UK: Cambridge University Press, 2008) provides some background on Zehe.

119. Arthur, "President's Report."

120. Arthur, "President's Report."

121. Rick Perlstein, *The Invisible Bridge: The Fall of Nixon and the Rise of Reagan* (New York: Simon & Schuster, 2014).

122. Lawrence Badash, "Science and McCarthyism," *Minerva* 38 (2000): 53–80. Other fields did the same: see, for instance, David H. Price, *Threatening Anthropology: McCarthyism and the FBI's Surveillance of Activist Anthropologists* (Durham: Duke University Press, 2004).

123. Memo from RA to FL, May 19, 1980, Folder 4, Box 180, Massachusetts Institute of Technology, Office of the President and Chancellor, Records of the President (Jerome B. Wiesner), 1960–1984, Archival Collection 8, Institute Archives and Special Collections, MIT Libraries.

124. Ibid.

125. Paul Gray, memo to file re: conversation with Dr. Robert E. Kahn, May 14, 1980, Folder 4, Box 180, Massachusetts Institute of Technology, Office of the President and Chancellor, Records of the President (Jerome B. Wiesner), 1960–1984, Archival Collection 8, Institute Archives and Special Collections, MIT Libraries.

126. John Krige, "Export Controls as Instruments to Regulate Knowledge Acquisition in a Globalizing Economy," in *How Knowledge Moves: Writing the Transnational History of Science and Technology*, ed. John Krige (University of Chicago Press, 2019): 62–94.

127. American Institute of Physics, "Free Dissemination of Scientific Information," in Folder 59, Box 2, American Vacuum Society Records, 1953–2007, NBL.

128. "Pentagon Goes on a Secrecy Tear," *Electronics* (November 17, 1983): 24. Italics in original.

129. Colin Norman, "To Catch a Spy," *Science* 222.4626 (November 25, 1983): 904–905.

130. Robert Jones, "Congress Shall Place No Restraint on Our Future," *Industrial Research and Development* (September, 1983): 11.

131. Kevles, *The Physicists* offers the clearest statement of this view, but there are hints of it in, for example, Leslie, *Cold War and American Science*, and Wisnioski, *Engineers for Change*. With specific respect to space history see Martin Collins, "The 1970s: Spaceflight and Historically Reinterpreting the In-Between Decade," in *Limiting Outer Space: Astroculture after Apollo*, ed. Alexander C. T. Geppert (New York: Palgrave Macmillan, 2018): 29–49.

132. For example, Hoddeson, Kolb, and Westfall, *Fermilab*.

Chapter 2

1. Kairn A. Klieman, "U.S. Oil Companies, the Nigerian Civil War, and the Origins of Opacity in the Nigerian Oil Industry," *Journal of American History* 99.1 (2012): 155–165; Itty Abraham, "Rare Earths: The Cold War in the Annals of Travancore,"

in *Entangled Geographies: Empire and Technopolitics in the Global Cold War*, ed. Gabrielle Hecht (Cambridge, MA: MIT Press, 2011): 101–124.

2. Bruce J. Schulman, *From Cotton Belt to Sunbelt: Federal Policy, Economic Development, and the Transformation of the South, 1938–1980* (New York: Oxford University Press, 1991); Jennifer Light, *From Warfare to Welfare: Defense Intellectuals and Urban Problems in Cold War America* (Baltimore: Johns Hopkins University Press, 2003); Manuel Castells and Peter Hall, *Technopoles of the World: The Making of Twenty-First-Century Industrial Complexes* (Milton Park, UK: Routledge, 1994); Steven Klepper, *Experimental Capitalism: The Nanoeconomics of American High-Tech Industries* (Princeton: Princeton University Press, 2016).

3. Peter J. Westwick (ed.), *Blue Sky Metropolis: The Aerospace Century in Southern California* (Berkeley: University of California Press, 2012); AnnaLee Saxenian, *Regional Advantage: Culture and Competition in Silicon Valley and Route 128* (Cambridge, MA: Harvard University Press, 1994); Martin Kenney (ed.), *Understanding Silicon Valley: The Anatomy of an Entrepreneurial Region* (Stanford: Stanford University Press, 2000).

4. Margaret Pugh O'Mara, *Cities of Knowledge: Cold War Science and the Search for the Next Silicon Valley* (Princeton: Princeton University Press, 2005); Stuart W. Leslie, "Regional Disadvantage: Replicating Silicon Valley in New York's Capital Region," *Technology and Culture* 42.2 (2001): 236–264; Paul Ceruzzi, *Internet Alley: High Technology in Tysons Corner, 1945–2005* (Cambridge, MA: MIT Press, 2008); Thomas J. Misa, *Digital State: The Story of Minnesota's Computing Industry* (Minneapolis: University of Minnesota Press, 2013); Martin Kenney and David C. Mowery (eds.), *Public Universities and Regional Growth: Insights from the University of California* (Stanford: Stanford University Press, 2014).

5. On Michigan/Ohio see Klepper, *Experimental Capitalism*, and Lillian Hoddeson and Peter Garrett, *The Man Who Saw Tomorrow: The Life and Inventions of Stanford R. Ovshinsky* (Cambridge, MA: MIT Press, 2018). On Dallas/Houston and other energy capitals, see Joseph A. Pratt, Martin V. Melosi, and Kathleen A. Brosnan (eds.), *Energy Capitals: Local Impact, Global Influence* (Pittsburgh: University of Pittsburgh Press, 2014). In the history of science and technology there are many classic studies of "high-tech" firms, universities, and other organizations in the Philadelphia–New York (really, the Wilmington–Schenectady) corridor, though it has received less focus *qua* region than places like Silicon Valley. Firms anchoring the corridor included DuPont, RCA, AT&T, IBM, and GE. Studies of R&D at those anchors include David A. Hounshell and John Kenly Smith Jr., *Science and*

Corporate Strategy: DuPont R&D, 1902–1980 (Cambridge, UK: Cambridge University Press, 1988); Benjamin H. Gross, *TVs of Tomorrow: How RCA's Flat-Screen Dreams Led to the First LCDs* (Chicago: University of Chicago Press, 2018); Jon Gertner, *The Idea Factory: Bell Labs and the Great Age of American Innovation* (London: Penguin, 2012); James W. Cortada, *IBM: The Rise and Fall and Reinvention of a Global Icon* (Cambridge, MA: MIT Press, 2019); George Wise, *Willis R. Whitney, General Electric, and the Origins of U.S. Industrial Research* (New York: Columbia University Press, 1985).

6. The military–industrial connection to biotech regions needs further examination, but it can be glimpsed by combining works on biotech regions with works on biotech and the military. For the former, see Joseph Cortright and Heike Mayer, *Signs of Life: The Growth of Biotechnology in the U.S.* (Washington, DC: Brooking Institution, 2002) and Martin Kenney, *Biotechnology: The University-Industrial Complex* (New Haven: Yale University Press, 1988). For the latter see Shelley L. Hurt, "The Military's Hidden Hand: Examining the Dual-Use Origins of Biotechnology in the American Context, 1969–1972," in *State of Innovation: The U.S. Government's Role in Technology Development*, ed. Frederick L. Block and Matthew R. Keller (New York: Routledge, 2016): 31–56; I take the latter as useful in understanding the biotech region that emerged near Washington, DC, in particular.

7. Turner, *Counterculture to Cyberculture*; Vettel, *Biotech*; John Markoff, *What the Dormouse Said: How the 60s Counterculture Shaped the Personal Computer Industry* (New York: Penguin, 2005).

8. Heather Paxson, "When Chèvre was Weird: Hippie Taste, Technoscience, and the Revival of American Artisanal Food Making," in *Groovy Science: Knowledge, Innovation, and American Counterculture*, ed. David Kaiser and W. Patrick McCray (Chicago: University of Chicago Press, 2016); Pat Kehde, "The Changing Face of Downtown Lawrence, 1970–1985," mailer put out by the Watkins Museum of History, 2018.

9. James R. Lehning, "Technological Innovation, Commercialization, and Regional Development: Computer Graphics in Utah, 1965–1978," *Information & Culture* 51.4 (2016): 479–499.

10. Westwick, *Blue Sky Metropolis*; Matthew H. Hersch, "Equitable growth and Southern California's aerospace industry." White paper for Washington Center for Equitable Growth, 2015.

11. "In Memory of Glen Culler," *Engineering Today* 36 (2003), http://www.engineering.ucsb.edu/coe/pubs/noted/MemoriamF03.pdf. Culler deserves more

attention than I can give him here, but his story closely parallels that of his friends Virgil Elings and David Phillips. Notably, Culler had somewhat more contact with student protesters than they did: his computer center was briefly occupied in an almost uniquely successful protest by the UCSB Black Student Union. Sean Gilleran, "'We Have Your Mechanical Brain': An Internet History of Malcolm X Hall," unpublished term paper, UCSB, 2018.

12. David Hounshell, "The Cold War, RAND, and the Generation of Knowledge, 1946–1962," *Historical Studies in the Physical and Biological Sciences* 27.2 (1997): 237–267.

13. Paul Erickson *et al.*, *How Reason Almost Lost Its Mind: The Strange Career of Cold War Rationality* (Chicago: University of Chicago Press, 2013).

14. Sharon Ghamari-Tabrizi, *The World of Herman Kahn: The Intuitive Science of Thermonuclear War* (Cambridge, MA: Harvard University Press, 2009).

15. Oreskes and Conway, *Merchants of Doubt.*

16. On management consulting, see Christopher D. McKenna, *The World's Newest Profession: Management Consulting in the Twentieth Century* (Cambridge, UK: Cambridge University Press, 2006). As this book was nearing completion, the gap in the literature on think tanks started to fill somewhat. See Pamela Lee, *Think Tank Aesthetics: Midcentury Modernism, the Cold War, and the Neoliberal Present* (Cambridge, MA: MIT Press, 2020); Stuart W. Leslie, "Cold War Suburbs: Thinking about the Unthinkable in Style," *Southern California Quarterly* 102.1 (2020): 24-56.

17. Stephen B. Adams, "Growing Where You Are Planted: Exogenous Firms and the Seeding of Silicon Valley," *Research Policy* 40.3 (2011): 368–379.

18. *TEMPO Santa Barbara* pamphlet (approx. 1961), quoted in Special Collections staff, "Guide to the G.E. Tempo Collection, 1957–1969," MSS 139 (Santa Barbara: University of California at Santa Barbara Davidson Library, 1999).

19. Bretton Fosbrook, "How scenarios became corporate strategies: Alternative futures and uncertainty in strategic management," PhD dissertation, York University, 2017 discusses the migration of scenario planning from defense to corporate contexts, but with a focus on Herman Kahn and the Hudson Institute. There is surely a wider story about that migration that incorporates places like TEMPO.

20. *TEMPO Santa Barbara.*

21. Interview with Peter Zavattaro by Mary Palevsky, Las Vegas, NV, May 31, 2005.

22. "Finalists Vie for APS Industrial Physics Prize," *APS News*, 17.7 (2008): 1, 5.

23. The quote is from an interview with Philip Wyatt conducted by Joe Anderson and Orville Butler, Santa Barbara, CA, March 10, 2011. I also draw on an interview with Philip Wyatt conducted by the author, Santa Barbara, CA, October 13, 2011.

24. David T. Phillips, "Evolution of a Light Scattering Photometer," *BioScience* 26.16 (August 15, 1971): 865–867.

25. Jennifer S. Light, *From Warfare to Welfare: Defense Intellectuals and Urban Problems in Cold War America* (Baltimore: Johns Hopkins University Press, 2003). Similarly, Thomas Turnbull, From paradox to policy, chapter 6 looks at RAND's interest in energy conservation, which emerged from a military context but civilianized in the 1970s.

26. Belanger, *Enabling American Innovation*.

27. Paul Sabin, "Crisis and Continuity in U.S. Oil Politics, 1965–1980," *Journal of American History* 99.1 (2012): 177–186.

28. Robin Wolfe Scheffler, *A Contagious Cause: The American Hunt for Cancer Viruses and the Rise of Molecular Medicine* (Chicago: University of Chicago Press, 2019).

29. Chet Holcombe, "Science Spectrum Receives Contract," *Santa Barbara News-Press* May 26, 1970, and "Subsidies Are Urged in Pollution Fight," *Santa Barbara News-Press*, January 15, 1971.

30. Ibid., "Subsidies Are Urged."

31. "Harry Brown President of Science Spectrum," *Santa Barbara News-Press* July 14, 1972.

32. Philip J. Wyatt, David T. Phillips, and Edward H. Allen, "Laser Light Scattering Bioassay for Veterinary Drug Residues in Food Producing Animals. 1. Dose-Response Results for Milk, Serum, Urine, and Bile," *Journal of Agricultural and Food Chemistry* 24.5 (1976): 984–988.

33. Philip J. Wyatt *et al.*, "Laser Differential Light-Scattering Bioassay for Methotrexate (NSC-740)," *Cancer Treatment Reports* 60 (1976): 225–233.

34. Vettel, *Biotech*.

35. John Hillaby, "Anti-Matter Rain on Earth Hinted: Florida Physicist [i.e., Wyatt] Suggests Some Craters Resulted from Such Assaults," *New York Times* April 26,

1958; David J. Shayler and Colin Burgess, *NASA Scientist-Astronauts* (New York: Springer, 2007). Also Wyatt interview with Anderson and Butler.

36. Wyatt interview with author; "Harry Brown," *Santa Barbara Free-Press*; Value Line Development Capital Corporation, *Annual Report 1973*, Elgin Groseclose Papers, University of Oregon Special Collections and University Archives, Box 29, Folder 3.

37. Phillips, "Evolution."

38. Special feature on the IR 100 Competition and Awards, *Industrial Research*, 14 (August 1972): 28.

39. Wyatt Technology, *2009 American Physical Society Prize for Industrial Application of Physics*, 2009. http://www.wyatt.com/events/news/2009-american-physical-society-prize-for-industrial-application-of-physics.html, accessed June 12, 2019.

40. O'Mara, *Cities of Knowledge*.

41. Paul Barrett, departmental history manuscript, 1987, found in UCSB Physics Department basement "archive."

42. Jong-Jean Kim, "Raman scattering from ferroelectric crystals of KDP family," PhD dissertation, UCSB, 1970.

43. David T. Phillips and John West, "The Poor Man's Nitrogen Laser," *American Journal of Physics* 38.5 (1970): 655–657.

44. Barrett departmental history, 30.

45. Letter from Charles Tart to Lila Gatlin *et al.*, November 2, 1976, Folder 20, Box 1, Santa Barbara Parapsychology Collection (henceforth SBPC), ARC Mss 48, Department of Special Collections, Davidson Library, University of California at Santa Barbara (henceforth UCSB Special Collections). Tart, Puthoff, and Rauscher figure prominently in Kaiser, *How the Hippies Saved Physics*.

46. Jeffrey J. Kripal, *Esalen: America and the Religion of No Religion* (Chicago: University of Chicago Press, 2007).

47. "Esalen Presents Psychic Phenomena," flyer for symposium starting May 7, 1974, SBPC, Box 4, Folder 20.

48. William H. Brock, *William Crookes (1832–1919) and the Commercialization of Science* (Aldershot, UK: Aldgate Press, 2008) examines an establishment figure

who was deeply interested in parapsychology. As we'll see, Crookes resembles Phillips as well in commercializing his interest in the paranormal.

49. Harry Collins, *Gravity's Shadow: The Search for Gravitational Waves* (Chicago: University of Chicago Press, 2004), 110–112.

50. David Phillips, memo to Robert Morris, re: Independent Evaluation of Research Program and Technique, 1975?, Folder 32, Box 1, SBPC.

51. David Phillips, letter to Anne Francis, January 16, 1983, stapled into notebook "Audio Circuits 1982 to June 1984," Box 1, SBPC.

52. Interview with Linda and Dan Phillips conducted by the author, June 26, 2013.

53. David Phillips to Arthur Hastings, August 11, 1974, Folder 16, Box 1; David Phillips to Barry Singer, October 20, 1980, Folder 19, Box 1; both SBPC.

54. David Phillips to Zepha Bogert, Folder 13, Box 1, SBPC.

55. Glen Phillips, "Garage Lyrics," available at http://www.stlyrics.com/songs/g/glenphillips25809/garage2036108.html, accessed June 12, 2019.

56. Sonali Shah, "From Innovation to Firm Formation: Contributions by Sports Enthusiasts to the Windsurfing, Snowboarding, and Skateboarding Industries," in *The Engineering of Sport 6*, ed. Eckehard Mortiz and Steve Haake (New York: Springer, 2006): 29–34.

57. Letter from Phillips to Davis M. Peck, Research Chairman, Spiritual Frontiers Fellowship, Rockville, MD, September 4, 1974, Folder 18, Box 1; letter from Phillips to Bill Welch, Encino, CA, July 5, 1974, Folder 20, Box 1; both SBPC.

58. David Phillips, letter to Winifred Babcock, Santa Barbara, CA, February 6, 1974, SBPC, Box 1, Folder 13; letter to Irving Laucks, Santa Barbara, CA, June 11, 1975.

59. Harry Nelson, "Psychic Medicine—Is It Real Science or Is It Hokum?: 'New Type of Energy' Attracting Attention of Experts World Over," *Los Angeles Times* July 30, 1972; J. W. White, "Acupuncture: The World's Oldest System of Medicine," *Psychic* 3.6 (1972): 12–18.

60. David Phillips, letter to editors of JAMA, December 1972, Box 1, Folder 16, SBPC; advertisements for Acupointer ran in *Today's Chiropractic, Anesthesia and Analgesia*, and *The Digest of Chiropractic Economics*.

61. Walt Halley, "Parapsychologists Seek Public Acceptance of Their Research," *Daily Nexus* 59.50 (November 21, 1978).

62. David Phillips and Robert Morris, "Internal State Variables in Non-Causal Correlation Studies," [draft of NSF proposal], 1975; Robert Morris, letter to editors of *Science*, September 27, 1978; both Folder 32, Box 1, SBPC.

63. "Clairvoyance Test Planned Saturday," *Santa Barbara News-Press* February 28, 1973.

64. David Phillips, letter to Walter Uphoff, April 1, 1973, Folder 21, Box 1, SBPC. Croiset claimed that Phillips unexpectedly changed the parameters of his test and threw off his concentration. Willem Oltmans, *Memoires 1973* (Breda: Papieren Tijger, 2004), 57.

65. I. Fatt, E. L. Hahn, and J. D. Jackson, "Evaluation of the Graduate Program of the Department of Physics, University of California, Santa Barbara," June 11, 1974, Folder 9, Box 88, UArch 13 Academic Senate Records, UCSB Special Collections.

66. Ibid.

67. UCSB Technology Management Program, 2008, interview with Virgil Elings, available at http://www.youtube.com/watch?v=H9aQBF7rIg8, accessed June 12, 2019; also interview with Virgil Elings conducted by the author, Santa Barbara, CA, May 9, 2012.

68. Fatt, Hahn, and Jackson, "Evaluation."

69. Information here and elsewhere on the dates of UCSB physicists' employment and promotions is taken from inspection of university directories available in UCSB Special Collections.

70. Barrett departmental history, 4.

71. H. A. Neal, T. L. Smith, and J. B. McCormick, *Beyond Sputnik: U.S. Science Policy in the 21st Century* (Ann Arbor: University of Michigan Press, 2008); National Science Foundation, *National Patterns of R&D Resources: 1994* (Washington, DC: National Science Foundation, 1994).

72. National Science Foundation, *The Twenty-First Annual Report of the National Science Foundation*, NSF 72-1 (Washington, DC: National Science Foundation, 1972), 57–63. RANN is the kind of organization that pops up briefly in a number of histories of the period but has not been the focus of a dedicated study. The most extensive account is possibly Turnbull, "From paradox to policy," chapter 5.

73. UCSB Contracts and Grants Office, "Annual Report on Activities Financed from Extramural Funds," 1972, Folder "Annual Report on Activities Financed by Extramural Funds, Sept. 1972,", Box 1, UArch 87 Office of Research Collection, UCSB Special Collections.

74. Davd Kaiser, "Cold War requisitions, scientific manpower, and the production of American physicists after World War II," *Historical Studies in the Physical Sciences* 33 (2002): 131–159.

75. Ibid.

76. UCSB Physics, Undergraduate brochure, undated (late 1970s or very early 1980s), Folder "Brochures, c. 1970s,", Box 1, UArch 122, UCSB Dept. of Physics Collection, UCSB Special Collections.

77. UCSB Physics, Physics 13 Spring 1976 "Environmental Physics," Folder "Course Flyers, 1976–1990," Box 1, UArch 122, UCSB Dept. of Physics Collection, UCSB Special Collections.

78. UCSB Office of Research and Development, "Organized Research Units," Folder "Organized Research Units, UCSB 1973–1974, Nov. 1973," Box 1, UArch 87 Office of Research Collection, UCSB Special Collections.

79. UCSB Physics, "Undergraduate brochure."

80. David Phillips, letter to Anne Francis, Director of La Bergerie/Inner Space Foundation, Santa Barbara, CA, January 16, 1983, Folder 3, Box 2, SBPC.

81. Fatt, Hahn, and Jackson, "Evaluation," 21.

82. "Proposal for a Master of Science Degree Program," UCSB Physics Department basement archive. The same phrase was used to describe the MSI program in several years' "Graduate Study in Physics" booklets, Folder "'Graduate Study in Physics' Booklets, c. 1970s," Box 1, UArch 122, UCSB Dept. of Physics Collection, UCSB Special Collections.

83. Advertisement, *Daily Nexus* 53.115 (April 27, 1973), 11. It is likely that the same ad appeared in Berkeley's *Daily Californian*; UC Riverside's *The Highlander*; California StateUniversity, Los Angeles's *The University Times*; and California Polytechnic State University's *Mustang*. The much cheaper rate for in-state tuition meant that Elings focused his recruitments efforts on California campuses, which supplied most of the program's students.

84. "Oil Company Gives Computer to UCSB," unknown date, author, and publication (although most likely the *Santa Barbara News-Press*). Found in UCSB Physics Department basement archive.

85. Ibid.

86. "Two UCSB Students Develop Device to Aid Blind with Sound Meters," *Santa Barbara News-Press* March 30, 1975.

87. "UCSB Students to Describe Undersea Monitoring Device," undated but probably 1975, unknown author and publication (but most likely the *Santa Barbara News-Press*). Found in UCSB Physics Department basement archive.

88. UCSB Office of Public Information, "Physics Student Wins DuPont Fellowship," October 12, 1973, Folder "Elings, Virgil," Box 11, UArch 11, Public Information Office Biographical Files, UCSB Special Collections.

89. Bob English, "*It's Called a Chromophone*: Invention Enables Deaf to 'See' Sounds on Color TV," UCSB Office of Public Information press release, February 12, 1973, Folder "Physics—Scientific Instrumentation 1973–1975," Box 56, UArch 12, Public Information Office Subject Files, UCSB Special Collections. Emphasis in original. This press release was taken up by papers at UCSB and Goleta and even the *LA Times*. See George Alexander, "'OO' is Blue: Deaf Children to 'Read' Sounds by Color on TV Set," *Los Angeles Times*, February 13, 1973.

90. V. Elings and D. Phillips, "An Interdisciplinary Graduate Curriculum in Scientific Instrumentation," *American Journal of Physics* 41 (1973): 570–573.

91. Ibid.

92. Robert English, "Scientific Instrumentation: 'Practical Scientists' Produced by New UCSB Program," November 7, 1972, Folder "Physics—Scientific Instrumentation, 1973–1975," Box 56, UArch 12 Public Information Office Subject Files, UCSB Special Collections.

93. Arnold A. Strassenburg, "Preparing Students for Physics-Related Jobs," *Physics Today* (October, 1973): 23–29.

94. Untitled, unsigned (but must be by Buchin), undated (but probably early 1974) report, probably intended to help UCSB's press office write a press release about the thermodilution computer. Emphasis and punctuation in original. Folder "Physics—Scientific Instrumentation 1973–1975," Box 56, UArch 12 Office of Public Information Subject Files, UCSB Special Collections. In the same folder,

immediately preceding Buchin's report, is the press release it was probably written for: Bob English, "A Student Project: Simple System Tells How Much Blood Heart Pumps per Minute," May 24, 1974.

95. David Phillips, letter to Marian Nester, Director of Education, American Society for Psychical Research, September 10, 1972, Folder 1, Box 4, SBPC.

96. John Placer, Robert L. Morris, and David T. Phillips, "MCTS: A Modular Communications Testing System," in *Research in Parapsychology 1976*, ed. J. D. Morris, W. G. Roll, and R. L. Morris (Metuchen, NJ: The Scarecrow Press, 1977): 38–40.

97. Notebook "UCSB Parapsychology 1974 Plethysmograph Agi Ban/D.T. Phillips," Folder 1, Box 2, SBPC.

98. Elings and Phillips, "Interdisciplinary Graduate Curriculum."

99. David F. Nicoli, Paul H. Barrett, and Virgil B. Elings, "Masters in Instrumentation," *Physics Today* 31.9 (1978): 9.

100. Christophe Lécuyer, *Making Silicon Valley: Innovation and the Growth of High Tech, 1930–1970* (Cambridge, MA: MIT Press, 2005).

101. Kristen Haring, *Ham Radio's Technical Culture* (Cambridge, MA: MIT Press, 2006).

102. Turner, *From Counterculture to Cyberculture*; Walter Isaacson, *Steve Jobs* (New York: Simon & Schuster, 2011); Markoff, *What the Dormouse Said.*

103. I've interviewed several scientists who've commented on this. Probably the clearest in describing his joy at finding that he could order operational amplifiers from a catalog (and the leap in experimental capability that resulted) was Clayton Teague, Gaithersburg, MD, June 28, 2002.

104. Nicoli, Barrett, and Elings, "Masters."

105. Ibid.

106. Virgil Elings, review of *The Art of Electronics* by Horowitz and Hill, *Physics Today* (March, 1981): 68–70.

107. Virgil Elings, "'Invent or Die' Is the Key to Success in Science," *R&D Magazine* (March, 1995): 21.

108. Moore, *Disrupting Science*; Henry Trim, "A Quest for Permanence: The Ecological Visioneering of John Todd and the New Alchemy Institute," in *Groovy*

Science: Knowledge, Innovation and American Counterculture (Chicago: University of Chicago Press), 142–171.

109. "UCSB Professors Supporting the October 15 Vietnam Moratorium," *El Gaucho* 50.15 (October 14, 1969): 3; "UCSB Faculty Are Supporting the Environmental Slate Crawford-Mylod-Fulks for Goleta Water Board," *Daily Nexus* 68.41 (November 3, 1987): 4.

110. David Turner Phillips, "Intensity correlation spectroscopy," PhD dissertation, University of California, Berkeley, 1966; Virgil Elings biographical data form for application for UCSB position, Folder "Elings, Virgil," Box 11, UArch 11 Public Information Office Biographical Files, UCSB Special Collections.

111. Ibid., Elings biographical data.

112. Letter from Advertising Department of Rainbow Enterprises to Virgil Elings, April 9, 1973, re: Display advertising in Probe the Unknown magazine; followed by draft ad, "Electronic Instrument Detects Centers of Psychic Energy"; both Folder 3, Box 1, SBPC.

113. "SPOT: The Simple Speech Display," advertisement, Folder "Elings, Virgil," Box 11, UArch 11 Public Information Office Biographical Files, UCSB Special Collections; note summarizing call from Jay Alan Barker, teacher at Utah School for Deaf and Blind, to Phillips residence, June 14, 1981, Folder 2, Box 2, SBPC. I have also found a mostly blank notebook (that must have been Phillips's) entitled "Santa Barbara Technology Model 1700 Parts/Order/Production Planning," implying that Phillips was working on SBT's product line; Folder 7, Box 2, SBPC.

114. I've heard versions of this story from Elings's later employees, but not from Elings himself. It's possible those employees may have heard it not from Elings but from web and newspaper accounts sourced to Michael Levin, the founder of Optigone. See "Mirage® Mania in the Media," https://optigone.com/index.php/mirage-in-the-media/, accessed June 12, 2019. That site quotes at length Tom Carter, "Mirage Relies on Light to Create Illusion," *Lexington Herald-Leader* November 24, 1992.

115. Virgil B. Elings and Caliste J. Landry, "Optical Display Device," US patent no. 3,647,284, granted March 7, 1972.

116. Virgil B. Elings and David T. Phillips, "Apparatus and Method for Measuring Cardiac Output," US patent no. 4,015,593, granted April 5, 1977.

117. Interview with Glenn Schiferl conducted by the author, Santa Barbara, CA, October 13, 2011.

118. Barrett departmental history, 28–29.

119. Vettel, *Biotech*; Galison, *Image and Logic.*

120. V. B. Elings, G. E. Jahn, and J. H. Vogel, "A theoretical model of regionally ischemic myocardium," *Circulation Research* 41 (1977): 722–729.

121. Fatt, Hahn, and Jackson., "Evaluation," 16.

122. "David Turner Phillips," obituary posted by cantorjoeocho, http://boards.ancestry.com/topics.obits/66031/mb.ashx?pnt=1, accessed January 23, 2012.

123. Wyatt interview with the author.

124. Philip J. Wyatt, "Days of Wine and Lasers," *Science* 212.4500 (June 12, 1981): 1212–1214.

125. Philip J. Wyatt, "The Taste of Things to Come," *Applied Optics* 21.14 (July 15, 1982): 2471–2472; "Oenological Laser," *Popular Mechanics* 156.4 (October, 1981): 194.

126. Phillips notebook, July 1984–November 1985, Folder 4, Box 2, SBPC. Phillips may have been participating in the emergence of "financial astrology" as a popular and professional practice: Caley Horan, "Investing in the Stars: The Rise of Financial Astrology in 1970s America" (Milan: Society for the History of Technology Annual Meeting, October 25, 2019).

127. Mody, *Instrumental Community.*

128. For a statement of faith in the transformative economic and social consequences of such start-ups, see, for instance, Richard L. Florida, *Rise of the Creative Class: And How It's Transforming Work, Leisure, and Everyday Life* (New York: Basic Books, 2004).

129. For a case study of some of the difficulties in reaping that dividend, see Schrank, "Green Capitalists in a Purple State."

130. Mirowski, *Science-Mart* and Forman, "The Primacy of Science in Modernity.

131. Mowery, Nelson, Sampat, and Ziedonis: *Ivory Tower and Industrial Innovation*; Elizabeth Popp Berman, "Why Did Universities Start Patenting?: Institution-Building and the Road to the Bayh-Dole Act," *Social Studies of Science* 38.6 (2008):

835–871; Mark Peter Jones, "Biotech's perfect climate: The Hybritech story," PhD dissertation, UC San Diego, 2005; Sally Smith Hughes, *Genentech: The Beginnings of Biotech* (Chicago: University of Chicago Press, 2011); Doogab Yi, *The Recombinant University: Genetic Engineering and the Emergence of Stanford Biotechnology* (Chicago: University of Chicago Press, 2015); Jeannette Colyvas, "From divergent meanings to common practices: Institutionalization processes and the commercialization of university research," PhD dissertation, Stanford University, 2007.

Chapter 3

1. Creso M. Sá, "'Interdisciplinary Strategies' in U.S. Research Universities," *Higher Education* 55.5 (2008): 537–552; Julie T. Klein, "Evaluation of Interdisciplinary and Transdisciplinary Research: A Literature Review," *American Journal of Preventive Medicine* 35.2 supplement (2008): S116–S123.

2. Scott Frickel, Mathieu Albert, and Barbara Prainsack (eds.), *Investigating Interdisciplinary Collaboration: Theory and Practice across Disciplines* (New Brunswick: Rutgers University Press, 2017); Harvey J. Graff, *Undisciplining Knowledge: Interdisciplinarity in the Twentieth Century* (Baltimore: Johns Hopkins University Press, 2015); Jerry Jacobs, *In Defense of Disciplines: Interdisciplinarity and the Research University* (Chicago: University of Chicago Press, 2013); Sarah Kaplan, Jonathan Milde, and Ruth Schwarz Cowan, "Symbiont Practices in Boundary Spanning: Bridging the Cognitive and Political Divides in Interdisciplinary Research," *Academy of Management Journal* 60.4 (2016): 1387–1414; Andrew Barry and Georgina Born (eds.), *Interdisciplinarity: Reconfigurations of the Social and Natural Sciences* (Milton Park, UK: Routledge, 2013).

3. The notion of keywords comes from Raymond Williams, but here I borrow on John Patrick Leary, *Keywords: The New Language of Capitalism* (Chicago: Haymarket Books, 2018) for the application of the idea to the discourse of university administrators (though, oddly, "interdisciplinarity" is not one of Leary's keywords).

4. The inappropriateness—yet ubiquity—of Stanford as a model for higher education in general is particularly remarked upon in Steven Shapin, *The Scientific Life: A Moral History of a Late Modern Vocation* (Chicago: University of Chicago Press, 2008).

5. See, among others, Lowen, *Creating the Cold War University*; Leslie, *The Cold War and American Science*; Stuart W. Leslie and Robert H. Kargon, "Selling Silicon Valley: Frederick Terman's Model for Regional Advantage," *The Business History*

Review 70.4 (Winter 1996): 435–472; Christophe Lécuyer, *Making Silicon Valley*; O'Mara, *Cities of Knowledge*; Timothy J. Sturgeon, "How Silicon Valley Came to Be," in *Understanding Silicon Valley: The Anatomy of an Entrepreneurial Region* (Stanford: Stanford University Press, 2000), 15–47.

6. Colyvas, "From divergent meanings to common practices"; Andrew J. Nelson, *The Sound of Innovation: Stanford and the Computer Music Revolution* (Cambridge, MA: MIT Press, 2015); Sally Smith Hughes, "Making Dollars out of DNA: The First Major Patent in Biotechnology and the Commercialization of Molecular Biology, 1974–1980," *Isis* 92 (2001): 541–575; Doogab Yi, *The Recombinant University*; Tim Lenoir and Eric Giannella, "The emergence and diffusion of DNA microarray technology," *Journal of Biomedical Discovery and Collaboration* 1 (2006): 11; Kenney, *Biotechnology*.

7. Two partial exceptions are Vettel, *Biotech*, and Christophe Lécuyer, "What do Universities Really Owe Industry? The Case of Solid State Electronics at Stanford," *Minerva* 43 (2005): 51–71.

8. On MIT, see Leslie, *The Cold War and American Science*. For both Hopkins and MIT, see Michael Aaron Dennis, "'Our First Line of Defense': Two University Laboratories in the Postwar American State," *Isis*, 85.3 (1994): 427–455. For Cornell and MIT, see S. S. Schweber, "Big Science in Context: Cornell and MIT," in *Big Science: The Growth of Large-Scale Research* (Stanford: Stanford University Press, 1992): 149–183. For Cornell alone, see Cyrus C. M. Mody and Hyungsub Choi, "From Materials Science to Nanotechnology."

9. Stuart W. Leslie, "Playing the Education Game to Win: The Military and Interdisciplinary Research at Stanford," *Historical Studies in the Physical Sciences* 18.1 (1987): 55 88.

10. C. Stewart Gillmor, *Fred Terman at Stanford: Building a Discipline, a University, and Silicon Valley* (Stanford: Stanford University Press, 2004).

11. Lowen, *Creating the Cold War University*, 213.

12. Vettel, *Biotech*, ch. 4.

13. Linda Sargent Wood, *A More Perfect Union: Holistic Worldviews and the Transformation of American Culture after World War II* (New York: Oxford University Press, 2010) examines the rapid growth in antireductionist views in the 1960s. This surely underlies the turn toward interdisciplinarity at places like Stanford.

14. Interview conducted by the author. Some details changed to preserve anonymity.

15. Rudolf Kompfner, "Generalities about a national science policy," undated (probably 1971), Folder 1, Box 1, Rudolf Kompfner papers, SC 194 ACCN 85–074, SUA.

16. John Baldeschwieler memo to Hu Heffner, July 9, 1971, Folder "Domestic Council Study Group on NTO," Box 114, David papers.

17. "Values, Technology, and Society Included in Experimental Program," *Campus Report* May 19, 1971.

18. Ibid.

19. "Dr. Miller: Society Should 'Strike Balance' in Research Activities," *Campus Report* November 25, 1970.

20. Memo from Rambo to Peter C. Allen, re: Engineering pieces for General Secretary [asking for a rephrasing of a paragraph about Rambo in Allen's report], September 14, 1971, Folder 1, Box 6, William Rambo papers, SC 132 ACCN 97–093, SUA.

21. Untitled speech, Folder "Pres. Pitzer Dinner July 23, 1970," Box 19, William Rambo papers, SC 132 ACCN 97–093, SUA.

22. Simon Schaffer, "How Disciplines Look," in *Interdisciplinarity: Reconfigurations of the Social and Natural Sciences* (Milton Park, UK: Routledge, 2013): 57-81.

23. Cyrus C. M. Mody and Joseph D. Martin, "Materials Science," *Encyclopedia of the History of Science* (June 2020), accessed June 16, 2020. https://doi.org/10.34758/6afy-w006.

24. R. J. Wasilewski (section head, Materials Research Laboratory section, National Science Foundation), letter to Robert Huggins (director, Stanford Center for Materials Research), February 2, 1972. Folder "Center for Materials Research," Box 14, Stanford Sponsored Projects Collection, SUA.

25. Mody and Choi, "From Materials Science to Nanotechnology."

26. Memorandum of understanding between Stanford and NASA, April 1968, Folder "National Aeronautics and Space Administration," Box 125, Lyman Papers.

27. Ibid.

28. Particularly in Leslie, *Cold War and American Science*, chapter 9.

29. "A Message from the Stanford Sit-In," Folder "AEL Sit-In April 1969," Box 10, Stanford University Sponsored Projects Office collection, SC 344, SUA.

30. Richard W. Lyman, *Stanford in Turmoil: Campus Unrest, 1966–1972* (Stanford: Stanford General Books, 2009), 145–147.

31. Engineering Students for a Fair Hearing, "Demonstrations at Stanford Autumn 1965–Spring 1971," Folder "Demonstrations at Stanford Autumn 1965-Spring 1971," Box 1, Peter C. Allen Student Demonstration Papers, SC 157, SUA. The tie between defenses of academic freedom and, well, defense was a deliberate feature of Cold War strategy: see Wolfe, *Freedom's Laboratory*.

32. BM (Research Engineer, Systems Techniques Laboratory), letter to Kenneth Pitzer, April 15, 1969. In Folder 2 Disruption Documents, Box 1, Peter C. Allen Student Demonstration Papers, SC 157, SUA.

33. Lyman was an historian of the British Labor Party of the 1920s. Sharon Traweek suggests in conversation that this was key to his success: his research showed him how Labor politicians navigated between radicals bent on revolution and an establishment intent on crushing it (with armed force if necessary).

34. Stanton A. Glantz, "Comments about Engineers for Engineering by an Engineer," *Grindstone* 1.1 (November 1970), collection ARCH 3009 The Grindstone, SUA.

35. Stanton A. Glantz and Norm V. Albers, "Department of Defense R&D in the University," *Science* 186.4165 (November 22, 1974): 706–711.

36. "Professor Urges Ideology Tempered with Concern," *Campus Report* February 12, 1969.

37. Memo from C. J. Shoens to "all administrative officers of the university, all members of the Academic Senate and Academic Council, research staff, administrative assistants, secretaries, technicians, students of all political persuasions," re: SDS March on STL 11-5-68, November 6, 1968. In Folder 9, Box 6, William Rambo papers, SC 132 ACCN 97–093, SUA.

38. "Welcome Pitzer!!," December, 1968, Folder 10, Box 6, William Rambo papers, SC 132 ACCN 97–093, SUA. Underlining in original.

39. Stanford SDS, *Through the Looking Glass: A Radical Guide to Stanford*, Folder 9, Box 6, William Rambo papers, SC 132 ACCN 97–093, SUA.

40. "First SES Reports Released: Continuous Process Hoped for," *Campus Report* November 26, 1968; Lyman, *Stanford in Turmoil*, 126–128.

41. "New Engineering Curriculum Follows Specific SES Proposal," *Campus Report* February 12, 1969.

42. "Private Gifts during 1969–70 Top $30 Million; 19,000 Donors," *Campus Report* September 23, 1970.

43. William F. Massy, draft proposal for a Center for Interdisciplinary Research, 1972, Folder 9, Box 14, Sponsored Projects Office Records, SC-344 ACCN 1986–200, SUA.

44. "Educational Improvement Plan Possible under Ford Grant," *Campus Report* September 25, 1968.

45. Lyman, *Stanford in Turmoil*, 161–162.

46. "Revolt against Reason Can Have 'Ominous Consequences'," *Campus Report* January 27, 1971.

47. John G. Linvill, "Research and Development of Tactile Facsimile Reading Aid for the Blind," Final Report for Grant No. OEG-0-8-071112-2995, March 1973, page 5. My thanks to Mara Mills for sending me a copy of this report and for further discussion of this topic.

48. Memo from John G. Linvill to Committee for the Study of Stanford's Educational Program, re: Comments of Electrical Engineering on Stanford's Educational Program and Objectives, January 26, 1967, binder "Electrical Engineering," Calvin Quate papers SC 347 ACCN 1987 (henceforth Quate papers), SUA.

49. Linvill, Final Report, page 10.

50. Stanford Electronics Laboratories, *Stanford University Electronics Research Review 11*, August 12, 1969, Collection 3120/4 STAN, SUA.

51. "Research Exposition Is Planned; Organizational Meeting Tomorrow," *Campus Report* December 9, 1970.

52. "Research Exposition Is Planned," *Campus Report* January 27, 1971.

53. Linvill, "Final Report," 14.

54. The evidence and arguments of this section are given in greater detail in Cyrus C. M. Mody, "Conversions: Sound and Sight, Military and Civilian," in *Oxford Handbook of Sound Studies*, ed. Trevor Pinch and Karin Bijsterveld (New York: Oxford University Press, 2011): 224–248.

55. National Science Foundation, "The Promise of Acoustic Microscopy," *Mosaic* (March/April 1978): 35–41.

56. Rudolf Kompfner, "Recent Advances in Acoustical Microscopy," *British Journal of Radiology* 47 (1975): 615–627.

57. Gilbert B. Devey, "Ultrasonic Microscope," *Radio-Electronic Engineering* (February 1953): 8–9.

58. John Linvill memo to Calvin Quate, December 16, 1966, Binder "Electrical Engineering," Quate papers.

59. G. C. Coburn letter to Calvin Quate, May 18, 1976, Folder "Contract—Hartford Foundation, Inc., Correspondence—Proposals (1970–1981)," Box 1, Quate papers.

60. Calvin F. Quate, "JSEP Progress Report for the Acoustic Microscope," November 1977, Folder "ONR-632 JSEP," Box 1, Quate papers.

61. Rüdiger Graf, "Claiming Sovereignty in the Oil Crisis: 'Project Independence' and Global Interdependence in the United States, 1973/74," *Historical Social Research* 39.4 (2014): 43–69.

62. Turnbull, "From paradox to policy," chapter 5.

63. See the contents of Folder "Contract #GI 37518 NSF-RANN," Box 1, Quate papers.

64. Calvin F. Quate, Abdullah Atalar, and H. K. Wickramasinghe, "Acoustic Microscopy with Mechanical Scanning—A Review," *Proceedings of the IEEE* 67 (1979): 1092–1114.

65. Calvin F. Quate, "Acoustic Microscopic Study of Materials," Ginzton Lab Report No. 2890, November 1978, Folder "AF 77–3455," Box 2, Quate papers.

66. Mody, *The Long Arm of Moore's Law.*

67. Calvin Quate letter to John Dimmock (Office of Naval Research), February 9, 1977, Folder "Contract #N00014–75-C-0632 ONR (JSEP) Correspondence 1974–77," Box 1, Quate papers.

68. Calvin Quate letter to Capt. Stephen Wax (Air Force Office of Scientific Research), circa February 1977, Folder "AFOSR 98 Correspondence 1978–82," Box 2, Quate papers.

69. Calvin F. Quate, "Innovative Measurement Technology for the Semiconductor Industry—The Acoustic Microscope—A New Instrument for Viewing Integrated

Circuits," Ginzton Lab Report No. 2820, June 1978, Folder "NBS 5–35899 Reports (9/1/75–4/2/79)," Box 2, Quate papers.

70. Calvin F. Quate memo to NSF/RANN, re: Acoustic Microscopy (attached to Quate letter to Gilbert B. Devey, NSF, December 19, 1974), Folder "Contract APR75–07317 National Science Foundation April 1, 1975," Box 2, Quate papers.

71. Calvin Quate letter to Dr. Richard C. Atkinson (NSF), November 12, 1979, Folder "Correspondence 1979 July-December," Box 3, Quate papers.

72. Quate notebook, April–June 1983, Box 2, Quate papers (2004 accession).

73. Jamie Cohen-Cole, *The Open Mind: Cold War Politics and the Sciences of Human Nature* (Chicago: University of Chicago Press, 2014).

74. Beatrice Cherrier and Aurélien Saïdi, "A Century of Economics and Engineering at Stanford," *History of Political Economy* 52.S1 (2020): 85–111.

75. Nancy Donham, Stanford University News Service release on the Center for Interdisciplinary Research, July 10, 1972, Folder "Center for Interdisciplinary Research (CIR), 1972–1974," Box 14, Stanford University Sponsored Projects Office collection SC 344, SUA.

76. Holt Ashley, "New Directions for Engineering Research," *Grindstone* 1.1 (November 1970), collection ARCH 3009 Grindstone, SUA.

77. Further argument and evidence regarding SAIL and CCRMA can be found in Cyrus C. M. Mody and Andrew J. Nelson, "'A Towering Virtue of Necessity': Interdisciplinarity and the Rise of Computer Music at Vietnam-Era Stanford," *Osiris* 28 (2013): 254–277.

78. Lester Earnest (ed.), "Final Report: The First Ten Years of Artificial Intelligence Research at Stanford," *Stanford Artificial Intelligence Laboratory Memo AIM-228* (July 1973); Bruce G. Buchanan, "Introduction to the COMTEX Microfiche Edition of Memos from the Stanford University Artificial Intelligence Laboratory" *AI Magazine* 4.4 (Winter 1983): 37–42.

79. John M. Chowning, Leland C. Smith, and Albert Cohen. "The Computer Music Facility: A New Medium" proposal submitted to the National Endowment for the Arts in conjunction with the proposal "Computer Simulation of Music Instrument Tones in Reverberant Spaces" submitted to the National Science Foundation. June 18, 1974, CCRMA papers ACCN 2001–262, SUA.

80. Letter from William Miller, provost, to Joseph Danek, program manager, Institutional Grants for Science, NSF, May 13, 1975, Folder "National Science Foundation, Misc.," Box 23, Lyman Papers Series 3.

81. Letter from James Angell to Pat Devaney, re: RDF report, January 14, 1976, Folder "National Science Foundation, Misc.," Box 23, Lyman Papers Series 3.

82. Nelson, *The Sound of Innovation.*

83. See Paul Lehrman, "A Talk with John Chowning," *Mix*, Part I, February 1, 2005, http://www.mixonline.com/mag/audio_talk_john_chowning/, and Part II, March 1, 2005, http://www.mixonline.com/mag/audio_talk_john_chowning_2/.

84. Nelson, *The Sound of Innovation*, 48.

85. C. F. Quate, record of invention for "Electronic Device for Converting Written Music to Audible Sound," No. S74–09, Hansen Lab Docket ML 316, January 16, 1974. In Box 7, CCRMA papers, ACCN 2001–262, SUA.

86. Joel N. Shurkin, *Broken Genius: The Rise and Fall of William Shockley, Creator of the Electronic Age* (London: Macmillan, 2006).

87. John Pierce letter to Calvin Quate, October 19, 1970, blue binder "Correspondence, External August 1969-March 1971," Box 5, John R. Pierce Papers, Huntington Library.

88. Lécuyer, "What Do Universities Really Owe Industry?"

89. My thanks to Rebecca Slayton for conversation on this point. See also Slayton, "From a 'Dead Albatross' to Lincoln Labs."

90. For example, Sally Smith Hughes, "Making Dollars out of DNA: The First Major Patent in Biotechnology and the Commercialization of Molecular Biology, 1974–1980," *Isis*, 92.3 (2001): 541–575; Kaushik Sunder Rajan, "Introduction," in *Lively Capital: Biotechnologies, Ethics, and Governance in Global Markets*, ed. Sunder Rajan (Durham: Duke University Press, 2012).

91. For example, Doogab Yi, *The Recombinant University*, and Jeannette Colyvas, "From divergent meanings to common practices." The partial exception, which I lean on heavily, is Eric Vettel's *Biotech*, which makes the connection between the protests and the turn toward "relevance" among Stanford's molecular biologists but is more speculative on Stanford's transformation into a model of academic capitalism.

92. Quoted in Colyvas, "From divergent meanings to common practices," 69–70.

93. Robert Lamar, Stanford University News press release, January 15, 1975, Folder "American Optical Corporation," Box 1, Quate papers.

94. Hughes, "Making Dollars out of DNA."

95. Donald Kennedy, "Health Research: Can Utility and Quality Co-Exist?," draft of speech (at University of Pennsylvania?), December 6, 1980, Folder 10, Box 1, Stanford University Sponsored Projects Office collection, SC 344, SUA.

96. Memo from James D. Meindl (chair, Committee on Research) to Senate of the Academic Council, re: Report of the 1982–83 Committee on Research, September 29, 1983, Folder 14, Box 1, Stanford University Sponsored Projects Office collection, SC 344, SUA.

97. Brent D. Maher, "Technically Allowed: Federal Scrutiny of Stanford University's Indirect Cost Expenditures and the Changing Context for Research Universities in the Post-Cold War Era," *History of Education Quarterly* 59.1 (2019): 97–127.

Chapter 4

1. Evening in the contiguous 48 US states, that is.

2. Blount and Molina, "The Distance from the Ghetto to the Moon"; Asif A. Siddiqi, "Competing Technologies, National(ist) Narratives, and Universal Claims: Towards a Global History of Space Exploration," *Technology and Culture* 51.2 (2010): 425–443.

3. Matthew D. Tribbe, *No Requiem for the Space Age: The Apollo Moon Landings and American Culture* (New York: Oxford University Press, 2014); Roger D. Launius, "Public Opinion Polls and Perceptions of US Human Spaceflight," *Space Policy* 19.3 (2003): 163–175.

4. Denis Cosgrove, "Contested Global Visions: *One-World, Whole Earth,* and the Apollo Space Photographs," *Annals of the Association of American Geographers* 84.2 (1994): 270–294.

5. Roger D. Launius, *Reaching for the Moon: A Short History of the Space Race* (New Haven: Yale University Press, 2019).

6. John F. Kennedy, speech delivered September 12, 1962, in Houston, Texas. Transcripts of the speech can be found several places. JSC itself prominently displays

it here: "John F. Kennedy Moon Speech—Rice Stadium," https://er.jsc.nasa.gov/seh/ricetalk.htm, accessed June 14, 2019.

7. These figures are an example of the kind of information that Wikipedia is good at collecting and presenting: see "Budget of NASA," https://en.wikipedia.org/wiki/Budget_of_NASA, accessed June 14, 2019.

8. George M. Low, statement before Subcommittee on NASA Oversight, Committee on Science and Astronautics, US House of Representatives, 1970 (?), in Box 6, Center Series, Section 89, Poindexter Reference Files, Johnson Space Center History Collection, University of Houston, Clear Lake Archives, Alfred R. Neumann Library, UHCL (henceforth JSC collection).

9. For example, Wisnioski, *Engineers for Change*, 36.

10. For example, David P. D. Munns, *Engineering the Environment: Phytotrons and the Quest for Climate Control in the Cold War* (Pittsburgh: University of Pittsburgh Press, 2017); David A. Mindell, *Digital Apollo: Human and Machine in Spaceflight* (Cambridge, MA: MIT Press, 2008); Erik Conway, *Atmospheric Science at NASA: A History* (Baltimore: Johns Hopkins University Press, 2008); Joseph Tatarewicz, *Space Technology and Planetary Astronomy* (Bloomington: Indiana University Press, 1990).

11. Trevor Pinch and Frank Trocco, *Analog Days: The Invention and Impact of the Moog Synthesizer* (Cambridge, MA: Harvard University Press, 2002).

12. My exploration of existential success could be seen as a contribution to the study of "projects," that is, time-bounded efforts with a defined goal. See Philip Scranton, "Projects as a Focus for Historical Analysis: Surveying the Landscape," *History and Technology* 30.4 (2014): 354–373.

13. Scheffler, *A Contagious Cause*.

14. Space Task Group, "The Post-Apollo Space Program: Directions for the Future," August 27, 1969, in Folder "White House Space Task Force 1969–1," Box 35, David papers.

15. John Krige, Angelina Long Callahan, and Ashok Maharaj, *NASA in the World: Fifty Years of International Collaboration in Space* (New York: Palgrave Macmillan, 2013).

16. Wolfe, *Competing with the Soviets*.

17. Hamblin, *Arming Mother Nature*; Scott Kirsch, *Proving Grounds: Project Plowshare and the Unrealized Dream of Nuclear Earthmoving* (New Brunswick: Rutgers University Press, 2005).

18. Mark D. Bowles, *Science in Flux: NASA's Nuclear Program at Plum Brook Station, 1955–2005* (Washington, DC: Government Printing Office, 2006); Olof Hallonsten, *Big Science Transformed: Science, Politics, and Organization in Europe and the United States* (New York: Palgrave Macmillan, 2016).

19. Quoted in Loren Graham, "Science and Computers in Soviet Society," *Proceedings of the Academy of Political Science* 35.3 (1984): 124–134.

20. Macrakis, *Seduced by Secrets*, looks into East German attempts to gain access to Western and Japanese semiconductor equipment.

21. Michael Conrad, "The Lure of Molecular Computing," *IEEE Spectrum* 23.10 (October, 1986): 55–60; Konstantin K. Liharev, "Superconductor Digital Electronics," *Physica C* 482 (November 20, 2012): 6–18.

22. Kieran Downes, "From enthusiasm to practice: Users, systems, and technology in high-end audio," PhD dissertation, MIT, 2009.

23. Vacuum tube revenues peaked in 1954 and were lower than transistor revenues from 1958 onward (and lower than integrated circuit revenues from 1965 onward). Unni Pillai, "The Origins of the Tools Suppliers in the Semiconductor Industry," *Business History* (2020), DOI: 10.1080/00076791.2020.1844666.

24. Zachary, *Endless Frontier*; Allan A. Needell, *Science, Cold War, and the American State* (Milton Park, UK: Routledge, 2013).

25. Hyungsub Choi and Brit Shields, "A Place for Materials Science: Laboratory Buildings and Interdisciplinary Research at the University of Pennsylvania," *Minerva*, 53.1 (2015): 21–42.

26. Office of Physical Sciences of the National Research Council, *Revitalizing Laboratory Instrumentation* (Washington, DC: National Academy Press, 1982).

27. Michael J. Neufeld, "The 'Von Braun Paradigm' and NASA's Long-Term Planning for Human Spaceflight," in *NASA's First 50 Years: Historical Perspectives*, ed. Steven J. Dick (Washington, DC: Government Printing Office, 2012): chapter 13.

28. Valerie A. Olson, "American extreme: An ethnography of astronautical visions and ecologies," PhD dissertation, Rice University, 2010.

29. McCray, *The Visioneers.*

30. John Logsdon, *After Apollo: Richard Nixon and the American Space Program* (New York: Palgrave Macmillan, 2015).

31. Matthew H. Hersch, *Inventing the American Astronaut* (New York: Palgrave Macmillan, 2012).

32. Raymond Waldmann letter to Ed Harper and Arnie Weber, re: The L.A. Unemployment Situation, February 12, 1971, Folder "White House Manpower 1970–72," Box 12, David papers.

33. Richard K. Cook (Special Assistant to the President) report to Congressman Bill Archer, "Summary of Actions Taken by the Federal Government to Assist Unemployed Scientists and Engineers," March 2, 1971, Folder "White House Manpower 1970–72," Box 12, David papers.

34. Robert J. Bazell, "NSF: Is Applied Research at the Take Off Point?," *Science* 172.3990 (June 25, 1971): 1315–1317.

35. Lawrence A. Goldmuntz memo to Hubert Heffner, July 7, 1971, Folder "Domestic Council Study Group on NTO," Box 114, David papers.

36. Robert R. Gilruth (Director of the Manned Spaceflight Center) letter to Wernher von Braun (Deputy Associate Administrator, NASA), July 28, 1971, Folder "MIUS Correspondence 1971," Box 1, Center Series, Section 97, JSC collection.

37. Andrew J. Butrica, "The Electronics Research Center–NASA's little known venture in aerospace electronics," AIAA 2002–1138 for the Fortieth AIAA Aerospace Sciences Meeting & Exhibit (Reno, NV: 2002).

38. Glenn E. Bugos, *Atmosphere of Freedom: Sixty Years at the NASA Ames Research Center* (Washington, DC: NASA History Office, 2000); Felicity D. Scott, "Securing Adjustable Climate," in *Climates: Architecture and the Planetary Imaginary*, ed. James Graham (Zürich: Lars Müller Publishers, 2016): 90–105.

39. Bowles, *Science in Flux.*

40. Virginia P. Dawson, *Engines and Innovation: Lewis Laboratory and American Propulsion Technology*, NASA Document SP-4306 (Washington, DC: NASA, 1991), 205.

41. Logsdon, *After Apollo.*

42. "Overview of Program and Directorate Briefing," September 26–27, 1974, Folder "Overview Program and Directorate Briefing, September 26, 1974," Box 4, Center Series, Section 103 JSC collection.

43. Howard Haglund, interview conducted by Casani-Clauser, October 22, 1975, Ad Hoc Study Group on JPL/Campus Interactions, 1975–1976 collection, Caltech Archives.

44. Peter J. Westwick, *Into the Black: JPL and the American Space Program, 1976–2004* (New Haven: Yale University Press, 2007).

45. National Aeronautics and Space Act of 1958, Public Law 85–568, 72 Stat., 426–438, July 29, 1958.

46. Bruce Seely, "NASA and Technology Transfer in Historical Perspective," *Comparative Technology Transfer* 6.1 (2008): 1–16.

47. "Technology Applications Progress Report for the period 1 January 1972–31 December 1972," Technology Utilization Office, December 1972.

48. Ibid.

49. "Technology Utilization Program Report," December 1973, NASA SP-5119.

50. "Technology Applications Progress Report for the period 1 January 1972–31 December 1972."

51. Goldmuntz to Heffner.

52. Clare F. Farley (NASA Deputy Assistant Administrator for Technology Utilization) statement before the Committee on Aeronautical and Space Sciences, United States Senate, 1972, Box 6, Center Series, Section 89 Poindexter Files, JSC collection.

53. Sam McFarland, "Application of Features of the NASA Lunar Rover to Vehicle Control for Paralyzed Drivers," Final Report for contract NAS 9–14473, August 1975, report No. NASA-CR-151358, 67.

54. *Meal System for the Elderly: Conventional Food in Novel Form*, Lyndon B. Johnson School of Public Affairs Policy Research Project Report No. 16, 1977.

55. Abner S. Salant (Director, Food Engineering Laboratory, Natick R&D Command) letter to Richard S. Johnston (JSC Director of Space and Life Science), June 16, 1977, Box 7, Center Series, Section 89 Poindexter Files, JSC collection.

56. Ibid., 9.

57. Jurgen Schmandt (ed.), *Shelf Stable Meals for Public Sector Uses: Report of a Conference on the NASA Meal System* (Austin: Lyndon B. Johnson School of Public Affairs, 1977): 36.

58. Ibid.

59. Advertisement, Oregon Freeze Dry Foods, Inc., undated, Folder "Meal Systems for the Elderly 1971–1977," Box 7, Center Series, Section 89, Poindexter Reference Files, JSC collection.

60. Mark D. Bowles, *The 'Apollo' of Aeronautics: NASA's Aircraft Energy Efficiency Program, 1973–1987* (Washington, DC: NASA, 2010).

61. Roger D. Launius, "Managing the Unmanageable: Apollo, Space Age Management, and American Social Problems," *Space Policy* 24 (2008): 158–165.

62. R. S. Johnston, George G. Armstrong, and Sam L. Pool, "A Concept for an Expanded Health Services System," October 29, 1971, Folder XX0D00548, Box LS84, Medical Operations, NASA Space Life Sciences Archive, McGovern Historical Center, Texas Medical Center Library.

63. Jeremy A. Greene, Victor Braitberg, and Gabriella Maya Bernadett, "Innovation on the Reservation: Information Technology and Health Systems Research among the Papago Tribe of Arizona, 1965–1980," *Isis* 111.3 (2020): 443–470; Lisa Cartwright, "Reach out and Heal Someone: Telemedicine and the Globalization of Health Care," *Health* 4.3 (2000): 347–377; Rashid L. Bashshur and Gary W. Shannon, *History of Telemedicine: Evolution, Context, and Transformation* (New Rochelle: Mary Ann Liebert, Inc., 2007): 195–202; Andrew T. Simpson, Charles R. Doarn, and Stephen Garber, "Interagency Cooperation in the Twilight of the Great Society," *Journal of Policy History* 32.1 (2020): 25-51; Victor Braitberg, "Inventing Telemedicine on the Reservation" (unpublished paper).

64. Johnston, Armstrong, and Pool, "A Concept for an Expanded Health Services System." Ellipses in the original.

65. William J. Bifano, Anthony F. Ratajczak, and William J. Ice, "Design and Fabrication of a Photovoltaic Power System for the Papago Indian Village of Schuchuli (Gunsight), Arizona," Thirteenth Photovoltaic Specialists Conference (June 5-8, 1978), reprinted as DOE/NASA/1022-78/39 and NASA TM-78948.

66. "A Remote Area Health Services Research, Demonstration, and Evaluation Project," vol. 2 "Implementation Plan," contract HSM 110-69-243, State of New Mexico Health and Human Services Department, Folder XX0D00346, Box LS84; National Academy of Engineering Subcommittee on Technology and Systems Transfer, "Study of Some Aspects of Aerospace Technology Utilization," April 1, 1972, Folder XX0D00506, Box LS84, 19–21; Permian Basin Regional Planning Commission, "Application for Grant for Permian Basin Region Emergency Medical Services," March 31, 1975, Folder XX0D00742, Box LS79; all in NASA Space Life Sciences Collection, McGovern Historical Center, Texas Medical Center Library.

67. Johnston, Armstrong, and Pool, "A Concept for an Expanded Health Services System."

68. "A Remote Area Health Services Research, Demonstration, and Evaluation Project."

69. Robert R. Ebert, *Flxible: A History of the Bus and the Company* (Yellow Springs, OH: Antique Power, 2001).

70. Robert M. Anderson *et al.*, *Divided Loyalties: Whistle-Blowing at BART* (West Lafayette: Purdue University, 1980); Robert Lindsey, "Mass Transit Woes Bring Aerospace Down to Earth," *New York Times*, April 21, 1974.

71. Adi Robertson, "The Road Not Taken: How We Found (and Lost) the Dream of Personal Rapid Transit," *The Verge*, February 24, 2016, https://www.theverge.com/2016/2/24/11094524/prt-transit-history-self-driving-cars-alden-starrcar-tomorrowland-1960s, accessed July 2, 2020.

72. Wendell E. Pritchett, "Which Urban Crisis? Regionalism, Race, and Urban Policy, 1960–1974," *Journal of Urban History* 34.2 (2008): 266–286.

73. Alondra Nelson, *Body and Soul*; Blount and Molina, "The Distance from the Ghetto to the Moon."

74. Rick Perlstein, *Nixonland: The Rise of a President and the Fracturing of America* (New York: Scribner, 2008).

75. R. L. Sohn *et al.*, "Application of Mobile Digital Communications in Law Enforcement," JPL-SP-43-6-REV-1, 1974; B. D. L. Mulhall, "FBI Fingerprint Identification Automation Study," JPL Publication 80–79, November 15, 1980. There also

appears to have been a bullet ballistics identification project in 1974; Technology Utilization Program Report 1974 (Washington, DC: NASA, 1975): 52.

76. James J. Haggerty, "Aid for the Medical Laboratory," *Spinoff 1986* (Washington, DC: NASA, 1986): 60–61.

77. John Darnton, "NASA Consultant to the City is Finding Down-to-Earth Uses for Space Technology," *New York Times*, October 13, 1973.

78. H. L. Macomber and James H. Wilson, "California Four Cities Program, 1971–1973," JPL-SP-43–4, May 15, 1974.

79. *Aerospace Technology: Creating Social Progress* (Washington, DC: Aerospace Industries Association, undated but probably 1970).

80. The most extensive secondary source on the USPO is Neil M. Maher, *Apollo in the Age of Aquarius* (Cambridge, MA: Harvard University Press, 2017).

81. John Noble Wilford, "Nixon Men Map New Aid to Technology," *New York Times*, October 31, 1971: 1.

82. Edward L. Hays and H. Nugent Myrick, "Environmental Control in Manned Spacecraft: Its Potential Contribution to Pollution Control on Earth," October 25–28, 1971, AIAA Paper No. 71–1012, in box 22, folder 5 "AIAA Paper," Center Series, Section 117, JSC collection.

83. Jeffrey Craig Sanders, *Seattle and the Roots of Urban Sustainability: Inventing Ecotopia* (Pittsburgh: University of Pittsburgh Press, 2012).

84. Sabine Höhler, *Spaceship Earth in the Environmental Age, 1960–1990* (Milton Park, UK: Routledge, 2015); Peder Anker, "The Closed World of Ecological Architecture," *The Journal of Architecture* 13.1 (2008): 527–552.

85. "NASA Report on Urban Development v. 1 A Technical Proposal for a Total Systems Engineering and Integration Approach to Urban Development," August 24, 1971, Folder "Community Development 2," Box 83, David papers. My thanks to Sara Stevens for the comparison of the home on pages 7–47 and 7–48 to John Lautner's architecture.

86. By "relatively quickly" I mean within about three weeks. See the drawings in "NASA Report on Housing Development Final Report v. 2 Illustration of the Application of NASA Technology to Housing Problems," September 16, 1971, Folder 1, Box 2, Center Series, Section 97, JSC collection.

87. Ronald D. Scott (ET/Technology Applications Programs Division) memo to EZ/Manager, Urban Systems Projects Office, March 15, 1973, Folder 4, Box 1, Center Series, Section 97, JSC collection.

88. Charles W. Mathews (Associate Administrator for Applications) letter to Christopher Kraft Jr. (Director, JSC), August 28, 1973, Folder 5, Box 1, Center Series, Section 97, JSC collection.

89. Gerald S. Leighton (Utility Systems Program Manager, HUD Office of the Assistant Secretary for Research and Technology) letter to Arthur J. Miller (Coordinator of HUD Programs, Oak Ridge National Laboratory Reactor Division), September 21, 1972, Folder 3, Box 1, Center Series, Section 97, JSC collection.

90. Ben Fulbright memo to Ted Hays, re: ASHRAE Conference June 24–28, 1973, July 17, 1973, Folder 5, Box 1, Center Series, Section 97, JSC collection.

91. Office of Management and Budget, "Review of New Technology Initiatives—Community Development—Proposal #3: Modular Integrated Utility System," undated but probably 1971, Folder "Domestic Council Study Group on NTO," Box 114, David papers.

92. EZ/Manager (Urban Systems Projects Office) memo to EA2 (Assistant Director for Mechanical and Chemical Systems), re: Alaskan Mini-MIUS Feasibility Study; Status and Organization, June 19, 1972, Folder "MIUS Correspondence Jan-June 1972," Box 1, Center Series, Section 97, JSC collection.

93. Edward Harper memo to Edward David and William Magruder, December 27, 1971, re: Modular Integrated Utilities System Experiment, Folder "MIUS Correspondence 1971," Box 1, Center Series, Section 97, JSC collection.

94. Diane Vaughan, *The Challenger Launch Decision: Risky Technology, Culture, and Deviance at NASA* (Chicago: University of Chicago Press, 1996).

95. Lyn Gordon-Winkler, "NASA and MIUS—A Study in Bureaucratic Failure," term paper for Government 5534.01 (possibly at University of Houston), Folder "Winkler 1981," Box 2, Center Series, Section 97, JSC collection.

96. Memo from Chris Kraft Jr. to Rep. Mike McCormack (Chair, Subcommittee on Energy) and Rep. James W. Symington (Chair, Subcommittee on Space Science and Applications), August 2, 1973, Folder 5, Box 1, Center Series, Section 97, JSC collection.

97. Winkler, "NASA and MIUS," 14.

98. "Evaluation of the Housing Development Program According to Domestic Council Criteria," undated but must be 1971, Folder XX0D00266, Box LS105A, NASA Life Sciences Collection, McGovern Historical Center, Texas Medical Center Library.

99. "NASA Report on Urban Development v. 1 A Technical Proposal for a Total Systems Engineering and Integration Approach to Urban Development."

100. For example, Donald W. Kurtz and Bain Dayman Jr., "Experimental Aerodynamic Characteristics of Vehicles Traveling in Tubes," JPL Technical Memorandum 33–731, July 15, 1975.

101. D. G. Bradshaw (Deputy Manager, Boeing—Houston), "Foreword," Clear Creek Basin Water Quality Study 1974, Folder "Water Quality Study 1975–77," Box 7, Center Series, Section 89, Poindexter Reference Files, JSC collection.

102. Bill O'Donnell and Dennis Williams, "Space Technology Aids Community Water Monitoring Systems," NASA News press release 76–57, March 26, 1976, Folder "Water Quality Study 1975–1977," Box 7, Center Series, Section 89, Poindexter Reference Files, JSC collection.

103. NASA, "Program Plan Solar Heating and Cooling," SCH-1002, January 31, 1975, Folder "Solar Heating, January 31, 1975," Box 184, Center Series, Section NIO, JSC collection.

104. Pamela E. Mack, *Viewing the Earth: The Social Construction of the Landsat Satellite System* (Cambridge, MA: MIT Press, 1990); Brian Jirout, "One Space Age Development for the World: The American Landsat Civil Remote Sensing Program in Use, 1953–2003," PhD dissertation, Georgia Tech, 2016; Kim McQuaid, "Earthly Environmentalism and the Space Exploration Movement, 1960–1990," *Space Policy*, 26 (2010): 163–173.

105. Howard Allaway (NASA News), "Satellite Pictures Help Alaskan Natives Choose Best Land," Release No. 76–63, April 1, 1976; Dick McCormack and Mack Herring (NASA News), "Landsat Satellite to Inventory Navajo Nation's Resources," Release No. 77–223, October 18, 1977; both Box 28, Center Series, Section 89, Poindexter Reference Files, JSC collection.

106. M. Rush and S. Vernon, "Remote Sensing for Urban Public Health," *Photogrammetric Engineering and Remote Sensing* 41 (1975): 1148–1155; C. M. Barnes *et al.*, "Applications of Remote Sensing in Public Health," in *International Symposium*

on Remote Sensing of Environment, October 2–6, 1972 (Ann Arbor: Environmental Research Institute of Michigan, 1973): 677–684.

107. Jeff Womack, "Pipe Dreams for Powering Paradise: Solar Power Satellites and the Energy Crisis," in *American Energy Policy in the 1970s*, ed. Robert Lifset (Norman: University of Oklahoma Press, 2014): 203–220. Glaser's proposal reached Magruder by at least December, 1971; see William Magruder letter to General James M. Gavin (chair, Arthur D. Little), December 11, 1971, Folder "Miscellaneous Memos 1971–1972," Box 79, David papers.

108. McCray, *The Visioneers*.

109. Boeing Aerospace Company Research and Engineering Division, "Solar Power Satellite System Definition Study v. 1 Technical and Management Proposal D180-20147-1," October 1976, Folder "Boeing SPS October 1976," Box 184, Center Series, Section NIO, JSC collection.

110. Womack, "Pipe Dreams."

111. DOE/NASA Satellite Power System Concept Development and Evaluation Program, "Survey and Documentation of Emerging Technologies for the Satellite Power System (SPS)," April 1981, Folder "DOE-ER0097 Survey of Emerging Technologies for the SPS," Box 7, Center Series, Section 108 SPS, JSC collection.

112. Jet Propulsion Laboratory Annual Report 1971, available from "JPL Annual Reports," https://www.jpl.nasa.gov/about/reports.php, accessed June 19, 2019.

113. Jet Propulsion Laboratory Annual Report 1978, available from "JPL Annual Reports," https://www.jpl.nasa.gov/about/reports.php, accessed June 19, 2019.

114. Theodore W. Price, Thomas W. Shain, and James A. Bryant, "Vehicle Test Report: South Coast Technology Electric Conversion of a Volkswagen Rabbit," NASA-CR-164532 (JPL Publication 81–28), February 15, 1981.

115. "Electricity from Photovoltaic Solar Cells Flat-Plate Solar Array Project—10 Years of Progress," JPL 400–279, October 1985; "Mirror on the Sun," JPL 400–10, October 1980.

116. Bruce Murray, "Director's Note," Jet Propulsion Laboratory Annual Report 1978: 3.

117. Richard P. O'Toole and A. L. Walton, "Intergenerational Equity and Conservation," JPL 80–49 (June 15, 1980).

118. R. Rhoads Stephenson (PI), "Should We Have a New Engine?," JPL SP 43–17 (August 1975).

119. A. Nash Williams, "Compendium of Critiques of JPL Report SP-43–17," JPL Pub 77–40 (July 18, 1977).

120. Amy Sue Bix, *Girls Coming to Tech! A History of American Engineering Education for Women* (Cambridge, MA: MIT Press, 2013).

121. Ad Hoc Study Group on JPL/Campus Interactions, "A Study of Relations between the Jet Propulsion Laboratory and the Campus of the California Institute of Technology," April 2, 1975, Folder 35, JPL 173 Harris Schurmeier papers, JPL Archives.

122. "A Study of Relations between the Jet Propulsion Laboratory and the Campus of the California Institute of Technology," part 1, 1976, page 6. Section II, Box 13, Folder 17, The Frank J. Malina Collection, Caltech Archives.

123. RVM, "Reflections on JPL/Caltech Relations," January 23, 1976, Ad Hoc Study Group on JPL/Campus Interaction, 1975–1976 collection, Caltech Archives.

124. John Pierce letter to Bruce Murray, February 7, 1977, Folder 6, JPL 142 Office of the Director collection, JPL Archives.

125. Bruce Murray, "Solar Energy: True God or False Prophet?," *Engineering & Science* (March-April 1978): 4–36; "Supporting Material for Executive Council Retreat, 1979," November 20, 1979, Folder 2, JPL 229 Executive Council Retreat, JPL Archives.

126. "Civil Systems Retitled to Reflect Energy Goal," *Universe* (an in-house JPL newspaper), 8.17 (March 17, 1978), JPL Archives.

127. Interview with B. D. Martin, Biomedical Systems Manager, Civil Systems Program Office, October 27, 1975; interview with Norman Brooks, Environmental Engineering and Science department, Caltech, October 29, 1975; both from Ad Hoc Study Group on JPL/Campus Interactions, 1975-76, Notes from Bourke/Noll Interviews, Caltech Archives.

128. Jet Propulsion Laboratory, *Annual Report*, JPL 400-236 (June, 1984).

129. Bruce Murray memo to management committee, re: E/C [Executive Council] retreat, January 18, 1982, Folder 8, JPL 229 Executive Council Retreat, JPL Archives.

130. Amy E. Foster, *Integrating Women into the Astronaut Corps: Politics and Logistics at NASA, 1972–2004* (Baltimore: Johns Hopkins University Press, 2011); John Krige, "A Victory for Clean Interfaces: Europe's Participation in the Space Shuttle Program," in *Space Shuttle Legacy: How We Did It and What We Learned*, ed. Roger D. Launius, John Krige, and James I. Craig (Washington: AIAA, 2013): 265–282; Richard Paul and Steven Moss, *We Could Not Fail: The First African Americans in the Space Program* (Austin: University of Texas Press, 2015).

131. Maher, *Apollo in the Age of Aquarius.*

132. *Goddard News*, 22.12 (March 1975), in Folder "Technology Spin-Offs 1967–1979," Box 7, Center Series, Section 89, Poindexter Reference Files, JSC collection. There was a brief surge in the number of blind programmers from the mid-1960s to the mid-1970s, partly due to calls from the Association for Computing Machinery: see Elizabeth R. Petrick, *Making Computers Accessible: Disability Rights and Digital Technology* (Baltimore: Johns Hopkins University Press, 2015).

133. These shifts in NASA's mission and philosophy are presented—sometimes explicitly but sometimes between the lines—in Logsdon, *After Apollo*; Krige, Callahan, and Maharaj, *NASA in the World*; and Vaughan, *The Challenger Launch Decision.*

Chapter 5

1. Christopher E. Johnson, "'Turn on the sunshine': A history of the solar future," PhD dissertation, University of Washington, 2015.

2. Ed Millis, *Jack St. Clair Kilby: A Man of Few Words* (Dallas: Ed Millis Books, 2008).

3. Because some readers might question my math: Kilby was only two when my father was born, and ten when my mother was.

4. Millis, *Jack St. Clair Kilby*, 74.

5. The story is told several places but probably best in Michael Riordan and Lillian Hoddeson, *Crystal Fire: The Invention of the Transistor and the Birth of the Information Age* (New York: W. W. Norton, 1998).

6. For more on the tyranny of numbers, see Hyungsub Choi, "Manufacturing knowledge in transit: Technical practice, organizational change, and the rise of

the semiconductor industry in the United States and Japan, 1948–1960," PhD dissertation, Johns Hopkins University, 2007.

7. Christophe Lécuyer and David C. Brock, *Makers of the Microchip: A Documentary History of Fairchild Semiconductor* (Cambridge, MA: MIT Press, 2010).

8. "Demonstration of the first working integrated circuit," Historical Marker No. 6709, located at 13500 N. Central Expressway, Dallas, TX; "Site of invention of the first commercially practicable integrated circuit," California Historic Marker No. 1000, 844 E. Charleston Rd., Palo Alto, CA.

9. Leslie Berlin, *The Man behind the Microchip: Robert Noyce and the Invention of Silicon Valley* (New York: Oxford University Press, 2005).

10. T. R. Reid, *The Chip: How Two Americans Invented the Microchip and Launched a Revolution* (New York: Random House, 1985).

11. Millis, *Jack St. Clair Kilby*, 65.

12. Eric S. Hintz, "The Post-Heroic Generation: American Independent Inventors, 1900–1950," PhD dissertation, University of Pennsylvania, 2010.

13. Cyrus C. M. Mody, "Academic Centers and/as Industrial Consortia in American Microelectronics Research," *Management and Organizational History* 12.3 (2017): 285–303.

14. Matthew Wisnioski, *Every American an Innovator: How Innovation Became a Way of Life* (Cambridge, MA: MIT Press, forthcoming).

15. Most of these disclosures were abandoned, but Jack S. Kilby, "Reliable Flashlight," US patent no. 3,898,450, filed November 1, 1973, issued August 5, 1975, received a patent. The full list of the disclosures that TI made with Kilby as inventor is at Melvin Sharp (TI, vice president corporate staff and general counsel), letter to Jack Kilby, October 31, 1983, Folder 2, Box 100, Jack S. Kilby papers, MSS 85317, Library of Congress (henceforth Kilby LoC).

16. Jack S. Kilby, Robert F. Schweitzer, and John McCrady, "Electronic Check Writer," US patent no. 3,920,979, filed October 19, 1973, issued November 18, 1975.

17. Jack Kilby letter to Raymond Matichard (Sintra), April 25, 1974, Folder 16, Box 5, Kilby LoC.

18. Millis, *Jack St. Clair Kilby*, 66.

19. Turner, *From Counterculture to Cyberculture.*

20. Thierry Bardini, *Bootstrapping: Douglas Engelbart, Coevolution, and the Origins of Personal Computing* (Stanford: Stanford University Press, 2000).

21. Joy Lisi Rankin, *A People's History of Computing* (Cambridge, MA: Harvard University Press, 2018); Morgan G. Ames, *The Charisma Machine: The Life, Death, and Legacy of One Laptop per Child* (Cambridge, MA: MIT Press, 2019).

22. Lécuyer, *Making Silicon Valley*, 202; James D. Finn, *Extending Education through Technology*, ed. Ronald J. McBeath (Greenwich, CT: Information Age Publishing, 1972): 287.

23. Barbara Hof, "The Turtle and the Mouse: How Constructivist Learning Theory Shaped Artificial Intelligence and Educational Technology in the 1960s," *History of Education* 50.1 (2021): 93-111.

24. For the mention of Magruder and various companies that Kilby contacted about his teaching meachine, see Kilby's monthly reports to Fred Bucy (then a vice president at Texas Instruments) for 1971 and 1972, Folder 4, Box 100, Kilby LoC.

25. Jay W. Lathrop interview conducted by David Morton, May 1, 1996, interview #265 for the Center for the History of Electrical Engineering, IEEE.

26. Integrated circuit revenues overtook discrete transistor revenues in 1970; by 1976 IC revenues were twice those of discrete transistors (with the divergence between the two rising sharply). Pillai, "The Origins of the Tools Suppliers."

27. Mody, "Academic Centers and/as Industrial Consortia"; Berman, *Creating the Market University.*

28. Jack Kilby letter to Ray Warner, October 17, 1983 (in response to Warner's letter of August 22), in Folder 16, Box 11, Kilby LoC.

29. P. R. Morris, *A History of the World Semiconductor Industry* (London: Peter Peregrinus, 1990).

30. Willis Adcock interview conducted by David Morton, July 27, 2000, interview #403 for the Center for the History of Electrical Engineering, IEEE; Jack S. Kilby interview conducted by Arthur L. Norberg, June 21, 1984, Charles Babbage Institute.

31. Lécuyer, *Making Silicon Valley.*

32. Jack Kilby report to Fred Bucy, November 4, 1978, Folder 7, Box 100, Kilby LoC.

33. Jack Kilby report to Fred Bucy, May 4, 1981, Folder 10, Box 100, Kilby LoC.

34. Jack Kilby report to Fred Bucy, July 2, 1980, Folder 9, Box 100, Kilby LoC.

35. Jack Kilby report to Fred Bucy, August 5, 1980, Folder 9, Box 100, Kilby LoC.

36. Jack Kilby, nomination of Gordon Moore for membership in the National Academy of Engineering, 1975, Folder 4, Box 3, Jack Kilby papers A2006.0023 Series 1: Manuscripts, DeGolyer Library Special Collections, Southern Methodist University (henceforth Kilby SMU). The extreme squareness of its subject is made plain in Thackray, Brock, and Jones, *Moore's Law*. Moore and Kilby seem to have had rather similar personalities, especially in being "men of few words," so I find it unsurprising that Kilby would support Moore's nomination.

37. Jack Kilby report to Fred Bucy, February 6, 1981, Folder 10, Box 100, Kilby LoC.

38. Jack Kilby report to Fred Bucy, November 6, 1980, Folder 9, Box 100, Kilby LoC.

39. See Timothy J. Mitchell, "The Resources of Economics: Making the 1973 Oil Crisis," *Journal of Cultural Economy* 3.2 (2010): 189–204 for a contrarian view of the reality of the oil crisis. For a more conventional summary of the public's experience, see Juliet E. Carlisle *et al.*, *The Politics of Energy Crises* (New York: Oxford University Press, 2017).

40. Millis says Kilby's notebooks contain the first entry on the solar energy system on April 26, 1974. I've found no indication in his correspondence that he was working on it any earlier, and indeed some evidence that it had not crossed his mind as late as February 22, 1974.

41. The fee was $2,400 per month—about $12,000 per month in 2019 dollars. Jack Kilby letter to Jay Lathrop, May 10, 1974, Folder 1, Box 6, Kilby LoC.

42. Wilbur Arthur Porter, "The fundamental nature and control of dislocations in single crystal silicon," PhD dissertation, Texas A&M University, 1970. W. Arthur Porter, *The Knowledge Seekers: Creating Centers for the Performing Sciences* (Austin: IC^2 Institute, 1998).

43. That size estimate and one of many pictures of pellets on the head of a pin can be found in "TI/DOE Joint Press Release," approx. July 7, 1973, in Folder 35, Box 3, Kilby SMU.

44. Keen eyes may notice that there is a missing step: half the pellets would be p-doped and half n-doped for photons to generate potentials among the pellets.

45. Jack Kilby report to Fred Bucy, November 4, 1974, Folder 4, Box 100, Kilby LoC.

46. Jack Kilby report to Fred Bucy, April 5, 1975, Folder 4, Box 100, Kilby LoC.

47. "Proposal to grant TI exclusive patent rights from the TI/Kilby entity under the solar energy converter development," July 23, 1975, Folder 1, Box 100, Kilby LoC.

48. Jack Kilby letter to Hal Levine (TI, assistant vice president and general patent counsel), April 5, 1976, Folder 1, Box 100, Kilby LoC.

49. The analysis that follows is strongly influenced by the picture of a fragmented, politically diverse solar community in Frank N. Laird, "Constructing the Future: Advocating Energy Technologies in the Cold War," *Technology and Culture* 44 (2003): 27–49.

50. Carroll Pursell, "The Rise and Fall of the Appropriate Technology Movement in the United States, 1965–1985," *Technology and Culture*, 34.3 (1993): 629–637.

51. Trim, "A Quest for Permanence."

52. Adam Rome, "The Genius of Earth Day," *Environmental History* 15.2 (2010): 194–205.

53. Lillian Hoddeson and Peter Garrett, *The Man Who Saw Tomorrow: The Life and Inventions of Stanford R. Ovshinsky* (Cambridge, MA: MIT Press, 2018).

54. Jack Kilby report to Fred Bucy, June 5, 1981, Folder 10, Box 100, Kilby LoC.

55. For example, Jack Kilby memo to Morris Chang and Jim Fischer, re: the Illinois Project, January 25, 1977, Folder 1, Box 53, Kilby LoC.

56. News from McDonnell Douglas press release, June 1, 1982, given to the author by Anna-Fay Williams.

57. For lists of firms see, among others, Anthony W. Adler, "Stock Listings of Solar-Related Companies," *Solar Engineering* 2.1 (January 1977): 12. On oil and solar see, for instance, Mark Etherides, "The U.S. Solar Energy Industries and the Role of Petroleum Firms," Research Study #018, March 12, 1980, American Petroleum Institute.

58. Booz Allen Hamilton, "Texas Instruments Solar Energy System Business Plan Briefing," undated but must be late 1982 or early 1983, Folder 24, Box 3, Kilby SMU.

59. Booz Allen Hamilton, "Progress Meeting: Preliminary Marketing Analysis of the Texas Instruments Solar Energy System," October 21, 1982, unfoldered bound material, Box 4, Kilby SMU.

60. Ibid.

61. Sheila Jasanoff and Sang-Hyun Kim, "Sociotechnical Imaginaries and National Energy Policies," *Science as Culture* 22.2 (2013): 189–196.

62. Again, this is a point made with respect to solar energy in the 1970s by Laird, "Constructing the Future," though without drawing on the sociotechnical imaginaries concept.

63. Unknown author, "Development with Government Support," and "Texas Instruments Takes a Partner," undated (probably 1977), Box 2, Record Group 4, Texas Instruments Records, A2005.0025, DeGolyer Library Special Collections, Southern Methodist University; and "Texas Instruments Alone," Box 3, Kilby SMU.

64. Fosbrook, "How scenarios became corporate strategies."

65. "TI Takes a Partner," 6. The iceberg episode opens Toby Craig Jones, *Desert Kingdom: How Oil and Water Forged Modern Saudi Arabia* (Cambridge, MA: Harvard University Press, 2010); Jones paints Faisal as the kind of high-modernist technophile whom TI could plausibly have approached.

66. "Development with Government Support," page 5 of "Conclusions."

67. "Development with Government Support," 3.

68. William D. Metz, "MIT Chemist, Schlesinger Ally Assumes Energy Research Post," *Science* 198.4322 (December 16, 1977): 1125–1126. Deutch would later serve as Deputy Secretary of Defense and then Director of Central Intelligence under Bill Clinton; obviously his future appointments were unknowable at the time, but they show us something about his self-identification with the national security state.

69. Heilmeier was one of the leaders in RCA's development of liquid crystal displays: Gross, *TVs of Tomorrow.*

70. George Heilmeier memo to Jim Fischer (copied to Kilby), re: Solar Energy Discussion with Department of Energy Official, January 31, 1978, Folder 13, Box

53, Kilby LoC. All text and formatting in original except for the words between square brackets.

71. Mike Rice memo to Jack Kilby, Pete Johnson, and Bob Stratton, re: TI Solar Energy Program, July 2, 1979, Folder 13, Box 53, Kilby LoC.

72. Ibid.

73. For example, Jules Levine memo to Pete Johnson, re: funding possibilities for dry arrays in 1983–84, February 2, 1983, Folder 8, Box 53, Kilby LoC.

74. Pete Johnson memo to Bob Stratton, re: TI Solar Energy Program, July 23, 1979, Folder 13, Box 53, Kilby LoC.

75. Matthew Eisler, *Overpotential: Fuel Cells, Futurism, and the Making of a Power Panacea* (New Brunswick: Rutgers University Press, 2012).

76. Unknown SERI employee to unknown TI employees, re: Texas Instruments, Inc. proposal regarding the photovoltaic chemical cell, May 12, 1978, Folder 13, Box 53, Kilby LoC.

77. Ibid.

78. "Texas Instruments Closes Residential Solar-H2 Project," *Hydrogen Times* 4 (Winter 1984): 1, 3, 6.

79. Jack Kilby report to Fred Bucy, December 5, 1978, Folder 7, Box 100, Kilby LoC.

80. Jack Kilby report to Fred Bucy, November 3, 1979, Folder 8, Box 100, Kilby LoC.

81. Jack Kilby report to Fred Bucy, December 1, 1979, Folder 8, Box 100, Kilby LoC.

82. "Hirsch Casts Doubt on Many Aspects of ERDA's Solar Electric Program," *The Energy Daily* October 21, 1976.

83. Ray Reece, *The Sun Betrayed: A Report on the Corporate Seizure of U.S. Solar Energy Development* (Boston: South End Press, 1979): 111.

84. Pat Haggerty press interview conducted by Buzz Selby, September 18, 1979, Folder 2, Box 58, Kilby LoC.

85. "Unsolicited Proposal to Develop Solar Energy System, Volume 1—Technical Proposal," Proposal No. 23-R78, Folder 4, Box 54, Kilby LoC.

86. Millis, *Jack St. Clair Kilby*, 91. Brackets in original. Note, though, that in Millis's anecdote Kilby is brought around by Clinton's charisma and comes to express his somewhat grudging admiration.

87. Jack Kilby report to Fred Bucy, December 9, 1980, Folder 9, Box 100, Kilby LoC.

88. Denis Hayes, "Washington Decrees a Solar Eclipse," *New York Times* August 12, 1981: A27.

89. Ibid.

90. Scott Bennett, "Tempest on the High Tech Seas," *D Magazine* (March 1986).

91. Ibid.

92. Booz Allen Hamilton, "Preliminary Marketing Analysis of the Texas Instruments Solar Energy System," July 1, 1982, unfoldered bound material, Box 4, Kilby SMU.

93. Pete Johnson, "OPC review, Project Illinois," May 17, 1982, Folder 4 Solar Energy Reviews 1978–1982, Box 68, Kilby SMU.

94. See Duccio Basosi, "A Small Window: The Opportunities for Renewable Energies from Shock to Counter-Shock," in *Counter-Shock: The Oil Counter-Revolution of the 1980s*, ed. Duccio Basosi, Giuliano Garavini, and Massimiliano Trentin (London: I. B. Tauris, 2018): 336–356 for a review and criticism of that conventional wisdom. The figures that follow are from Basosi, Garavini, and Trentin's introduction to that volume.

95. Booz Allen Hamilton, "Preliminary Marketing Analysis of the Texas Instruments Solar Energy System."

96. This list is from the documents in Box 1, Brown Williams papers, Hagley Museum and Library.

97. Melvin Sharp to Kilby, October 31, 1983.

98. Jack Kilby to Ray Warner, October 17, 1983.

Chapter 6

1. See, for example, Gary Helm Darden, "The New Empire in the 'New South': Jim Crow in the Global Frontier of High Imperialism and Decolonization," *Southern Quarterly* 46.3 (2009). See Moore, *Disrupting Science*, for discussion of New Left scientists' trips to places like North Vietnam and the People's Republic of China as manifestations of solidarity. No doubt many American scientists of color—square and otherwise—also perceived the global character of the civil rights movement.

2. As detailed from a global perspective in Odd Arne Westad, *The Global Cold War* (Cambridge, UK: Cambridge University Press, 2007).

3. Robert Z. Lawrence, *Can America Compete* (Washington, DC: Brookings Institution, 1984).

4. John Emerson Mohr, "Globalization and the auto industry in the South," PhD dissertation, Auburn University, 2018.

5. Timothy Mitchell, *Carbon Democracy: Political Power in the Age of Oil* (London: Verso, 2011).

6. Jessica Smith Rolston, *Mining Coal and Undermining Gender: Rhythms of Work and Family in the American West* (New Brunswick: Rutgers University Press, 2014), chapter 2.

7. Jeffrey T. Macher, David C. Mowery, and David A. Hodges, "Reversal of Fortune? The Recovery of the U.S. Semiconductor Industry," *California Management Review* 41.1 (1998): 107–136.

8. Bassett, *To the Digital Age*; Lécuyer, *Making Silicon Valley*.

9. For an overview of related approaches and the turn from diplomatic history toward transnational perspectives, see Michael J. Hogan (ed.), *America in the World: The Historiography of American Foreign Relations since 1945* (Cambridge, UK: Cambridge University Press, 1995).

10. Bylinsky, "How Intel Won Its Bet on Memory Chips."

11. Stewart Brand interview with Jeffrey Inaba, *Volume* 24 [Counterculture] (2011).

12. There are subtle and less subtle versions of this view; for the former see the essays in Kenney, *Understanding Silicon Valley*.

13. Klepper, *Experimental Capitalism*.

14. Shurkin, *Broken Genius*; Riordan and Hoddeson, *Crystal Fire*; Lécuyer, *Making Silicon Valley*.

15. Lionel E. Kattner, "Signetics History," undated but before 2008, lot no. X7847.2017, Information Technology Corporate Histories Collection, Computer History Museum.

16. The thumbnail history of Signetics's first few years that follows is taken from Lionel Kattner interview conducted by Craig Addison, lot no. X4363.2008,

Semiconductor Equipment and Materials International (SEMI) oral histories and company videos, Computer History Museum.

17. Ibid.

18. Ibid.

19. "Big First for Signetics," *The Current Probe* 4.3 (May-June 1969), Folder 12, Box 13, Don Liddie papers on Signetics, lot no. X4609.2008, Computer History Museum (henceforth Liddie papers).

20. The events and employment and revenue figures that follow are taken from Donald F. Liddie, "Signetics Corporation Key Event Chronology," December 28, 2006, lot no. X7847.2017, Computer History Museum.

21. Charles (Chuck) Harwood interview conducted by Duane Wadsworth, October 29, 2009, oral history collection, Computer History Museum. Words in brackets in original.

22. "Harwood's Innovative Leadership New Product Data Base," December 14, 1984, Folder "History—Signetics and Philips, 1971–1998 3 of 3," Box 14, Liddie papers.

23. Kattner interview.

24. Margaret B. W. Graham and Alec T. Shuldiner, *Corning and the Craft of Innovation* (New York: Oxford University Press, 2001), 272

25. Ibid., 273.

26. Ibid., 306.

27. Harwood interview.

28. "Signetics's Halter Reveals 'Dutch Connection'," *Electronic Business* (February 1977): 22–23.

29. Dick Wittenberg, "Topman Signetics: Philips schuld van neergang bedrijf," *NRC Handelsblad* (April 4, 1990), 13. The full quote is "Volgens [Signetics president James] Dykes is het met de samenwerking tussen Philips and Signetics al vijftien jaar geleden misgegaan. Signetics ging ervaan uit dat Philips met de acquisitie wel een bedoeling zou hebben en met plannen zou komen. 'Dat is niet zo vreemd,' zegt Dykes. De onderneming stelde zich dus heel afwachtend op, maar Philips nam maar niet de leiding, waardoor een vacuüm ontstond."

30. Jones, "Biotech's perfect climate."

31. Thackray, Brock, and Jones, *Moore's Law*, chapter 5, depicts this milieu vividly.

32. Napoleone Cavlan and Ronald Cline interview conducted by Andrew Haines, September 15, 2009, oral history collection, Computer History Museum. Cavlan is the one speaking in this excerpt.

33. "Don Hoefler, Writer Who Coined the Term 'Silicon Valley'," *San Jose Mercury News* April 16, 1986: 8B.

34. Cavlan and Cline interview.

35. Marie Hicks, "Why Tech's Gender Problem Is Nothing New," *The Guardian* October 12, 2018.

36. Glenna Matthews, *Silicon Valley, Women, and the California Dream: Gender, Class, and Opportunity in the Twentieth Century* (Stanford: Stanford University Press, 2003). For an electronics manufacturing district that absorbed a workforce of women from the textile industry, see Stuart W. Leslie, "Blue Collar Science: Bringing the Transistor to Life in the Lehigh Valley," *Historical Studies in the Physical and Biological Sciences*, 32.1 (2001): 71–113.

37. For example, "Congratulations Girls," *The Current Probe* 5.5 (October–November 1969) [but also "Cupid's Arrow" from the same issue]; "Louise Crain Trains Scottish Lasses," *The Probe*, 7.5 (December 1972), Folder "Signetics History Probe Publications 1967–1970," Box 27, Liddie papers.

38. "Eventful Queen of Outer Space Search Starts Soon," *Lab-Oratory* 19.3 (October 1969), JPL Archives.

39. For an ethnographic look at different modes of objectification of women in engineering, see Wendy Faulkner, "Doing Gender in Engineering Workplace Cultures. 1. Observations from the Field," *Engineering Studies* 1.1 (2009): 3–18. See Bix, *Women Coming to Tech!* (especially chapter 5) for an historical view of women's fraught participation in engineering organizations similar to the sites examined in this book. Many studies of the objectification of women in technoscientific cultures look back in some way to Carolyn Merchant's argument that the form of rationality and experimentation fostered by the scientific revolution—and perpetuated in Western technoscience ever since—was inherently sexist. See Carolyn Merchant, *The Death of Nature: Women, Ecology, and the Scientific Revolution* (New York: Harper & Row, 1980).

40. "Orem Plant Report: We Are In! . . . Well, Sorta," *Spec Sheet* II.1 (January 1969), Folder "1970 Spec Sheet Orem," Box 27, Liddie papers.

41. For example, a story about the promotions of Ann La Voto and Paula Kerley: "Stepping up to Foreman," *The Probe* 7.4 (June 1972), Folder "Signetics History Probe Publications 1967–1970," Box 27, Liddie papers.

42. "Some Thoughts on Our Past," *Probe* 16.8 (1981), Folder "Signetics History Probe Publications 1981," Box 27, Liddie papers.

43. Matthews, *Silicon Valley, Women, and the California Dream*, 141–142.

44. Jim Riley, "General Manager's Message," *The Current Probe* 4.5 (September 1968), Folder "Signetics History Probe Publications 1967–1970," Box 27, Liddie papers.

45. Jim Riley, "General Manager's Message," *The Current Probe* 5.5 (October–November 1969), Folder "Signetics History Probe Publications 1967–1970," Box 27, Liddie papers. First ellipsis in original.

46. "How Will Historians View the Sixties?," *The Current Probe* 6.1 (March 1970), Folder "Signetics History Probe Publications 1967–1970," Box 27, Liddie papers.

47. "Big First for Signetics."

48. Bill Gilchrist, "Plant Manager's Message," *The Spec Sheet* III.6 (June 1970), Folder "1970 Spec Sheet Orem," Box 27, Liddie papers.

49. Harwood interview.

50. Lécuyer, *Making Silicon Valley*, 204–208. Hong Kong was at the time a British colony, and some aspects of the rule of law prevailed there; although some democratic reforms were introduced in the 1980s and 1990s, when Fairchild arrived the territory was largely ruled by decree. Signetics was also following Fairchild's lead in placing plants in the US Southwest. The latter founded a plant in Shiprock, New Mexico (on the territory of the Navajo Nation), in 1965. Lisa Nakamura, "Indigenous Circuits: Navajo Women and the Racialization of Early Electronic Manufacture," *American Quarterly* 66.4 (2014): 919–941.

51. Don Liddie (president, Signetics International Corporation) letter to Charles LaFollette (Sovirel), February 11, 1969, Folder "Outgoing Memoranda and Correspondence 1969," Box 1, Liddie papers.

52. Steven C. McKay, *Silicon Islands or Satanic Mills: The Politics of High-Tech Production in the Philippines* (Ithaca: ILR Press, 2006).

53. D. F. Liddie (manager, International Facilities) letter to Ezra G. Hester (Corning Glass Works), December 18, 1968, Folder "Outgoing Memoranda and Correspondence 1968," Box 1, Liddie papers.

54. J. F. Riley and D. F. Liddie memo re: Site Selection—Signetics's United Kingdom Plant, December 19, 1968, Folder "Outgoing Memoranda and Correspondence 1968," Box 1, Liddie papers.

55. D. Liddie memo to C. Harwood, re: Union, April 8, 1974, Folder "Outgoing Memoranda and Correspondence 1974," Box 1, Liddie papers.

56. Press Release, October 8, 1976, Folder "Outgoing Memoranda and Correspondence 1976," Box 1, Liddie papers.

57. D. F. Liddie memo to R. F. Kruger, re: Plant Site Survey, April 4, 1968, Folder "Outgoing Memoranda and Correspondence 1968–2," Box 1, Liddie papers.

58. "Harwood's Innovative Leadership New Product Data Base." Note that Harwood lists the year of closing as 1973, but this is clearly incorrect; Liddie's timeline says that the Portugal plant closed in April 1976—after the Philips takeover, and at the same time as the Linlithgow (Scotland) plant's closure. Charles Downs, *Revolution at the Grassroots: Community Organizations in the Portuguese Revolution* (Albany: State University of New York Press, 1989): 36–38 gives some flavor of the strife between multinationals (including Philips/Signetics) and workers in the region in the year prior to the shutdown. Note that the above-mentioned Fairchild plant in New Mexico closed down in 1975, reputedly because the American Indian Movement planned to unionize the workforce. See Nakamura, "Indigenous Circuits."

59. Harwood interview. "Signetics Corp. Will Close Its Manufacturing Plant in the Philippines," UPI, January 28, 1983.

60. Michael S. Malone, "Signetics Finds the Right Buyer Was the Key to Its Future Here," *San Jose Mercury News* June 24, 1979.

61. John Gosch, "Giant Philips Flexes Semiconductor Muscle," *Electronics* (June 24, 1976): 72.

62. Ibid.

63. "Ranking within Industry," March 14, 1980, Folder "Financial Prospectus, 1980-04-07," Box 24, Liddie papers.

64. Mila Davids and Geert Verbong, "Intraorganizational Alignment and Innovation Processes: Philips and Transistor Technology," *Business History Review* 80.4 (2006): 657–688.

65. "Ontslagen Philips niet uitgesloten," *Het Parool* (April 4, 1975): 11.

66. "Signetics's Halter."

67. Ibid.

68. D. F. Liddie memo to C. C. Harwood, re: Meeting with Tony Opstelten, May 2, 1975, May 15, 1975, Folder "Outgoing Memoranda and Correspondence 1975," Box 1, Liddie papers.

69. "Philips Enters U.S. Semiconductor Production Market," *Electronic Business* (December 1979).

70. René Raaijmakers, *De Architecten van ASML* (Nijmegen: Techwatch Books, 2017), 140.

71. Ibid., 153–154.

72. Bassett, *To the Digital Age*; Lécuyer, *Making Silicon Valley*; Tim Jackson, *Inside Intel: How Andy Grove Built the World's Most Successful Chip Company* (New York: HarperCollins, 1998).

73. Paul Gerven and René Raaijmakers, *Natlab: Kraamkamer van ASML, NXP en de CD* (Nijmegen: Techwatch Books, 2018).

74. Casper Schuuring, "Computer dringt gezin binnen," *Amigoe di Curacao* (October 26, 1976): 9. The Briarcliff Manor lab was about 54 miles from Bell Labs' main complex, roughly eighty miles to RCA's Princeton lab, and approximately 130 miles to GE's main lab.

75. Evan Herbert, "Japanese R&D in the United States," *Research Technology Management* 32.6 (1989): 11–20.

76. "New Research Building Planned for 1981," *Probe* 2.1 (January 1980), Folder "Signetics History Probe Publications 1975–1980," Box 27, Liddie papers.

77. "Signetics Bootstraps $250 Million MOS Offensive with Philips-Wide Resources," January 21, 1981, Folder "Old Signetics Information 2," Box 13, Liddie papers.

78. "New Research Building." See also "Signetics: Opens Major Facility to Conduct Research and Development on ICs," *Business Wire* October 5, 1984.

79. Ruud Horeman, "Philips verschuift werk van Europa naar VS," *De Volkskrant* April 14, 1977: 2.

80. "Philips poogt positie op Amerikaanse markt te versterken," *De Waarheid* July 30, 1977: 4.

81. Horeman, "Philips verschuift werk van Europa naar VS."

82. Harwood interview. First brackets original, latter two are mine.

83. Willem Beusekamp, "Amerikaanse chipmakers overheersen," *De Volkskrant* October 10, 1978: 19; Mark Blackburn, "Europe Chips Away at U.S. High-Technology Lead," *Oakland Tribune* February 8, 1981: B-1.

84. Olivier Blanchard and Thomas Philippon, "The Quality of Labor Relations and Unemployment," NBER Working Paper Series No. 10590.

85. "Lichte winststijging Philips," *De Volkskrant* May 20, 1983: 2, specifically credits Signetics's increased revenues for blazing the trail back toward profit for all of Philips.

86. Here again I rely on Philip Jenkins's argument that setbacks for the United States on the international stage and the erosion of a domestic social order with white, straight men at the top created conditions in which the media depicted almost every news event in terms of decline, disorder, and moral panic. Jenkins cites the emergence of new high-tech industries such as biotech and personal computing as almost the only events treated with any optimism. Jenkins, *Decade of Nightmares*, 190–192.

87. AnnaLee Saxenian, *Regional Advantage: Culture and Competition in Silicon Valley and Route 128* (Cambridge, MA: Harvard University Press, 1994).

88. AnnaLee Saxenian, "The Urban Contradictions of Silicon Valley: Regional Growth and the Restructuring of the Semiconductor Industry," *International Journal of Urban and Regional Research* 7.2 (1983): 237–262.

89. Foucault, *The Birth of Biopolitics*; Huber, *Lifeblood*.

90. Gina Neff, *Venture Labor: Work and the Burden of Risk in Innovative Industries* (Cambridge, MA: MIT Press, 2012). I take this to be one of the underappreciated points of Turner, *From Counterculture to Cyberculture*.

91. Allen Boraiko, "The Chip: Electronic Mini-Marvel That Is Changing Your Life," and Moira Johnston, "High Tech, High Risk, and High Life in Silicon Valley," *National Geographic* (October 1982): 419–457 and 458–476.

92. Michael S. Malone, "Going Underground in Silicon Valley," *New York Times* May 30, 1982.

93. For example, Alice Marwick and Rebecca Lewis, "Media Manipulation and Disinformation Online," report of the Data & Society Research Institute, 2017; Whitney Phillips, *This Is Why We Can't Have Nice Things: Mapping the Relationship between Online Trolling and Mainstream Culture* (Cambridge, MA: MIT Press, 2015).

94. Susan Benner, "Storm Clouds over Silicon Valley," *Inc.* September 1, 1982.

95. Ron Deutsch memo for distribution, re: Documentary, April 28, 1982, Folder "History—Signetics and Philips, 1971–1998 3 of 3," Box 14, Liddie papers.

96. "Arrests in Computer Chip Sting Operation," *United Press International* February 27, 1984.

97. Benner, "Storm Clouds."

98. Joyce Maupin (Union WAGE) letter to Charles Harwood, September 11, 1979, Folder 34, Box 9, DataCenter Records, larc.ms.0061, Labor Archives and Research Center, J. Paul Leonard Library, San Francisco State University; "County Candidate Visits Signetics," *The Probe* 20.4 (November 1984): 4, Folder "Signetics History Probe Publications 1983–1985," Box 27, Liddie papers.

99. Johnston, "Silicon Valley."

100. AnnaLee Saxenian, "In Search of Power: The Organization of Business Interests in Silicon Valley and Route 128," *Economy and Society* 18.1 (1989): 25–70.

101. "Semi Equipment Makers Not up to Par, Complain Top Signetics Execs," *Electronic Engineering Times* (November 10, 1980): 2.

102. William H. Dana memo to Robert Dettmer, esq. re: Political Action Committee (PAC), March 16, 1981, Folder "Community and Government Relations 1981," Box 6, Liddie papers.

103. William H. Dana memo to C. C. Harwood. re: Political Action Committee, April 24, 1981, Folder "Community and Government Relations 1981," Box 6, Liddie papers.

104. Mody, "Academic Centers and/as Industrial Consortia."

105. "Governmental Relations Program," unsigned and undated (but 1981, probably March) memo to Signetics upper management, Folder "Community and Government Relations 1981," Box 6, Liddie papers.

106. C. C. Harwood internal correspondence to all Signetics employees, re: Measure A, October 24, 1984, Folder "Community Affairs (1981–1983) 2," Box 6, Liddie papers.

107. Christophe Lécuyer, "From Clean Rooms to Dirty Water: Labor, Semiconductor Firms, and the Struggle over Pollution and Workplace Hazards in Silicon Valley," *Information and Culture* 52.3 (2017): 304–333.

108. Lécuyer examines the case in some depth. For some wider context see Ted Smith, David A. Sonnenfeld, and David Naguib Pellow, *Challenging the Chip: Labor Rights and Environmental Justice in the Global Electronics Industry* (Philadelphia: Temple University Press, 2006).

109. For example, Susan Yoachum and Michael Malone, "The Chemical Handlers," *San Jose Mercury News* April 6, 1980.

110. "IC Industry Gets Good Safety Rating by State Agency," *Probe* 17.2 (March 1982): 3, Folder "Signetics History Probe Publications 1983–1985," Box 27, Liddie papers.

111. Lécuyer and *Challenging the Chip* summarize efforts by both Ross and the SVTC. Ross is also prominently profiled in Benner, "Storm Clouds."

112. Dan Eastman memo to Don Liddie, re: Proposed 1979 Donations Budget, October 11, 1978, Folder "Charitable Contributions, 1976-2001-2," Box 6, Liddie papers.

113. Dan Eastman memo to Don Liddie, re: 1981 Donations/Contribution Budget, September 4, 1980, Folder "Charitable Contributions, 1976-2001-2," Box 6, Liddie papers.

114. "Public Television," unsigned and undated (but probably 1981) handwritten list of shows that Signetics could sponsor on the San Francisco, Sacramento, and Albuquerque public television stations, Folder "Charitable Contributions, 1976-2001-2," Box 6, Liddie papers. Scattered through the rest of the folder is correspondence with KQED (the San Francisco public radio/television station) about how sponsorship of shows could help with "heightened esteem in the eyes of customers, employees, and stockholders" and comparing Signetics's sponsorship with that of other semiconductor firms.

115. See O'Mara, *Cities of Knowledge*, for some of the earliest stirrings of opposition to Bay Area manufacturing on environmental grounds. There's little indication,

however, that groups such as the Sierra Club made any difference in the semiconductor industry's lax attitude toward pollution and occupational safety.

116. Other studies make this point more explicitly. For example, Aubrey Underwood, "The Invisible Issue: Southern Women and Nuclear Opposition in the Tennessee 'Nuclear' River Valley, 1974–2011," *Journal of the Southern Association for the History of Medicine and Science* 1.1 (2019): 27–44; Natasha Zaretsky, *Radiation Nation: Three Mile Island and the Political Transformation of the 1970s* (New York: Columbia University Press, 2018); Rich Newman, "Making Environmental Politics: Women and Love Canal Activism," *Women's Studies Quarterly* 29.1/2 (2001): 65–84.

117. For example, Joanna Weidler-Lewis, "Transformation and Stasis: An Exploration of LGBTQA Students Prefiguring the Social Practices of Engineering for Greater Inclusivity," *Engineering Studies* 12.2 (2020): 127-149.

118. *The Probe*, 19.3 (October 1984), Folder "Signetics History Probe Publications 1983–1985," Box 27, Liddie papers.

119. "Signetics and the Energy Crunch," *Probe*, 8.14 (December 14, 1973), 1, Folder "Signetics History Probe Publications 1971–1974;" "Vanpooling Comes to Signetics," *Probe*, 2.2 (February, 1980): 6; Folder "Signetics History Probe Publications 1975–1980," Box 27, both in Liddie papers; "Signetics Announces 10-Day Closing Will Be Spread over Summer Months," *Electronic Engineering Times* June 22, 1981.

120. Roger Williams, "Visit to Signetics, Sunnyvale," January 1989, Folder "Quality Improvement Process (QIP) 1982-1991-3," Box 14, Liddie papers.

121. Eefke Smit and Dick Wittenberg, "Het tegenoffensief: Monsterverbond europese chip-industrie tegen VS en Japan," *NRC Handelsblad* February 17, 1988: S5; "Philips verdwijnt dit jaar uit top-10 van chipsfabrikanten," *NRC Handelsblad* September 13, 1988: 13.

122. Wittenberg, "Topman Signetics."

123. Oedzge Atzema, Egbert Wever, and Marloes Krol, "Philips: A Global Electronics Firm Restructuring Its Home Base," in *International Business Geography: Case Studies of Corporate Firms*, ed. Piet Pallenberg and Egbert Wever (Milton Park, UK: Routledge, 2007): 186–215.

124. "Zeggenschap over VS bedrijven weer bij Philips: U.S. Philips heft zichzelf op," *NRC Handelsblad* December 19, 1986: 15.

125. "Philips brengt chips onder in aparte groep," *Trouw* November 10, 1990: 7.

126. Erik van Gruijthuijsen and Peter Junge, "Chipdivisie Philips bijna in orde: Signetics in Silicon Valley onverkoopbaar," *Het Parool* November 25, 1991: 8. "In deze ideal omgeving, omgeven door vestigingen van wereldmerken als Apple, Intel en Hewlett-Packard, duurt de lijdensweg van Signetics al meer dan vijftien jaar. Sinds Philips er in 1975 huisbaas werd, leed Signetics slechts verliezen. In de loop der jaren hebben die Philips zeker een miljard gulden gekost."

127. "CD-fabriek Philips in België geblokkeerd door boos personeel," *Trouw* January 17, 1992: 7.

128. Williams, "Visit to Signetics."

129. "Signetics: Quality Program Saves Millions; Reduces Defects Nearly 100 Percent," *Business Wire* October 25, 1983.

130. Louise Kehoe, "2nd Chip Firm Admits Errors," *The Globe and Mail* November 9, 1984.

131. Kathleen Burton, "Signetics to Cut Work Force," *Computerworld* March 11, 1985: 127.

132. Dick Wittenberg, "Mega-project commerciële flop voor Philips," *NRC Handelsblad* August 31, 1990: 11. See also Marcel Metze and Bart Vink, *Kortsluiting: Hoe Philips zijn talenten verspilde* (Nijmegen: Sun, 1992).

133. Chris Thompson, "Military Spending Boom Forecast by Signetics's Boss," *Sacramento Union* March 27, 1981.

Chapter 7

1. Steve Coll, *The Deal of the Century: The Breakup of AT&T* (New York: Atheneum, 1986).

2. Hall is not mentioned, but for studies of Picturephone and its demise see Kenneth Lipartito, "Picturephone and the Information Age: The Social Meaning of Failure," *Technology and Culture* 44.1 (2003): 50–81; A. Michael Noll, "Anatomy of a Failure: Picturephone Revisited," *Telecommunications Policy* 16.4 (1992): 307–316.

3. Arthur D. Hall III, "Highly Automated Agricultural Production System," US patent no. 4,015,366, filed April 11, 1975, issued April 5, 1977.

4. Neff, *Venture Labor.*

5. For RCA, see Gross, *TVs of Tomorrow*; for IBM, see Cortada, *IBM*. AT&T (especially Bell Labs) has garnered attention as a place where quite square engineers and managers embraced and to some extent coopted the counterculture. See Gertner, *The Idea Factory*; Matthew Wisnioski, "How the Industrial Scientist Got His Groove: Entrepreneurial Journalism and the Fashioning of Technoscientific Innovators," in *Groovy Science*, ed. David Kaiser and W. Patrick McCray (Chicago: University of Chicago Press, 2016); 337–365.

6. There is a scattered literature on the topic, but as far as I know no synthetic work. But see Schulman, *From Cotton Belt to Sunbelt*; O'Mara, *Cities of Knowledge*; and Caroline Rose Peyton, "Radioactive Dixie: A history of nuclear power and nuclear waste in the American South, 1950–1990," PhD dissertation, University of South Carolina, 2016; J. Merritt McKinney, "Air pollution, politics, and environmental reform in Birmingham, Alabama, 1940–1971," PhD dissertation, Rice University, 2011, for some examples of what a synthetic study might cover.

7. Akera, *Calculating a Natural World*; Amy Slaton, *Race, Rigor, and Selectivity in U.S. Engineering: The History of an Occupational Color Line* (Cambridge, MA: Harvard University Press, 2010); Joanna S. Ploeger, "The Art of Science at Fermi National Accelerator Laboratory: The Rhetoric of Aesthetics and Humanism in the National Laboratory System in the Late 1960s," *History and Technology* 18.1 (2002): 23–49.

8. Arthur D. Hall, "Qualifications Statement," undated but possibly 1983, Folder 8, Box 6, Arthur D. Hall III papers (Accession 2628), Hagley Museum and Library (henceforth Hall papers).

9. Ibid.

10. Advanced Decision Handling, "Experience Briefs," undated but early 1980s, Folder 8, Box 6; Arthur D. Hall III, speech on "Capitalizing Emerging Markets and Technology," given to the 1972 General Technical Advisory Committee meeting in St. Louis, February 15, 1972, Folder 12, Box 1; both in Hall papers.

11. Arthur D. Hall III, "Multiple Site Remote Sensing and Control by Automated Spectrometry," proposal to Defense Small Business Advanced Technology Program, August 20, 1981. In Folder 13, Box 7, Hall papers.

12. Michael Egan, *Barry Commoner and the Science of Survival: The Remaking of American Environmentalism* (Cambridge, MA: MIT Press, 2007).

13. For examples, see Mark Solovey and Hamilton Cravens (eds.), *Cold War Social Science* (New York: Palgrave Macmillan, 2012); or David H. Price, *Weaponizing*

Anthropology: Social Science in Service of the Militarized State (Chico, CA: AK Press, 2011).

14. Chris Knight, "Decoding Chomsky," *European Review*, 12.4 (2004): 581–603; also Janet Martin-Nielsen, "'It Was All Connected': Computers and Linguistics in Early Cold War America," in *Cold War Social Science*, ed. Mark Solovey and Hamilton Cravens (New York: Palgrave Macmillan, 2012): 63–78.

15. David Edgerton, *The Shock of the Old: Technology and Global History since 1900* (New York: Oxford University Press, 2007); Andrew Russell and Lee Vinsel, "Let's Get Excited about Maintenance!," *New York Times* July 23, 2017: SR5.

16. For example, Paul Harold Rubinson, "Containing science: The U.S. National security state and scientists' challenge to nuclear weapons during the Cold War," PhD dissertation, University of Texas, 2008.

17. Badash, *A Nuclear Winter's Tale*; Hamblin, *Arming Mother Nature*.

18. Rebecca Slayton, *Arguments That Count: Physics, Computing, and Missile Defense, 1949–2012* (Cambridge, MA: MIT Press, 2013). Bethe and to some extent York were at least partially visible to the public eye, but they hardly sought the limelight.

19. Quote is from Frank von Hippel and Tomoko Kurokawa, "Citizen Scientist: Frank von Hippel's Adventures in Nuclear Arms Control," *Journal for Peace and Nuclear Disarmament* 3[supl 1] (2019): 1–37. The interview with von Hippel contains a nice glimpse of Bethe's square suspicion of public performance: "I remember I went to a 'teach-in' in a big lecture hall with multiple speakers denouncing the war. I happened to sit next to Bethe. He saw that I was getting excited by one speaker and advised me quietly, 'He's a demagogue.' . . . He just didn't like the speaker's style." See also Elinor Langer, "National Teach-In: Professors Debating Viet Nam, Question Role of Scholarship in Policy-Making," *Science* 148.3673 (1965): 1075–1077.

20. Joel N. Shurkin, *True Genius: The Life and Work of Richard Garwin* (Buffalo, NY: Prometheus Books, 2017).

21. Harvey Brooks letter to Norman Hackerman, May 14, 1970, Folder 1, Box 3, Norman Hackerman personal and professional papers, 1956–1986, MS 407, Woodson Research Center, Fondren Library, Rice University.

22. Harvey Brooks, "Technology Assessment in Retrospect II," *Newsletter on Science, Technology & Human Values* 1.4 (1976): 17–29. Note that this was published

in the journal that became *Science, Technology & Human Values*, the journal of the main professional organization of the field of Science and Technology Studies. In the article, Brooks lists seven "lessons" from TA's recent past, of which #2 relates to inclusivity and the rest (particularly #3) to reflexivity.

23. Shanley, "Reimagining the Future through Revisiting the Past."

24. Oreskes and Conway, *Merchants of Doubt*.

25. Erik Conway, *High-Speed Dreams: NASA and the Technopolitics of Supersonic Transportation, 1945–1999* (Baltimore: Johns Hopkins University Press, 2005).

26. I base this observation on conversations with some of Meindl's longtime coworkers at the Center for Integrated Systems. But see also the picture spread "One Haircut, Three Decades," in Tekla S. Perry, "Wizard of Watts," *IEEE Spectrum* (June 2006): 32–37.

27. For example, Arthur D. Hall III, "Energy Self-Sufficiency on the Farm," proposal to the Department of Energy, January 1, 1980, Folder 14, Box 3, Hall papers.

28. Arthur D. Hall III, "A Syllabus for a New Course on Communications Technology and Public Policy," undated but probably 1973, Folder 8, Box 1, Hall papers.

29. Ibid.

30. Ibid.

31. I have David Kaiser's work on both high energy physics and parapsychology somewhat in mind here. That's not to single Kaiser out, but rather to argue that if even someone as perceptive as Kaiser has missed the big picture then we can be sure that the rest of the field hasn't done much better. My praise for Kaiser's contributions leavened by the critique that I'm offering here is rehearsed in Cyrus C. M. Mody, "Faster-than-Light Reading," *Social Studies of Science* 42 (2012): 159–164.

32. See Thomas Turnbull, "From paradox to policy," chapter 7, for a look at a similar group of US physicists.

33. Kenneth A. Fegley letter to George Gerbner, October 30, 1973, Folder 8, Box 1, Hall papers.

34. Arthur D. Hall III, "A Systematic Approach to the Total Enterprise," speech to the Senior Management Program of the American Management Association, undated but probably early 1970s, Folder 13, Box 1, Hall papers.

35. Fegley to Gerbner.

36. Ibid.

37. Nelson, *The Sound of Innovation*.

38. John Pierce letter to Wolfgang Kuhn, Acting Chairman, Stanford Department of Music, sometime in 1978. Folder Misc. Correspondence H-Z 1973–79, Box 5, John R. Pierce Papers, Huntington Library.

39. Ibid., 47–48.

40. For a lengthy discussion of Klüver's collaborations with a number of artists (and his role in bring about other intersections of engineering and art), see W. Patrick McCray, *Making Art Work: How Cold War Engineers and Artists Forged a New Creative Culture* (Cambridge, MA: MIT Press, 2020). McCray cites John Pierce's fervent support for Klüver but doesn't mention the much squarer Kompfner at all, even though Kompfner was Klüver's direct supervisor (i.e., Pierce was Klüver's boss's boss). See Julie Martin, oral history interview conducted by Brent Edwards, August 5 and 14, 2013, Rauschenberg Oral History Project, Columbia Center for Oral History.

41. Douglas Kahn, "From Brainwaves to Outer Space: John Cage and Karl Jansky," in *Earth Sound Earth Signal: Energies and Earth Magnitude in the Arts* (Berkeley: University of California Press, 2013).

42. Calvin F. Quate, "Electronic Device for Converting Written Music to Audible Sound," Stanford University Record of Invention No. S74–09, January 16, 1974. An accompanying letter indicates that Niels Reimers was "to present [Quate's idea] to Yamaha on occasion of their next visit." Both documents given to author by Andrew J. Nelson.

43. Rudolf Kompfner, manuscript on music, Folder "Music, personal reminiscences," Box 1, Rudolf Kompfner Papers SC 194 ACCN 86–125, SUA. See also J. R. Pierce, "Rudolf Kompfner, 1909–1977: A Biographical Memoir," in *National Academy of Science Biographical Memoir* (Washington, DC: National Academy of Sciences, 1983).

44. Arthur D. Hall III, "A Proposal for Exploratory Development of a Highly Automated Agricultural Production System (AUTOFARM)," Folder 16, Box 2, Hall papers.

45. Arthur D. Hall III, "First Monthly Report to the U.S. Department of Justice Antitrust Division on Contract #J-42432," April 10, 1976, Folder 15, Box 1, Hall papers.

46. ADH "Experience Briefs."

47. Sean F. Johnston, *Techno-Fixers: Origins and Implications of Technological Faith* (Montreal: McGill-Queen's University Press, 2020).

48. Hall discusses GE's "Geniponics" system in Arthur D. Hall III, letter to Robert Rodale (chairman, Rodale Press), December 25, 1981, Folder 15, Box 2. And he made inquiries to Grumman about the Sailwing Wind Generator; see Maureen Emmerich, Grumman Aerospace Corporation, New Business Operations, Energy Programs, letter to Hall, June 13, 1974, Folder 16, Box 2. Both documents in Hall papers. Geniponics was a fairly direct spin-off from the space program; after a few years of development GE sold it to Control Data Corporation. See Erin E. Murphy and Jerome J. Suran, "Whatever Happened to Geniponics?," *IEEE Spectrum* 24.11 (November 1987): 17.

49. Arthur D. Hall III, "AUTOFARM: A Vision for Computerized Agriculture," December 18, 1989, Folder 5, Box 3, Hall papers.

50. Arthur D. Hall III, "Preliminary Design and Evaluation of an Audience Research System," July 15, 1968, Folder 11, Box 1, Hall papers.

51. Arthur D. Hall III, "Small Businesses from Biomass Gasification," proposal to US Department of Energy Appropriate Energy Technology Small Grants Program, 1981, April 16, 1981. In Folder 15, Box 2, Hall papers.

52. David P. D. Munns and Kärin Nickelsen, "To Live among the Stars: Artificial Environments in the Early Space Age," *History and Technology* 33.3 (2017): 272–299.

53. Sabine Höhler, "The Environment as a Life Support System: The Case of Biosphere 2," *History and Technology* 26.1 (2010): 39–58.

54. Arthur D. Hall III, letter to Robert H. Roy (chair, Hazardous Waste Facilities Siting Board, Maryland Environmental Service), July 13, 1981, Folder 2, Box 3, Hall papers.

55. Arthur D. Hall III, "A Prospectus for AUTOFARM," undated but approxmately 1990, Folder 13, Box 3, Hall papers.

56. A common view of what Oreskes and Conway call "cornucopianism." See also Schleper, "Life on Earth," for ties between cornucopian technooptimists and the idea of sustainable development.

57. Arthur D. Hall, "Thomas Jefferson, Freedom, and the SMC Society," *Conference Proceedings 1991 IEEE International Conference on Systems, Man, and Cybernetics* (New York: IEEE, 1991); Arthur D. Hall III, "A Broken Bell Tolls for New Freedom," draft letter to the editors of *Wall Street Journal* January 7, 1985, Folder 9, Box 7, Hall papers.

58. Hall, "Thomas Jefferson."

59. I. W. Edminster letter to Arthur D. Hall III, November 5, 1974, Folder 16, Box 2, Hall papers.

60. Quoted in Hall to Rodale. Note on underlining in Hall's quotation: I'm doubtful that the underlining was in the original review.

61. G. E. Ferguson (John Bean Division, FMC Corporation) letter to W. S. Thompson, re: Mr. Arthur D. Hall, September 1, 1964, Folder 16, Box 2, Hall papers.

62. Arthur D. Hall III, letter to Alex Schwartzhoff (NSF—RANN), November 7, 1975, Folder 17, Box 2, Hall papers.

63. Arthur D. Hall III, letter to Charles R. Fisher, Anaconda Telecommunications, January 15, 1980, Folder 18, Box 2, Hall papers.

64. Andrew J. Roman letter to Arthur Hall, III, February 20, 1978, Folder 18, Box 2; Arthur D. Hall III, letter to Donald McGrath (president, Tyler Corporation), March 20, 1993, Folder 19, Box 2; Arthur D. Hall III, "A Proposal for Israel to Develop the First Highly Automated Agricultural Production System," April 5, 1979, Folder 8, Box 3; Arthur D. Hall III, "A Proposal for Egypt to Develop the First Highly Automated Agricultural Production System," March 7, 1979, Folder 9, Box 3; Arthur D. Hall III, "A Proposal for Saudi Arabia to Develop the First Highly Automated Agricultural Production System," June 20, 1977, Folder 18, Box 7; all Hall papers.

65. Bowles, *Science in Flux*; Akera, *Calculating a Natural World*; Peter J. Westwick, "From the Club of Rome to Star Wars: The Era of Limits, Space Colonization, and the Origins of SDI," in *Limiting Outer Space: Astroculture after Apollo*, ed. Alexander C. T. Geppert (New York: Palgrave Macmillan, 2018): 283–302; Cyrus C. M. Mody, "How I Learned to Stop Worrying and Love the Bomb, the Nuclear Reactor, the

Computer, Ham Radio, and Recombinant DNA," *Historical Studies in the Natural Sciences* 38.3 (2008): 451–461.

66. The exception being what Lillian Hoddeson, Adrienne Kolb, and Catherine Westfall call "megascience"—that is, a shift from lots of giant "Big Science" projects to one single megaproject requiring enormous coalitions pooling their resources. Hoddeson, Kolb, and Westfall, *Fermilab.*

67. Jones, *Desert Kingdom.*

68. Arthur D. Hall III, letter to The Hon. Robert E. Bauman, November 21, 1980, Folder 17, Box 2, Hall papers.

69. Arthur D. Hall III, letter to The Hon. Charles McC. Mathias Jr., November 21, 1980, Folder 17, Box 2, Hall papers.

70. Steven M. Flipse, Maarten C. A. van der Sanden, and Patricia Osseweijer, "Midstream Modulation in Biotechnology Industry: Redefining What Is 'Part of the Job' of Researchers in Industry," *Science and Engineering Ethics* 19.3 (2013): 1141–1164.

71. David H. Guston and Kenneth Keniston, "Introduction: The Social Contract for Science," in *The Fragile Contract: University Science and the Federal Government*, ed. Guston and Keniston (Cambridge, MA: MIT Press, 1994): 1–41.

72. Harro van Lente, Tsjalling Swierstra, and Pierre-Benoît Joly, "Responsible Innovation as a Critique of Technology Assessment;" Rinie van Est, "Responsible Innovation as a Source of Inspiration for Technology Assessment and vice versa: The Common Challenge of Responsibility, Representation, Issue Identification, and Orientation;" and Pierre Delvenne, "Responsible Research and Innovation as a Travesty of Technology Assessment?;" all three in *Journal of Responsible Innovation*, 4.2 (2017): 254–261, 268–277, and 278–288 respectively.

73. Van Lente, Swierstra, and Joly, "Responsible innovation as a critique."

74. Bruce Bimber, *The Politics of Expertise in Congress: The Rise and Fall of the Office of Technology Assessment* (Albany: State University of New York Press, 1996).

75. Ibid.

76. Daniel Lee Kleinman and Mark Solovey, "Hot Science/Cold War: The National Science Foundation after World War II," *Radical History Review* 63 (1995): 110–139.

77. Hall, "Capitalizing Emerging Markets and Technology."

78. Scott Frickel and Neil Gross, "A General Theory of Scientific/Intellectual Movements," *American Sociological Review* 70.2 (2005): 204–232.

79. Jessica M. Smith, *Extracting Accountability: Engineers and Corporate Social Responsibility* (Cambridge, MA: MIT Press, 2021).

80. Stephen Macekura, *Of Limits and Growth*; Danielle Shanley, "Looking Forward, Looking Back."

81. Anthony P. Pontello (Appropriate Technology Program Manager) letter to Advanced Decision Handling, Inc., September 15, 1981, Folder 19, Box 2, Hall papers.

82. From van Lente, Swierstra, and Joly. Full quote is "The original model of TA institutionalized in the US through the creation of [the] Office of Technology Assessment in 1972 did not include an emphasis on public or stakeholder participation." The authors aren't really wrong, though, except perhaps in their use of the word *original*. Shanley shows that TA was initially more open, but as it institutionalized it became more technocratic and less public facing.

83. Michael Almon, "Report to the Lawrence City Commission on Energy Conservation and Renewable Energy Systems," undated but probably fall 1979, Folder County Energy 1979–1980; and *Kansas Natural Guard Newsletter*, issue 2 (undated but about 1981), Folder Kansas Natural Guard Newsletter; both Box 1, Center for Appropriate Technology Collection (RH MS 821), Kenneth Spencer Research Library, University of Kansas.

84. Dee Tolar, unofficial notes from County Energy Study meeting, February 20, 1980, Folder "County Energy 1982," Box 1, Center for Appropriate Technology Collection (RH MS 821), Kenneth Spencer Research Library, University of Kansas.

85. "Energy and the Community," *Small Farm Energy* 35 (March/April 1982): 10–11.

86. Hall to Rodale.

87. Arthur D. Hall III, letters to W. C. Doud (vice president, IBM), May 3, 1977; Harold Ransom (procurement director, Campbell Soup Company), October 15, 1980; Robert Howell (Control Data Corporation); R. C. Cosgrove (chairman, Green Giant Company), July 2, 1980; all Folder 18, Box 2; Sylvia Hadowanetz (unsolicited proposals coordinator, US Army Armament Research and Development Command) letter to Arthur D. Hall III, September 23, 1981, Folder 19, Box 2; Thomas McClure letter to Jordan Lewis (Battelle Development Corporation), September 28, 1971, Folder 7, Box 3; all Hall papers.

88. Jack Stilgoe, Richard Owen, and Phil Macnaghten, "Developing a Framework for Responsible Innovation," *Research Policy* 42.9 (2013): 1568–1580.

89. Winner's essay "Do Artifacts Have Politics?" has passed into the timeless canon of STS, so it is quite easy for readers today to forget that it was written at the tail end of the long 1970s and had that decade's developments in science and technology in its sights. The "artifacts" that Winner discusses are largely products of square physical and engineering science. Indeed, one of his cases—admittedly, the one that is usually neglected in classroom discussions—was an automated tomato-picker that closely resembled some aspects of Hall's Autofarm. I would even argue that Winner's question presupposes squareness in many cases. That is, artifacts that arise from ambivalence and reticence are more likely to have a politics obscure enough to ask whether they have politics at all. Langdon Winner, "Do Artifacts Have Politics?," *Daedalus* 109.1 (1980): 121–136.

90. Arthur D. Hall III, "Capitalizing Emerging Markets and Technology," Report to Emerson Electronics, May 16, 1972, Folder 12, Box 1, Hall papers.

91. Ibid.

92. DOE/NASA Satellite Power System Concept Development and Evaluation Program, "A Methodology for the Comparative Assessment of the Satellite Power System (SPS) and Alternative Technologies," January 1980, DOE/ER-0051, box 7, folder "DOE-ER0051 Methodology for Comparative Assessment of SPS and Alternatives," Center Series, Section 108 SPS, JSC collection.

93. Hall, "A Broken Bell."

94. Stanton Glantz memo to Stanford Committee on Research, re: Interview with Engineering Professor, March 3, 1971, Folder 1, Box 1, Stanford University Sponsored Projects Office papers, SC 344 ACCN 1987–130, SUA.

95. Rebecca Onion, *Innocent Experiments: Childhood and the Culture of Popular Science in the United States* (Chapel Hill: University of North Carolina Press, 2016).

96. Chowning, Smith, and Cohen, "The Computer Music Facility."

97. For more on the ambivalence of inclusivity, see Steven Epstein, *Inclusion: The Politics of Difference in Medical Research* (Chicago: University of Chicago Press, 2007).

98. Hall, "Capitalizing Emerging Markets and Technology."

99. Ashley, "New Directions for Engineering Research."

100. Daniel Sarewitz, "Saving Science," *New Atlantis* (Spring/Summer 2016).

101. Erik Fisher, Roop L. Mahajan, and Carl Mitcham, "Midstream Modulation of Technology: Governance from Within," *Bulletin of Science, Technology & Society* 26.6 (December 2006): 485–496.

102. Gabrielle Hecht, *Being Nuclear: Africans and the Global Uranium Trade* (Cambridge, MA: MIT Press, 2012); Egan, *Barry Commoner and the Science of Survival*; Lindsey Dillon, "War's Remains: Slow Violence and the Urbanization of Military Bases in California," *Environmental Justice* 8.1 (2015); Lauren B. Hirshberg, "Targeting Kwajalein: U.S. Empire, Militarization, and Suburbanization and the Marshall Islands, 1944–1986," PhD dissertation, University of Michigan, 2011; Abraham, "Rare Earths'; Joseph Masco, *The Nuclear Borderlands: The Manhattan Project in Post-Cold War New Mexico* (Princeton: Princeton University Press, 2006).

103. For example, Weidler-Lewis, "Transformation and Stasis."

104. Arthur D. Hall III, letter to Pawel Warchol, General Manager of State Administration of Breeding Animal Farms, Warsaw, November 4, 1974, Folder 16, Box 2, Hall papers.

105. Kaiser, *How the Hippies Saved Physics.*

106. Thomas Heinze and Richard Münch, "Institutional Conditions for Progress and Renewal in Science," in *Innovation in Science and Organizational Renewal: Historical and Sociological Perspectives* (London: Palgrave Macmillan, 2016): 1–20.

107. Sean F. Johnston, *Techno-Fixers: Origins and Implications of Technological Faith* (Montreal: McGill-Queen's University Press, 2020), chapter 4.

108. Slaton, *Race, Rigor, and Selectivity*; "Meritocracy, Technocracy, Democracy: Understandings of Racial and Gender Equity in American Engineering Education," in *International Perspectives on Engineering Education*, ed. Steen Hyldgaard Christensen *et al.* (New York: Springer, 2015): 171–189.

109. Quoted in "Defense Research Will Shift If Forced out of Universities," *Campus Report* September 23, 1970.

INDEX

INSIDE TECHNOLOGY SERIES

Edited by Wiebe E. Bijker, Trevor Pinch, and Rebecca Slayton

Cyrus C. M. Mody, *The Squares: US Physical and Engineering Scientists in the Long 1970s*

Brice Laurent, *European Objects: The Troubled Dreams of Harmonization*

Florian Jaton, *The Constitution of Algorithms: Ground-Truthing, Programming, Formulating*

Kean Birch and Fabian Muniesa, *Assetization: Turning Things into Assets in Technoscientific Capitalism*

David Demortain, *The Science of Bureaucracy: Risk Decision-Making and the US Environmental Protection Agency*

Nancy Campbell, *OD: Naloxone and the Politics of Overdose*

Lukas Engelmann and Christos Lynteris, *Sulphuric Utopias: The History of Maritime Fumigation*

Zara Mirmalek, *Making Time on Mars*

Joeri Bruynincx, *Listening in the Field: Recording and the Science of Birdsong*

Edward Jones-Imhotep, *The Unreliable Nation: Hostile Nature and Technological Failure in the Cold War*

Jennifer L. Lieberman, *Power Lines: Electricity in American Life and Letters, 1882–1952*

Jess Bier, *Mapping Israel, Mapping Palestine: Occupied Landscapes of International Technoscience*

Benoît Godin, *Models of Innovation: The History of an Idea*

Stephen Hilgartner, *Reordering Life: Knowledge and Control in the Genomics Revolution*

Brice Laurent, *Democratic Experiments: Problematizing Nanotechnology and Democracy in Europe and the United States*

Cyrus C. M. Mody, *The Long Arm of Moore's Law: Microelectronics and American Science*

Tiago Saraiva, *Fascist Pigs: Technoscientific Organisms and the History of Fascism*

Teun Zuiderent-Jerak, *Situated Interventions: Sociological Experiments in Healthcare*

Basile Zimmermann, *Waves and Forms: Electronic Music Devices and Computer Encodings in China*

Andrew J. Nelson, *The Sound of Innovation: Stanford and the Computer Music Revolution*

Sonja D. Schmid, *Producing Power: The Pre-Chernobyl History of the Soviet Nuclear Industry*

Casey O'Donnell, *Developer's Dilemma: The Secret World of Videogame Creators*

Christina Dunbar-Hester, *Low Power to the People: Pirates, Protest, and Politics in FM Radio Activism*

Eden Medina, Ivan da Costa Marques, and Christina Holmes, editors, *Beyond Imported Magic: Essays on Science, Technology, and Society in Latin America*

Anique Hommels, Jessica Mesman, and Wiebe E. Bijker, editors, *Vulnerability in Technological Cultures: New Directions in Research and Governance*

Amit Prasad, *Imperial Technoscience: Transnational Histories of MRI in the United States, Britain, and India*

Charis Thompson, *Good Science: The Ethical Choreography of Stem Cell Research*

Tarleton Gillespie, Pablo J. Boczkowski, and Kirsten A. Foot, editors, *Media Technologies: Essays on Communication, Materiality, and Society*

Catelijne Coopmans, Janet Vertesi, Michael E. Lynch, and Steve Woolgar, editors, *Representation in Scientific Practice Revisited*

Rebecca Slayton, *Arguments that Count: Physics, Computing, and Missile Defense, 1949–2012*

Stathis Arapostathis and Graeme Gooday, *Patently Contestable: Electrical Technologies and Inventor Identities on Trial in Britain*

Jens Lachmund, *Greening Berlin: The Co-Production of Science, Politics, and Urban Nature*

Chikako Takeshita, *The Global Biopolitics of the IUD: How Science Constructs Contraceptive Users and Women's Bodies*

Cyrus C. M. Mody, *Instrumental Community: Probe Microscopy and the Path to Nanotechnology*

Morana Alač, *Handling Digital Brains: A Laboratory Study of Multimodal Semiotic Interaction in the Age of Computers*

Gabrielle Hecht, editor, *Entangled Geographies: Empire and Technopolitics in the Global Cold War*

Michael E. Gorman, editor, *Trading Zones and Interactional Expertise: Creating New Kinds of Collaboration*

Matthias Gross, *Ignorance and Surprise: Science, Society, and Ecological Design*

Andrew Feenberg, *Between Reason and Experience: Essays in Technology and Modernity*

Wiebe E. Bijker, Roland Bal, and Ruud Hendricks, *The Paradox of Scientific Authority: The Role of Scientific Advice in Democracies*

Park Doing, *Velvet Revolution at the Synchrotron: Biology, Physics, and Change in Science*

Gabrielle Hecht, *The Radiance of France: Nuclear Power and National Identity after World War II*

Richard Rottenburg, *Far-Fetched Facts: A Parable of Development Aid*

Michel Callon, Pierre Lascoumes, and Yannick Barthe, *Acting in an Uncertain World: An Essay on Technical Democracy*

Ruth Oldenziel and Karin Zachmann, editors, *Cold War Kitchen: Americanization, Technology, and European Users*

Deborah G. Johnson and Jameson W. Wetmore, editors, *Technology and Society: Building Our Sociotechnical Future*

Trevor Pinch and Richard Swedberg, editors, *Living in a Material World: Economic Sociology Meets Science and Technology Studies*

Christopher R. Henke, *Cultivating Science, Harvesting Power: Science and Industrial Agriculture in California*

Helga Nowotny, *Insatiable Curiosity: Innovation in a Fragile Future*

Karin Bijsterveld, *Mechanical Sound: Technology, Culture, and Public Problems of Noise in the Twentieth Century*

Peter D. Norton, *Fighting Traffic: The Dawn of the Motor Age in the American City*

Joshua M. Greenberg, *From Betamax to Blockbuster: Video Stores tand the Invention of Movies on Video*

Mikael Hård and Thomas J. Misa, editors, *Urban Machinery: Inside Modern European Cities*

Christine Hine, *Systematics as Cyberscience: Computers, Change, and Continuity in Science*

Wesley Shrum, Joel Genuth, and Ivan Chompalov, *Structures of Scientific Collaboration*